The Theory of Stochastic Processes

The Theory of
Stochastic Processes

D. R. COX
H. D. MILLER

CHAPMAN & HALL

London · Glasgow · New York · Tokyo · Melbourne · Madras

Published by Chapman & Hall, 2–6 Boundary Row, London SE1 8HN

Chapman & Hall, 2–6 Boundary Row, London SE1 8HN, UK

Blackie Academic & Professional, Wester Cleddens Road, Bishopbriggs, Glasgow G64 2NZ, UK

Chapman & Hall, 29 West 35th Street, New York NY10001, USA

Chapman & Hall Japan, Thomson Publishing Japan, Hirakawacho Nemoto Building, 6F, 1-7-11 Hirakawa-cho, Chiyoda-ku, Tokyo 102, Japan

Chapman & Hall Australia, Thomas Nelson Australia, 102 Dodds Street, South Melbourne, Victoria 3205, Australia

Chapman & Hall India, R. Seshadri, 32 Second Main Road, CIT East, Madras 600 035, India

First edition 1965
Reprinted as a paperback 1967
Reprinted 1968, 1970, 1972, 1980, 1984, 1987, 1990, 1992

© 1965 D.R. Cox and H.D. Miller

Printed in Great Britain by J. W. Arrowsmith Ltd, Bristol

ISBN 0 412 15170 7

Contents

Preface

This book is an introductory account of the mathematical analysis of stochastic processes, i.e. of systems that change in accordance with probabilistic laws. We have written for statisticians and applied mathematicians interested in methods for solving particular problems, rather than for pure mathematicians interested in general theorems. To stress the wide variety of applications for the theory, we have included in outline illustrations from a number of scientific and technological fields; we have, however, not attempted to give detailed realistic discussion of particular applications.

A good knowledge of elementary probability theory is essential in order to understand the book. The mathematical techniques used have been kept as elementary as possible and are mostly standard results in matrix algebra and advanced calculus, including the elementary theory of Laplace and Fourier transforms.

Three very important aspects of stochastic processes that are not dealt with, except marginally, are

(a) the construction, rather than the solution, of models;
(b) the solution of problems by simulation;
(c) the statistical analysis of data from stochastic processes.

The first of these, the choice of a mathematical model that is reasonably tractable and also represents the essential features of an applied problem, is of vital importance for successfully applying the theory. Isolated brief comments about the choice of models occur throughout the book.

The most direct way for obtaining numerical results for particular complex systems is often by simulation on an electronic or desk computer. Simulation is also often very useful as a device for getting the 'feel' of a problem as a preliminary to theoretical work and for suggesting or checking approximations made in theoretical work. We have tried to emphasize the importance of straightforward simulation in the introductory chapter and in the exercises, but have not attempted to describe any of the more subtle devices that can be used to increase the precision of simulation studies.

We have not discussed the statistical analysis of observations on stochastic processes. This would require a separate book.

The book is intended both as a reference book for research workers and as a text-book for students. We have tried to make the different chapters as self-contained as possible and each chapter contains material of varied difficulty. If the book is used as a basis for a course of lectures, it will be necessary to select material depending on the students' interests and mathematical ability. Thus it would be possible to take the more elementary sections from all the chapters, or to concentrate on a small number of chapters, for instance on the relatively thorough account of Markov chains in Chapter 3.

We have not tried to trace the history of the subject or to give detailed references for well-established results. References are in general given only where the treatment in the text is incomplete; at the end of the chapters and in Appendix 2 there are suggestions for further reading.

We thank the Syndics of the Cambridge University Press for permission to reproduce problems from the examinations for the Cambridge Diploma in Mathematical Statistics.

We are very grateful to Dr P. A. W. Lewis for helpful comments. It is a special pleasure to thank also our secretary, Miss Dorothy Wilson, for her unfailing efficiency.

D. R. COX

H. D. MILLER

Birkbeck College, London
August, 1964

CHAPTER ONE

Introduction

1.1. Examples of stochastic processes

The theory of stochastic processes deals with systems which develop in time or space in accordance with probabilistic laws. Applications of the theory can be made to a wide range of phenomena in many branches of science and technology. The present chapter is designed to illustrate, by simple examples, some of the ideas and problems with which the theory deals.

Example 1.1. *Simple random walk.* Figure 1.1 illustrates one of the simplest stochastic processes. Here X_n is a random variable denoting the position at time n of a moving particle ($n = 0, 1, 2, \ldots$). Initially the particle is at the origin, $X_0 = 0$. At $n = 1$, there is a jump of one step, upwards to position 1 with probability $\frac{1}{2}$, and downwards to position -1 with probability $\frac{1}{2}$. At $n = 2$ there is a further jump, again of one step, upwards or downwards with equal probability, the jumps at times $n = 1, 2$ being independent. In general,

$$X_n = X_{n-1} + Z_n, \tag{1}$$

where Z_n, the jump at the nth step, is such that the random variables $\{Z_1, Z_2, \ldots\}$ are mutually independent and all have the distribution

$$\text{prob}(Z_n = 1) = \text{prob}(Z_n = -1) = \tfrac{1}{2} \quad (n = 1, 2, \ldots), \tag{2}$$

where $\text{prob}(A)$ denotes the probability of the event A. Equation (1), taken with the initial condition $X_0 = 0$, is equivalent to

$$X_n = Z_1 + \ldots + Z_n. \tag{3}$$

The distribution of (3) for given n is closely related to the binomial distribution. For in n trials the probability that there are j downward jumps, and hence $n - j$ upward jumps, is

$$\binom{n}{j} 2^{-n}$$

Hence

$$\text{prob}(X_n = n - 2j) = \binom{n}{j} 2^{-n}.$$

For even n, the possible positions correspond to the even integers $(-n, \ldots, -2, 0, 2, \ldots, n)$, whereas for odd n the possible positions are the odd integers $(-n, \ldots, -1, 1, \ldots, n)$. In fact, if we were concerned solely with a single value of n, we would have a typical problem in elementary probability theory. When considering the system as a stochastic process, we think of position as a random function of the discrete time variable, n.

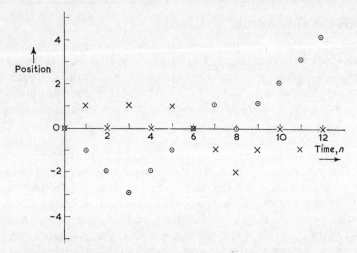

Fig. 1.1. Two independent realizations of a simple random walk.

There are very many generalizations of this process. For instance, (2) can be replaced by

$$\text{prob}(Z_n = 1) = p, \qquad \text{prob}(Z_n = -1) = q,$$
$$\text{prob}(Z_n = 0) = 1-p-q; \tag{4}$$

or, again, the steps Z_n may have many possible values. We shall use the term *random walk* for any stochastic process of the form (1) in which $\{Z_1, Z_2, \ldots\}$ are independent and identically distributed random variables. The term *simple random walk* will be used when each Z_n is restricted to the values 1, 0 and -1, so that (4) holds. The random walk in which the steps are equally likely to be one up or one down is thus a special case of the simple random walk.

We can use this example to illustrate a few important ideas and terms. First, the possible positions of the particle, corresponding to the integers, are often called the *states* of the system. Instead of saying that the particle is in position i, we then say that the state i is occupied. The set of all possible states is called the *state space*.

Next, we stress that the process is described in terms of random variables. If, by appropriately using tables of random numbers or some equivalent randomizing device, we assign particular numerical values to the random variables, we obtain a *realization* (or sample path) of the process. In fact, Fig. 1.1 shows two such independent realizations. Note that they are in detail quite different from one another. These detailed differences are, of course, typical of random phenomena.

In applications, we are often interested in the time at which a particular state, a say, is first entered. That is, for $a \neq 0$, we define N by the requirement that $X_N = a$, and that $X_r \neq a$ $(r < N)$. The random variable N is called the *first passage time* from the origin to state a. Note that in the realizations shown in Fig. 1.1, if we take $a = -2$, the observed values of N are 2 and 8.

There is another way of looking at first passage times that is often helpful. Suppose that an absorbing barrier is placed at state a. That is, the random walk continues in accordance with (1)–(3) until state a is first reached. The random walk then stops and the particle remains in state a thereafter. We now call a an *absorbing state* and the first passage time N the *time of absorption*.

Natural questions concern whether absorption is ultimately certain and the form of the probability distribution of the time, i.e. number of steps, to absorption. If there are two absorbing states, say at a and $-b$ $(a, b > 0)$, we will be interested in the probability that absorption occurs at the upper barrier, i.e. that in the unrestricted random walk the particle enters the state a before passing through the state $-b$. These problems, and others associated with random walks, are examined in Chapter 2.

Example 1.2. *A Markov chain with three states.* Consider a component, such as a valve, which is subject to failure. Let the component be inspected each day and classified as being in one of three states:

state 0, satisfactory;
state 1, unsatisfactory;
state 2, failed.

For example, we may say that the valve is unsatisfactory if, although it is still functioning, it does not pass a subsidiary test, thereby suggesting that it is likely to fail soon. We might then be interested in practice in whether it is worth replacing unsatisfactory components before they fail.

A possible very simplified model for the system is as follows. Suppose that at time n, the process is in state 0. Then let the probabilities of being, at time $n+1$, in states 0, 1, 2 be p_{00}, p_{01}, p_{02}, with $p_{00} + p_{01} + p_{02} = 1$, and let these probabilities not depend on n. Next, if the process is in state 1 at time n, let the probabilities of being, at time $n+1$, in states 0

1, 2 be 0, p_{11}, p_{12}, with $p_{11}+p_{12}=1$. That is, once the valve is unsatis-factory, it can never return to the satisfactory state. We suppose that p_{11}, p_{12} are constants independent of n and of the history of the process before n. Finally we suppose that if the process is in state 2 at time n, it is certain to be in state 2 at time $n+1$. That is, state 2, failed, is an absorbing state. The p_{ij} are called *transition probabilities* and are speci-fied by the 3×3 matrix $\mathbf{P} = (p_{ij})$. Note that the p_{ij} are probabilities taken conditionally on the state i. We can emphasize the meaning of $\mathbf{P}$ by displaying it in the following form:

$$\begin{array}{c} \text{\textit{Final state}} \\ \text{\textit{Initial}} \overbrace{\qquad\qquad} \\ \text{\textit{state}} \quad 0 \quad\; 1 \quad\; 2 \end{array}$$

$$\mathbf{P} \;=\; \begin{array}{c} 0 \\ 1 \\ 2 \end{array} \begin{bmatrix} p_{00} & p_{01} & p_{02} \\ 0 & p_{11} & p_{12} \\ 0 & 0 & 1 \end{bmatrix} \tag{5}$$

Usually, however, we omit the row and column labellings.

We have just described a special case of a type of process called a Markov chain. Chapter 3 deals with their theory.

An important special feature of the process, called the Markov property, is that the transition probabilities for the transition from time n to time $n+1$ depend on the state given to be occupied at time n and the final state at time $n+1$, but not in addition on what happened before time n. The very special nature of this property is best understood by constructing realizations of a special case. Table 1.1 shows ten such realizations for the system with

$$\mathbf{P} = \begin{bmatrix} \frac{1}{2} & \frac{2}{5} & \frac{1}{10} \\ 0 & \frac{1}{2} & \frac{1}{2} \\ 0 & 0 & 1 \end{bmatrix}; \tag{6}$$

it is assumed that at $n = 0$, the system is always in state 0. Note that as soon as state 2 is reached, the realization can be regarded as ended.

To construct a realization, we use a table of random digits. If the system is at some point in state 0, we read a single digit and the next state is determined as follows:

> digit 0–4: stay in state 0;
> digit 5–8: pass to state 1;
> digit 9: pass to state 2.

This is in accordance with the first row of (6). Similarly, if at some point the system is in state 1, the system remains in state 1 if the digit is 0–4, and moves to state 2 if the digit is 5–9. This is in accordance with the second row of (6).

Table 1.1. *Ten independent realizations of the three-state Markov chain with transition probability matrix* (6)

Realization No.	Time, n									
	0	1	2	3	4	5	6	7	8	9
1	0	0	1	1	2					
2	0	2								
3	0	0	0	0	0	0	2			
4	0	0	2							
5	0	1	1	2						
6	0	1	1	1	1	1	1	1	2	
7	0	1	1	1	2					
8	0	1	1	2						
9	0	0	0	0	0	1	1	2		
10	0	0	0	0	1	1	1	1	1	2

The meaning of the Markov property can be illustrated from Table 1.1. Note that we are in state 1 at $n = 8$ in realization 10. Then the probability that at $n = 9$ the system enters state 2 is $\frac{1}{2}$, in accordance with the second row of (6). This probability is quite unaffected by the occurrence of a previous long run of 0's and 1's, before $n = 8$. Similarly, at $n = 1$ in realization 5, state 1 is occupied. Again there is a probability $\frac{1}{2}$ that at $n = 2$, state 2 will be occupied, this probability being unaffected by the fact that state 1 has only recently been entered. The transition probabilities governing the change from time n to time $n+1$ are determined entirely by the state occupied at n. Physically, this is a very strong restriction on the process. We shall see in Chapter 3 that there is no difficulty, in principle, if we generalize to allow the transition probabilities to depend on the time, n. The essence of the Markov property is that, once the state occupied at n is given, the states occupied before n are not involved in the subsequent transition probabilities.

One aspect of the process that we shall often be particularly interested in is the distribution of time to failure, i.e. the distribution of first passage time from state 0 to state 2.

In Examples 1.1 and 1.2, the time coordinate n is restricted to the non-negative integers ($n = 0, 1, 2, \ldots$). We concern ourselves with the system at these time points only. When the system is defined just at a finite or an enumerably infinite set of time points, we say we have a problem in *discrete time*. If, on the other hand, the system is defined for a continuous range of times, for example, for all times t ($t \geqslant 0$), we have a system in *continuous time*. We now give a few examples of such processes.

Example 1.3. *The Poisson process.* The following very important stochastic process serves as a mathematical model for a wide range of empirical phenomena, including the arrival of calls at a telephone exchange, the emission of particles from a radioactive source, and the occurrence of serious coal-mining accidents. We consider point events occurring singly in time, and completely randomly in the following special sense. There is a constant ρ $(\rho > 0)$, with dimensions $[\text{time}]^{-1}$, which we call the *rate of occurrence*. Then, if $N(t, t + \Delta t)$ is the number of events in the interval $(t, t + \Delta t]$, we suppose that, as $\Delta t \to 0 +$,

$$\text{prob}\{N(t, t + \Delta t) = 0\} = 1 - \rho \Delta t + o(\Delta t), \tag{7}$$

$$\text{prob}\{N(t, t + \Delta t) = 1\} = \rho \Delta t + o(\Delta t), \tag{8}$$

so that

$$\text{prob}\{N(t, t + \Delta t) > 1\} = o(\Delta t), \tag{9}$$

where, in the usual notation, $o(\Delta t)$ denotes a function tending to zero more rapidly than Δt. Further, we suppose that $N(t, t + \Delta t)$ is independent of occurrences in $(0, t]$. That is, the probabilities (7)–(9) are unaltered, if taken conditionally on a complete specification of the occurrences in $(0, t]$. A stochastic process of point events satisfying these conditions is called a *Poisson process* of rate ρ.

The last condition is the precise condition for the randomness of the occurrences, whereas (9) is the condition that events occur singly. The constancy of ρ in (8) means that there are no time trends or other systematic variations in the rate of occurrence.

Properties of the Poisson process will be developed in Section 4.1. Two important ones are

(a) the number of events in any time interval $(t,\ t + h]$ has a Poisson distribution of mean ρh. In particular, ρ is the expected number of events per unit time;

(b) the interval from 0 up to the first event, and thereafter the intervals between successive events, are independently distributed with the exponential probability density function (p.d.f.)

$$\rho e^{-\rho x} \quad (x > 0). \tag{10}$$

To obtain realizations of the process, we could proceed directly from (7)–(9) as follows. Divide the time scale into convenient finite intervals Δt so that $\rho \Delta t \ll 1$, say so that $\rho \Delta t < 0.01$. Then, using a table of random digits, determine for each interval $(r\Delta t,\ (r + 1)\Delta t)$ whether

(a) an event occurs, with probability $\rho \Delta t$;

(b) no event occurs, with probability $1 - \rho \Delta t$.

The probability of multiple occurrences is excluded, and the choice of Δt would depend on the accuracy with which it is desired to reproduce the behaviour over very small time intervals. If (a) is realized, an event is regarded as having occurred at the centre of the time interval.

Fortunately, this rather cumbersome procedure is unnecessary, in view of (10). For let us generate quantities independently distributed with the p.d.f. (10). These quantities can be regarded as intervals between successive events and hence the realized series of events can be constructed. Quantities with the distribution (10) are easily produced from tables of random digits in view of the following easily proved result. If the random variable U is uniformly distributed over $(0,1)$, then $(-\log U)/\rho$ has the p.d.f. (10). That is, we can take sets of, say, three random digits, regarded as defining a decimal number in $(0,1)$, and a simple transformation induces the distribution (10). In fact, the Poisson process is of such importance that it is worth having a separate table of exponentially distributed quantities and a short table for $\rho = 1$ is given

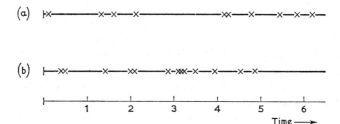

Fig. 1.2. Two independent realizations of a Poisson process.

in Appendix 1. Thus the first few values are 0·22, 2·46, 0·56, 1·01, ... Hence to construct a realization of a Poisson process with $\rho = 1$, we place events at the time points

$$0.22, 2\ 68, 3.24, 4.25, \ldots$$

If the rate ρ is different from 1, we simply multiply these values by $1/\rho$. Figure 1.2 shows two independent realizations of a Poisson process with $\rho = 2$ which are constructed in this way.

We normally call the coordinate t, the time. In the present example, t may equally be a space coordinate. For instance, we may consider the occurrence of faults along the length of a textile yarn. A further natural generalization is to take t as a vector and to consider a Poisson process in two, three or more dimensions.

The Poisson process can be specified in terms of random variables in various ways. One, already mentioned, is by specifying the joint distribution of intervals between successive events. A second is in terms of the random variables $N(0,t)$, the number of events in $(0,t]$. A realization of the process $N(0,t)$ is a step function with a jump of one at the values of t where events occur.

We shall examine in Chapter 4 the theory both of the Poisson process itself and of many processes intimately connected with the Poisson process.

Example 1.4. *A simple queueing process.* An important field of application of the theory of stochastic processes is the study of congestion, in such contexts as telephone engineering, traffic studies, and in a variety of industrial problems. One of the simplest models of a queueing process is the following. Let customers arrive at a service point in a Poisson process of rate α. Suppose that customers can be served only one at a time and that customers arriving to find the server busy queue up in order of arrival until their turn for service comes. Further, suppose that the length of time taken to serve a customer is a random variable with the exponential p.d.f.

$$\beta e^{-\beta x} \quad (x > 0). \tag{11}$$

In a generalization of the problem, the p.d.f. of service-time may be arbitrary. The exponential distribution is a reasonably good approximation for the duration of local telephone calls.

The special significance of the exponential distribution of service-time arises from its close connexion with the Poisson process, already noted in the discussion of Example 1.3. In fact the converse of (10) is that if the service of a customer is in progress at time t, and the p.d.f. of service-time is (11), then the probability that service is completed in the time $(t, t + \Delta t]$ is

$$\beta \Delta t + o(\Delta t). \tag{12}$$

This probability is independent of the behaviour of the process up to time t, and in particular of the length of time for which service of the customer has been in progress. This property holds for no other distribution of service-time.

For some purposes it is useful to define the state of the process by the number of customers in the system, including the one, if any, being served. The possible states are thus $\{0, 1, 2, \ldots\}$. Because the input is a Poisson process and the p.d.f. of service-time is exponential, the process now has the property that the probability distribution of the transition occurring in $(t, t + \Delta t]$ depends only on the state occupied at t. Thus processes of the present type have the Markov property. In fact, in the present example the probability distribution of the transition is, to the first order in Δt, the same for all states except the zero state 0. Since in the zero state there is no customer being served, transitions from the zero state must be upwards. If the distribution of service-time were not exponential, the Markov property would not hold.

We can distinguish by a simple argument between two types of behaviour of the system. The rate of arrival of customers is α and in a long time t_0 the number of customers arriving is approximately αt_0. Now the mean of the distribution (11) is $1/\beta$ so that, if the service of customers were to go on continuously, the number of customers served in time t_0 would be about βt_0. Hence, if $\alpha > \beta$, we can expect the queue of unserved customers to increase indefinitely; in a practical application the queue would increase until the laws governing the system change, for example because new customers are deterred from joining the queue.

On the other hand, if $\alpha < \beta$ the server needs to work only for a total time of about $\alpha t_0/\beta$ in order to serve the customers arriving in a long time t_0: that is, the server will be idle, i.e. the process will be in state 0, for about a proportion $1 - \alpha/\beta$ of the time. Further, it is very reasonable to expect the system to settle down to a stable statistical behaviour in which, for example, the proportions of a long time period spent in states 0, 1, 2, ... tend to limits. The system in fact has a stationary or *equilibrium probability distribution* $\{p_i\}$ $(i = 0, 1, 2,...)$ specifying equivalently

(a) the limiting proportion of a very long time period spent in state i; or

(b) the probability that the system is in state i at a particular time point a long way from the time origin, this probability being unaffected by the initial state of the process.

In many applications the equilibrium probability distribution is the aspect of the process of most interest.

The construction of realizations of this process is straightforward using the table of exponentially distributed quantities in Appendix 1. Table 1.2 gives a short section of a realization with $\alpha = 1$, $\beta = 1 \cdot 25$. The process starts empty at $t = 0$. Column 2 gives the exponentially distributed intervals between successive arrivals, from which the arrival instants are reconstructed in column 3. Column 4 gives the exponentially distributed service-times obtained by multiplying tabulated random quantities by $1/\beta = 0 \cdot 8$. From columns 3 and 4 the behaviour of the system is easily built up. A customer's queueing time is the time elapsing between arrival in the system and the start of service.

The realization illustrates well the very irregular behaviour of the system. Note particularly the run of long queueing times for customers 7–12.

The information in Table 1.2 is equivalent to a complete history of the process but does not give directly the number of customers in the system as a function of time. This is, however, easily found and is shown in Fig. 1.3.

Table 1.2. *A realization of a simple queueing process* ($\alpha = 1$, $\beta = 1.25$)

Customer No.	Interval between arrivals	Arrival time	Service time	Time service starts	Time service ends	Customer's queueing time	No. of customers in system on arrival
1	1·66	1·66	0·18	1·66	1·84	0	0
2	0·31	1·97	1·03	1·97	3·00	0	0
3	0·45	2·42	0·35	3·00	3·35	0·58	1
4	1·64	4·06	1·11	4·06	5·17	0	0
5	1·81	5·87	0·17	5·87	6·04	0	0
6	0·26	6·13	2·54	6·13	8·67	0	0
7	0·36	6·49	0·83	8·67	9·50	2·18	1
8	0·27	6·76	0·24	9·50	9·74	2·74	2
9	0·18	6·94	1·56	9·74	11·30	2·80	3
10	2·30	9·24	0·52	11·30	11·82	2·06	3
11	0·56	9·80	0·28	11·82	12·10	2·02	2
12	0·50	10·30	0·18	12·10	12·28	1·80	3
13	2·11	12·41	1·18	12·41	13·59	0	0
14	1·71	14·12	1·26	14·12	15·38	0	0
15	0·82	14·94	0·21	15·38	15·59	0·44	1
.	.	.	.	.	.	.	.
.	.	.	.	.	.	.	.
.	.	.	.	.	.	.	.

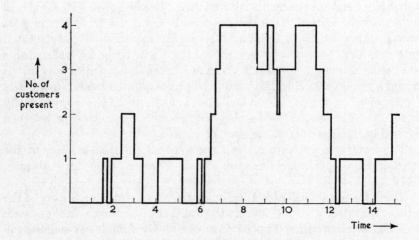

Fig. 1.3. Number of customers in system for realization of Table 1.2.

A few examples of properties of the process which may be of interest are:

(a) the equilibrium distribution, already described, of the number of customers in the system;

(b) the equilibrium probability distribution of customers' queueing time;

(c) the probability distribution of the length of a server's busy period and of the number of customers served during a busy period. A busy period is defined to start when a customer arrives to find the system empty and to end the next time a customer departs to leave the system empty. Thus in Table 1.2 there is a busy period during which customer 1 is served, another in which customers 2, 3 are served, and so on;

(d) the distribution of the time taken for the system first to become empty starting from a given congested state, i.e. the distribution of first passage time from a given state to state 0.

This example can be used to illustrate one further important idea. Suppose that we consider the process only at those time points where a change of state occurs, i.e. where a customer arrives or where the service of a customer is completed. These are the points of discontinuity in Fig. 1.3. We can label these time points $n = 1, 2, \ldots$ and think of n as a new discrete time variable. (Of course, equal 'time' intervals in terms of n correspond to unequal and random time intervals in terms of t.) We

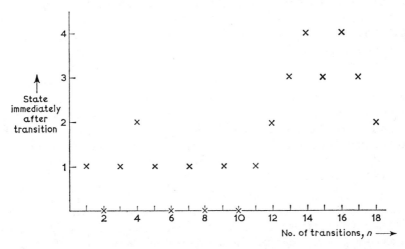

Fig. 1.4. Realization of imbedded process obtained by considering state immediately following a transition.

can define a new stochastic process in discrete time by defining the state of the process at time n to be that immediately following the nth transition in the old process. Figure 1.4 shows the realization of Table 1.2 and Fig. 1.3 represented in this new way. The new process in discrete time is said to be *imbedded* in the process in continuous time.

It is quite easy to show that if, at time n, the system is in a non-zero state, there is a probability $\alpha/(\alpha+\beta)$ that the next step will be one upwards and a probability $\beta/(\alpha+\beta)$ that it will be one downwards. All steps are independent. From the zero state, the next jump must be one upwards. Thus, except when the zero state is occupied, the imbedded process is a simple random walk (Example 1.1).

Other imbedded processes can be defined by considering the state of the system immediately before an arrival or immediately after the departure of a customer.

Examples 1.1–1.4 have been discussed in detail because they illustrate some important ideas in terms of very simple examples. The following further examples are treated more briefly.

Example 1.5. *Random walk.* In the simple random walk of Example 1.1 the state space is discrete. If, however, the individual steps are continuously distributed, the state space is continuous, being part or all of the real axis. A further generalization is to consider random walks occurring in two, three or more dimensions, when the state space is multidimensional.

Example 1.6. *A stationary process.* Figure 1.5 shows the variation of weight per unit length along a length of wool yarn. It is found empirically that the statistical properties of the variation are nearly constant over very long lengths of yarn.

This type of variation can be regarded as a realization of a *stationary* stochastic process, in this case in continuous time and with a continuous state space. The important property of stationarity will be defined precisely in Chapter 7. Roughly, however, it is that the probabilistic structure of different sections of the process is the same. Thus the probability distribution of thickness is the same at all times and the joint distribution of the thickness at pairs of times distant h apart is always the same, etc.

In applications of stochastic process, two distinct steps arise. The first step is, given some system in science or technology, to set up a formal stochastic process that will be a useful representation of the system. This is the step of *model building*. Almost inevitably some simplification of the real problem will be necessary. The second step is, given a precisely formulated stochastic process, to derive the properties of interest. This can be done either by simulation or by mathematical analysis.

In simulation, realizations of the system are usually constructed by using hand calculation in simple cases and an electronic computer for larger-scale studies. The advantage of simulation is that it can deal in a fairly routine way with problems that are too complex for mathematical analysis. The disadvantage is that very many simulations under a wide range of conditions may be necessary before general understanding of the system is obtained.

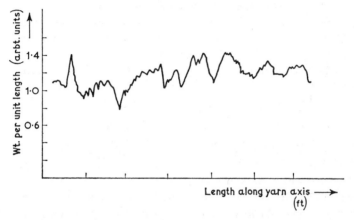

Fig. 1.5. Variation of weight per unit length along a length of wool yarn.

The present book is concerned very largely with the mathematical analysis of stochastic processes. As we shall see, useful and detailed mathematical results can ordinarily be obtained only for quite simple processes. A knowledge of the types of process that can be solved in simple terms is, of course, of great value in model building, isolated remarks about which occur throughout the book. In major investigations it will be profitable to combine mathematical analysis of simple models with simulation of more complex models.

1.2. Specification of stochastic processes

We shall in this book approach the theory very largely through the discussion of particular processes. The remaining sections of this chapter, are, however, intended to give a broad review of some of the more general definitions and ideas. Those completely unfamiliar with the subject should omit these sections on first reading.

First we specify the set of 'time' points for which the process is defined. This will usually be either the set of points with integer co-ordinates ($n = 0, 1, 2, \ldots$; or $n = 0, \pm 1, \pm 2, \ldots$) or an interval on the real axis, usually $(0 \leqslant t;$ or $-\infty < t < \infty)$. These two main cases will be referred to as *discrete time* and *continuous time*, respectively. More generally, we have a problem in discrete time if the set of time points is finite or enumerably infinite. Occasionally, as in Example 1.3, we can consider processes in multidimensional 'time'.

We shall not attempt here a formal general mathematical definition of a stochastic process. The usual situation, though, is to have a collection of random variables $\{X_n\}$ or $\{X(t)\}$, defined for all relevant values of n (discrete time) or t (continuous time).

The *state space* of the process is the set of possible values of an individual X_n or $X(t)$. The state space can be classified first as one-dimensional or multidimensional, simple processes usually being one-dimensional. Secondly, the state space is *discrete* if it contains a finite or enumerable infinity of points, and otherwise is *continuous*.

For one-dimensional processes, we can thus distinguish four cases, as follows:

discrete time, discrete state space: Examples 1.1, 1.2;
discrete time, continuous state space: Example 1.5;
continuous time, discrete state space: Example 1.4;
continuous time, continuous state space: Example 1.6.

In applications, we usually start from the way the process is built up, for example from the steps of the random walk (Example 1.1), the transitions of the Markov chain (Example 1.2) and the arrival instants and service-times of the customers in the queue (Example 1.4). The problem is to deduce various properties of the process from the initial specification.

The following are examples of properties which may be required. For simplicity, we deal with univariate processes in discrete time.

First, we may require, for an arbitrary n, the probability distribution of X_n. In complex cases we may have to be content with the first few moments of X_n and in particular with the mean and variance of X_n, denoted by $E(X_n)$ and $V(X_n)$ respectively. They will, in general, be functions of n. If the distribution of X_n is known to be normal, $E(X_n)$ and $V(X_n)$ determined the distribution. In random walks, defined by (1) or (3) with the Z_i independently and identically distributed, we have immediately that

$$E(X_n) = nE(Z_i), \qquad V(X_n) = nV(Z_i). \tag{13}$$

The next possibility is to consider for arbitrary m, n the joint distribution of the random variables X_m, X_n. If we are working only with first and second moments we consider, together with means and variances, the covariance of X_m, X_n, denoted by $C(X_m, X_n)$. This will be a function of m, n, called the *covariance function*.

More generally, we may consider the process at k arbitrary time points m, n, ... and examine the joint distribution of X_m, X_n, ... If all such distributions, for all k, are multivariate normal, the process is called *Gaussian*. Such processes are completely characterized by $E(X_n)$, $V(X_n)$, and the covariance function.

It can be shown that a stochastic process is in principle completely determined if we know for every k and every choice of m, n, ... the relevant joint distribution. This is the basis for a general definition of a stochastic process. In applications, however, we rarely work directly with the joint distribution except for rather small values of k.

Another important group of properties, already mentioned in Section 1.1, concern first passage times and related properties. These involve, instead of the random state occupied at a fixed time, the random time taken to reach a fixed state. Suppose that at $n = 0$ the system is in a given state, i say. Let A denote a set of states, possibly consisting of just one state, but not including i. Let the random variable N, or more explicitly N_{iA}, be defined as the first time at which the system enters the set of states A. It is called the *first passage time* from i to A. We may be interested in the probability distribution of N_{iA} and often especially in $\text{prob}(N_{iA} < \infty)$, the probability that the set A is ever entered. Problems about first passage times are often solved by defining a new stochastic process in which the states in the set A are absorbing; that is when a state in the set A is entered, the process remains in that state thereafter. Then, if X'_n denotes a random variable in the new process,

$$\text{prob}(N_{iA} \leqslant n) = \text{prob}(X'_n \in A) \cdot \qquad (14)$$

and, once the right-hand side has been found for each fixed n, properties of N_{iA} are easily derived. Examples of this argument occur in Sections 2.2 and 5.4.

A related idea, which is dealt with more precisely later, is of *recurrence time*, the first time at which the system is again in its initial state.

Examples of other properties of the process that might be of interest in particular applications are the largest state occupied by the process over a given time, the range of states (i.e. difference between largest and smallest states), the length of time spent in a specified state, etc. All these are random variables whose probability distribution may be required.

The general situation is that for each possible sample function, there is defined one or more new variables. Mathematically these are functionals on the space of sample functions. The object of the theory is to calculate the probability distribution of these variables from the basic properties of the stochastic process.

To return to the specification in terms of the joint distribution of the states occupied at k different times, suppose that $k = 3$ and that $l < m < n$

are three time points, X_l, X_m, X_n being the corresponding random variables. We describe a distribution by the probability density function (p.d.f.), it being understood that if the states are discrete the probability densities are ordinary probabilities. When, in later chapters, we discuss processes of particular types, it will be simpler to use rather different notations for the different types of process. For work with general processes, however, we denote the marginal p.d.f. of X_n by $p_{X_n}(x)$. Thus with a discrete state space

$$p_{X_n}(x) = \text{prob}(X_n = x),$$

and with a continuous state space

$$p_{X_n}(x) = \lim_{\Delta x \to 0+} \frac{\text{prob}(x < X_n < x + \Delta x)}{\Delta x}.$$

We can define various conditional p.d.f.'s. Thus $p_{X_n}(x|X_m = y)$ is the conditional p.d.f. at time n, given the state occupied at time m. Similarly we can define $p_{X_n}(x|X_m = y, X_l = z)$. Then, because one and only one of the possible values of X_m must occur, we have that

$$p_{X_n}(x|X_l = z) = \int_{-\infty}^{\infty} p_{X_m}(y|X_l = z)p_{X_n}(x|X_m = y, X_l = z)\,dy \quad (15)$$

where the integral, or sum, is taken over all possible states. This is an important equation. It is a version, in a possibly unfamiliar notation, of the following result in elementary probability theory:

$$\text{prob}(B) = \sum_i \text{prob}(B|C_i)\text{prob}(C_i),$$

where B is an arbitrary event and $\{C_i\}$ are a set of mutually exclusive and exhaustive events.

1.3. Markov processes

A large part of this book is concerned with the special class of processes called Markov processes and we have already discussed the Markov property informally in connexion with Examples 1.2–1.4. The general definition can be expressed in terms of the conditional p.d.f.'s of the previous section. The stochastic process is a Markov process if for arbitrary times $\ldots < l < m < n$,

$$p_{X_n}(x|X_m = y, X_l = z, \ldots) = p_{X_n}(x|X_m = y). \quad (16)$$

The right-hand side may depend on m and n. The essential point is that $X_m = y$, the condition on the right-hand side of (16), is the one referring to the *latest* of the time points $(m, l, \ldots)$. In a Markov process we need never consider p.d.f.'s conditional on values at two or more time points; they can always be reduced to the form of the right-hand side of (16).

Because of (16), equation (15) can be written for Markov processes in the form

$$p_{X_n}(x|X_l = z) = \int_{-\infty}^{\infty} p_{X_m}(y|X_l = z) p_{X_n}(x|X_m = y)\, dy, \qquad (17)$$

where $l < m < n$. This is called the Chapman–Kolmogorov equation. Its importance is that it enables us to build up the conditional p.d.f.'s over the 'long' time interval (l, n) from those over the 'short' time intervals (m, n) and (l, m).

Now equation (16), although it expresses the fundamental property of Markov processes in a natural form, is often too general a formulation to be directly useful in applications. Indeed we can, in discrete time, take $n = m + 1$ in (16), obtaining the special case that

$$p_{X_{m+1}}(x|X_m = y, X_l = z, \ldots) = p_{X_{m+1}}(x|X_m = y). \qquad (18)$$

The right-hand side of (18) is the conditional p.d.f. that one step of the process takes us from a given point y to x; it is called a one-step transition p.d.f. In special Markov processes arising in applications, it is usually this one-step transition p.d.f. that is known, and the Markov property (16) is given to hold only when $n = m + 1$, i.e. in the form (18). It is, however, easily deduced from (18) and a generalized form of (15) that (16) holds for all $n > m$. That is, the Markov property in discrete time can, without loss of generality, be taken in the restricted form (18).

The corresponding general definition in continuous time requires only the replacement of $\ldots, X_l, X_m, X_n$ by $X(t), X(u), X(v)$ for $t < u < v$. That is for $t < u < v$

$$p_{X(v)}\{x|X(t) = z\} = \int_{-\infty}^{\infty} p_{X(u)}\{y|X(t) = z\} p_{X(v)}\{x|X(u) = y\}\, dy. \qquad (19)$$

Just as in discrete time we are particularly concerned with $n = m + 1$, so in continuous time we are particularly concerned with the limiting cases in which $v = u + \Delta u$, and $\Delta u \to 0 +\cdot$ or, slightly less commonly, $u = t + \Delta t$, $\Delta t \to 0 +$. In the first case, the Markov condition (16) asserts that the probability distribution of the transition occurring in a small time interval $(u, u + \Delta u)$ depends only on the state occupied at u and not further on the states occupied before u.

We stress that the formulae of this section, and in particular the Chapman–Kolmogorov equation (17), will rarely be used in their general form. Nevertheless, the general idea of (17), that transition probabilities over long time intervals can be built up recursively from transition probabilities over shorter time intervals, is of fundamental importance, and will recur repeatedly, especially in Chapters 3–5.

Whether a particular system leads to a Markov process depends on how the random variables specifying the stochastic process are defined.

In fact the main method for dealing with a non-Markov process is to redefine the state space so that the Markov property does hold. This idea, and the general nature of the Markov property, may be clarified by taking two examples of non-Markov processes.

Example 1.7. *Modified model of failure.* In Example 1.2, a three-state model of failure was used to illustrate the Markov property in a very simple form. Suppose now that the assumptions are modified, so that if state 1 (unsatisfactory) is entered, the system remains there for exactly two time periods before passing to state 2. Let X_n denote the state occupied at time n. Then for the new process

$$\text{prob}(X_{n+1} = 1 | X_n = 1, \ X_{n-1} = 1) = 0,$$
$$\text{prob}(X_{n+1} = 1 | X_n = 1, \ X_{n-1} = 0) = 1. \tag{20}$$

Thus the Markov property (18) does not hold. The probability of a transition out of state 1 is not uniquely determined by the fact that state 1 is currently occupied.

A simple extension of the state space, however, converts the process into a Markov process. Suppose that we divide the original state 1 into two states, $(1,0)$ and $(1,1)$, where $(1,0)$ is the state corresponding to $X_n = 1, X_{n-1} = 0$ and $(1,1)$ the state corresponding to $X_n = X_{n-1} = 1$. The new process with four states is easily shown to be a Markov chain and its matrix of transition probabilities is, in the same notation as (5),

$$
\begin{array}{c}
\textit{Initial} \\
\textit{state} \\
\begin{array}{c}
0 \\
(1,0) \\
(1,1) \\
2
\end{array}
\end{array}
\overset{\textstyle \textit{Final state}}{
\begin{array}{cccc}
0 & (1,0) & (1,1) & 2 \\
\left[\begin{array}{cccc}
p_{00} & p_{01} & 0 & p_{02} \\
0 & 0 & 1 & 0 \\
0 & 0 & 0 & 1 \\
0 & 0 & 0 & 1
\end{array}\right]
\end{array}}. \tag{21}
$$

This example is typical of a very important set of initially non-Markov processes in which the transition probabilities depend on how long the current state has been occupied. One important general method for dealing with such processes is the one illustrated in (21); namely, the state space is augmented sufficiently to ensure that the current state determines the transition probabilities.

Example 1.8. *Self-avoiding random walk.* The following process is much more drastically non-Markovian than the process of Example 1.7. Let the state space be the lattice in the plane, i.e. the set of points (i, j), where i, j are integers. A particle starts at the origin. Suppose that at time n, the particle is at (i, j). At time $n + 1$, the particle moves to one of the points $(i, j + 1)$, $(i + 1, j)$, $(i - 1, j)$, $(i, j - 1)$ with the provisos that

(a) no point passed through at time $0, 1, \ldots, n-1$ may be revisited;
(b) each remaining point of the four is chosen with equal probability;
(c) if all four points have been visited before, the walk ends.

Thus the possible transitions from time n to time $n + 1$ may be affected by the states occupied at all previous times. Hence, not only do we have a non-Markov process, but also it cannot be converted into a Markov one by a simple extension of the state space.

Bibliographic Notes‡

The topics introduced in this chapter are nearly all dealt with in more detail in later chapters and references will be given there. Methods for simulation are discussed systematically in the books of Tocher (1963) and Hammersley and Handscomb (1964) and in the symposium edited by Meyer (1956). Bartlett (1953) has given a brief and elementary introduction to a wide range of applications of the theory of stochastic processes. Neyman (1960) has discussed, with examples, the general role of stochastic processes in science. Cramér (1964) has made a broad survey of model building with stochastic processes. A very useful collection of exercises on stochastic processes has been published by Takács (1960). In the present book we deal entirely with the theory of special types of stochastic process. A general definition and theory of stochastic processes involves measure–theoretic ideas beyond the scope of this book. For an introduction to the general theory see the books of Rosenblatt (1962) and Pitt (1963), and for a full account that of Doob (1953).

Exercises

1. For the special random walk defined by (2) and (3), show that the probability that the particle after $2m$ steps is again at the origin is equal to

$$\binom{2m}{m} 2^{-2m} \sim \frac{1}{\sqrt{(\pi m)}} \quad \text{as } m \to \infty.$$

Prove further that if the steps are either one up or one down, with probabilities p and $1 - p$, then if $p \neq \frac{1}{2}$, the corresponding probability is exponentially small as $m \to \infty$.

2. If X_n is the position after n steps in the random walk of Exercise 1, show that $E(X_n) = n(2p - 1)$, $V(X_n) = 4np(1 - p)$. Noting that by the central limit theorem X_n is asymptotically normally distributed, draw

‡ The references for the whole book are collected in Appendix 2.

rough sketches to show the distribution of X_n for large n, when (a) $p = \frac{1}{2}$, (b) $p > \frac{1}{2}$, (c) $p < \frac{1}{2}$. Use the sketches to explain qualitatively the results of Exercise 1.

3. Suppose that in the random walk of Exercises 1 and 2 there is an absorbing barrier at a $(a > 0)$ and let N_a be the time of absorption. Prove that

$$\text{prob}(N_a \leqslant n) \geqslant \text{prob}(X_n \geqslant a),$$

where the probability on the right refers to the unrestricted random walk. Hence, using Exercises 1 and 2, show that ultimate absorption is certain if $p > \frac{1}{2}$.

4. A random walk with steps distributed according to (4) starts at 0. Show that the random walk plus the following boundary conditions form Markov chains and write down the matrices of transition probabilities:

   (a) unrestricted process,
   (b) absorbing barriers at 2 and -2,
   (c) barriers at 2 and -2 such that if the system reaches a barrier at time n, the system is at time $n+1$ equally likely to occupy the states 1, 0, -1.

5. In the Markov chain of Example 1.2, state 0 is occupied at $n = 0$. Let p_{00} be the conditional probability, given that the state 0 is occupied at time n, that the same state is occupied at time $n+1$. Prove that Z, the time at which the system is first not in state 0, has a geometric distribution, with $\text{prob}(Z = r) = p_{00}^{r-1} (1 - p_{00})$. Find $E(Z)$ and $V(Z)$. Prove further that given that a transition out of state 0 occurs, the probability is $p_{01}/(p_{01} + p_{02})$ that the transition is to state 1.

6. Form twenty realizations of length 10 of the Markov chain with two states and transition probability matrix

$$\begin{bmatrix} \frac{3}{5} & \frac{2}{5} \\ \frac{2}{5} & \frac{3}{5} \end{bmatrix},$$

starting each realization from the same state. Is an equilibrium probability distribution, with the properties explained in Example 1.4, to be expected? If so, what do symmetry considerations indicate this distribution to be and to what extent are your realizations consistent with it?

7. Give two proofs, along the following lines, that in a Poisson process defined by (7)–(9), the interval X from 0 to the first event has p.d.f. $\rho e^{-\rho x}$:

   (a) Divide the interval $(0, x)$ into n intervals of length Δ, where $n\Delta = x$. Write down the probability that there are no events in the first $(n-1)$ intervals and an event in the last interval; take a limit as $n \to \infty$.

(b) Let $\Phi(x) = \mathrm{prob}(X > x)$. Show that

$$\Phi(x+\Delta x) = \Phi(x)(1-\rho\Delta x)+o(\Delta x)$$

and hence that $\Phi(x) = e^{-\rho x}$.

8. Prove, by the type of argument used in Exercise 7(a), that in a Poisson process the number of events in an interval of length h has a Poisson distribution of mean ρh.

9. Suppose that in the queueing system of Example 1.4 a customer is being served. Prove that the probability that his service is completed before a new arrival is $\beta/(\alpha+\beta)$.

CHAPTER TWO

The Random Walk

2.1. Introduction

The simple random walk was described briefly in Example 1.1. In the present section we introduce the general random walk and give some examples. Section 2.2 is devoted to a detailed discussion of the simple random walk and the material is referred to again in Chapter 3 in the larger context of Markov chains. It is therefore recommended that as a minimum Section 2.2 be studied before proceeding to Chapter 3, although not all detailed results are required. In Section 2.3 the general random walk is treated in detail and this section is not essential for a first reading. Some results, however, will be required in Sections 5.2 and 5.3.

Suppose that a particle is initially at a given point X_0 on the x-axis. At time $n = 1$ the particle undergoes a step or jump Z_1, where Z_1 is a random variable having a given distribution. At time $n = 2$ the particle undergoes a jump Z_2, where Z_2 is independent of Z_1 and with the same distribution, and so on. Thus the particle moves along a straight line and after one jump is at the position $X_0 + Z_1$, after two jumps at $X_0 + Z_1 + Z_2$ and, in general, after n jumps, i.e. at time n, the position of the particle is given by

$$X_n = X_0 + Z_1 + Z_2 + \ldots + Z_n, \tag{1}$$

where $\{Z_i\}$ is a sequence of mutually independent, identically distributed random variables. We say that the particle undergoes a general one-dimensional random walk or more briefly a random walk. Alternatively we may write (1) as

$$X_n = X_{n-1} + Z_n \quad (n = 1, 2, \ldots). \tag{2}$$

In the particular case where the steps Z_i can only take the values 1, 0 or -1 with the distribution

$$\text{prob}(Z_i = 1) = p, \quad \text{prob}(Z_i = 0) = 1 - p - q, \quad \text{prob}(Z_i = -1) = q,$$

we shall call the process a *simple random walk*. It is usual in the literature to define a simple random walk as one for which each jump is either $+1$ or -1 with $p + q = 1$. However, we shall assume that $p + q \leqslant 1$ with $1 - p - q$ as the probability of a zero jump. This extra generality is achieved with hardly any cost in analytical complication.

According to our definition, the random walk is a stochastic process in discrete time. The state space will be continuous if the steps $\{Z_i\}$ are continuous random variables and discrete if the steps are restricted to

22

integral values as in the simple random walk. Anticipating the more general terminology of stochastic processes we shall sometimes refer to the position of the particle as its state, e.g. if $X_n = x$ we say that the particle or process is in state x at time n. For simplicity we shall refer to the positive direction as *up* and the negative direction as *down*.

It should be emphasized that it is simply a matter of convenience to use the language and ideas associated with the motion of a particle. While the theory is of course applicable to the study of physical particles undergoing random motion of the type described above, it is also true that several other stochastic processes and physical systems, when viewed appropriately, can be seen to be equivalent to a random walk in all but terminology.

If the particle continues to move indefinitely according to equation (2) the random walk is said to be *unrestricted*. Frequently, however, we consider the motion of the particle to be restricted in some way, usually by the presence of *barriers*. For example, a random walk starting at $X_0 = 0$ may be restricted to within a distance a up and b down from origin in such a way that when the particle reaches or overreaches *either* of the points a or $-b$, the motion ceases. The points a and $-b$ are called *absorbing barriers* in this case and the states $x \geqslant a$ and $x \leqslant -b$ are all absorbing states. Other types of behaviour, such as reflection, are possible at a given barrier and we shall define these when we deal with them.

The following are some examples of stochastic processes arising in dissimilar physical situations which may be represented as random walks.

Example 2.1. Insurance risk. Consider an insurance company which starts at period 0 with a fixed capital, X_0. During periods 1, 2, ... it receives sum $Y_1, Y_2, ...$ in the form of premium and other income while it pays out sums $W_1, W_2, ...$ in the form of claims. Thus at time n, its capital is

$$X_n = X_0 + (Y_1 - W_1) + ... + (Y_n - W_n) \qquad (3)$$

and if at any time n, $X_n < 0$, then the company is ruined and cannot continue its operations. If we assume, perhaps unrealistically, that $\{Y_r\}$ and $\{W_r\}$ are two sequences of mutually independent and identically distributed random variables, then the capital X_n behaves like a random walk starting at X_0 and with jumps $Z_r = Y_r - W_r$ ($r = 1, 2, ...$). In addition, we have an absorbing barrier at the origin and so we may describe the process as a random walk on the half-line $(0, \infty)$ with an absorbing barrier at 0. The equations defining the process are for $n = 1, 2, ...$

$$X_n = \begin{array}{l} X_{n-1} + Z_n \quad (X_{n-1} > 0, \; X_{n-1} + Z_n > 0), \\ 0 \quad \text{(otherwise)}. \end{array}$$

2

A problem of interest here is to determine the probability of ruin for given X_0 and given statistical behaviour of claims and income.

Example 2.2. The content of a dam. Let X_n (in suitable units) represent the amount of water in a dam at the end of n time units; for the sake of argument we call the time unit a day. Suppose that during day r, Y_r units flow into the dam in the form of rainfall and supply from rivers, where Y_r has a statistical distribution over days. Water is released from the dam according to the following rule, by opening release gates at the beginning of each day. If the content at the end of day $r-1$ added to the inflow on day r exceeds a given quantity α then α units are released during day r; otherwise the dam becomes drained by the end of day r. If b represents the capacity of the dam, it will be seen that if $X_{r-1} + Y_r - \alpha > b$ then an overflow, of amount $(X_{r-1} + Y_r - \alpha) - b$, occurs on day r and such an overflow is presumed to be lost. The situation may be represented as follows:

$$X_{n-1} + Z_n \quad (0 < X_{n-1} + Z_n < b), \tag{4}$$
$$X_n = 0 \qquad (X_{n-1} + Z_n \leqslant 0), \tag{5}$$
$$b \qquad (X_{n-1} + Z_n \geqslant b), \tag{6}$$

where $Z_r = Y_r - \alpha$ is the change in the content of the dam on day r, provided such a change does not empty or fill the dam. If the dam is full it remains full until the first negative Z_r, i.e. until the first subsequent day when the amount released exceeds the inflow, while if empty it remains so until the first positive Z_r. If $\{Z_r\}$ is a sequence of mutually independent and identically distributed random variables then we may describe the process X_n as a random walk on the interval $[0,b]$ in the presence of *reflecting* barriers at 0 and b. A reflecting barrier is defined as a state which, when crossed in a given direction, say downwards, holds the particle until a positive jump occurs and allows the particle to move up and resume the random walk. For such a process some problems of interest are

(a) the determination of the long run or equilibrium probability distribution of X_n;

(b) the finding of the probability distribution of empty periods and of non-empty (wet) periods, i.e. periods during which a particle remains continuously on or continuously off a barrier.

The states of the system are the real numbers in $[0,b]$.

Example 2.3. Gambler's ruin. The gambler's ruin problem affords a classical illustration of the simple random walk with absorbing barriers. Consider two gamblers A and B who start off with a and b units of capital respectively. The game consists of a sequence of independent turns and at each turn A wins one unit of B's capital with probability q or B wins one unit of A's capital with probability p $(p+q=1)$. Let X_n denote B's cumulative gain at the end of n turns. Then provided $-b < X_n < a$ we have $X_n = Z_1 + \ldots + Z_n$, where the Z_i are mutually independent with the distribution

$$\mathrm{prob}(Z_i = 1) = p, \quad \mathrm{prob}(Z_i = -1) = q = 1-p.$$

If at any stage $X_n = a$, then B has gained all A's capital and A is ruined, while if $X_n = -b$, B is ruined. Thus X_n is a simple random walk starting at the origin with absorbing barriers at $-b$ and a. One problem of interest is to find the probability of ruin of each of the contestants, i.e. of absorption at each of the barriers a and $-b$, for given p, q, a and b.

Example 2.4. *The escape of comets from the solar system.* Kendall (1961a, b, c) has made an interesting application of the random walk to the theory of comets. We mention briefly one aspect of the theory. During one revolution round the earth the energy of a comet undergoes a change brought about by the disposition of the planets. In successive revolutions the changes in the energy of the comet are assumed to be independent and identically distributed random variables Z_1, Z_2, …. If initially the comet has positive energy X_0 then after n revolutions the energy will be

$$X_n = X_0 + Z_1 + \ldots + Z_n.$$

If at any stage the energy X_n becomes zero or negative, the comet escapes from the solar system. Thus the energy level of the comet undergoes a random walk starting at $X_0 > 0$ with an absorbing barrier at 0. Absorption corresponds to escape from the solar system. Problems of interest are to determine the probability of escape and the distribution of the time to escape. For further details the reader is referred to the above-mentioned papers.

It should be noted that the random walk as described in this chapter is a particular kind of Markov process of which more general kinds will be dealt with in later chapters. We may also observe that from one point of view an unrestricted random walk is a sum of independent random variables and we may thus expect an important role to be played by results in probability theory concerning sums of independent random variables, such as the laws of large numbers and the central limit theorem; see, for example, Parzen (1960).

2.2. The simple random walk

We now consider in more detail the simple random walk with independent jumps Z_1, Z_2, …, where for $i = 1, 2, \ldots$,

$$\mathrm{prob}(Z_i = 1) = p, \; \mathrm{prob}(Z_i = -1) = q, \; \mathrm{prob}(Z_i = 0) = 1-p-q. \quad (7)$$

(i) UNRESTRICTED

Suppose that the random walk starts at the origin and that the particle is free to move indefinitely in either direction. Then we have

$$X_n = \sum_{r=1}^{n} Z_r.$$

The possible positions of the particle at time n are $k = 0, \pm 1, \dots, \pm n$. In order to reach the point k at time n the particle has to make r_1 positive jumps, r_2 negative jumps and r_3 zero jumps where r_1, r_2, r_3 may be any non-negative integers satisfying the simultaneous equalities

$$r_1 - r_2 = k, \qquad r_3 = n - r_1 - r_2. \tag{8}$$

Hence the probability that $X_r = k$ is given by the summation of multinomial probabilities,

$$\text{prob}(X_n = k) = \sum \frac{n!}{r_1! \, r_2! \, r_3!} p^{r_1} (1 - p - q)^{r_3} \, q^{r_2}$$

over values of r_1, r_2 and r_3 satisfying (8). It should be noted that if $p + q = 1$ then $\text{prob}(X_n = k)$ vanishes for odd k when n is even and for even k when n is odd. The probability generating function (p.g.f.) of the jump Z_r is

$$G(z) = E(z^{Z_r}) = pz + (1 - p - q) + qz^{-1}$$

and hence that of X_n is

$$E(z^{X_n}) = \{G(z)\}^n.$$

Since $X_0 = 0$ we define $G_0(z) = 1$ and introduce a generating function

$$G(z, s) = \sum_{n=0}^{\infty} s^n \{G(z)\}^n = \frac{1}{1 - sG(z)} \qquad (|sG(z)| < 1)$$

$$= \frac{z}{-spz^2 + z\{1 - s(1 - p - q)\} - sq}.$$

Then $G(z, s)$ contains all the information about the process in the sense that $\text{prob}(X_n = k)$ is the coefficient of $z^k s^n$ in $G(z, s)$.

Let μ and σ^2 denote the mean and variance of a jump. Then $\mu = p - q$ and $\sigma^2 = p + q - (p - q)^2$ and hence

$$E(X_n) = n\mu, \qquad V(X_n) = n\sigma^2,$$

where $V(X)$ denotes the variance of the random variable X. It is of interest to calculate the probability that at time n the particle is found in one of the states $j, j+1, \dots, k$, where j and k are possible values of X_n ($j < k$). This may involve an inconvenient summation of a large number of multinomial probabilities and so we can resort to an approximation provided by the central limit theorem, according to which X_n

will be approximately normally distributed with mean $n\mu$ and variance $n\sigma^2$ for large n. Thus

$$\text{prob}(j \leqslant X_n \leqslant k) \simeq (2\pi\sigma^2 n)^{-\frac{1}{2}} \int_j^k \exp\left\{-\frac{(x-n\mu)^2}{2n\sigma^2}\right\} dx.$$

A still better approximation is obtained by employing a continuity correction, i.e. by using $j-c$ and $k+c$ as the limits of integration, where $c = \frac{1}{2}$ or $c = 1$ according as $p+q < 1$ or $p+q = 1$. We then have, after transforming the integral to standard form

$$\text{prob}(j \leqslant X_n \leqslant k) \simeq \Phi\left(\frac{k+c-n\mu}{\sigma\sqrt{n}}\right) - \Phi\left(\frac{j-c-n\mu}{\sigma\sqrt{n}}\right), \tag{9}$$

where

$$\Phi(y) = \frac{1}{\sqrt{(2\pi)}} \int_{-\infty}^y e^{-\frac{1}{2}x^2} dx$$

is the standard normal distribution function and is well tabulated.

We can regard (9) and subsequent similar results as asymptotic equivalences if, as $n \to \infty$, $j - \mu n \sim \alpha\sqrt{n}$, $k - \mu n \sim \beta\sqrt{n}$, for some fixed α, β $(\alpha < \beta)$.

Suppose now that $p > q$, i.e. the probability of a jump upwards is greater than that of a jump downwards, and let us investigate where the particle is likely to be found after a large number n of jumps. The mean jump μ is thus positive and again according to the central limit theorem X_n will, with high probability, be within say three standard deviations of its expected value. That is, $\text{prob}(n\mu - 3\sigma\sqrt{n} < X_n < n\mu + 3\sigma\sqrt{n})$ is nearly 1 and we can say that

$$X_n = n\mu + O(\sqrt{n}) = n\mu\{1 + O(n^{-\frac{1}{2}})\}$$

with high probability. Since $\mu > 0$ the probability that X_n is arbitrarily large becomes arbitrarily near one as n increases. More precisely, it can be deduced from (9) or from the weak law of large numbers that

$$\text{prob}(X_n > j) \to 1 \quad \text{as } n \to \infty$$

for any j. By using the strong law of large numbers we can make a stronger statement. For, according to the strong law we can say that for any $\epsilon > 0$ the probability that the particle remains in the region $n(\mu - \epsilon) < X_n < n(\mu + \epsilon)$ for all $n > n_0$ can be made as close to 1 as we please by choosing n_0 sufficiently large. Hence for any j we have that

$$\text{prob}(X_n > j, X_{n+1} > j, \dots) \to 1 \quad \text{as } n \to \infty,$$

so that with probability one the particle drifts off to $+\infty$. Similarly, if $p < q$ the particle drifts off to $-\infty$ with probability one. But if $p = q$ then the jumps are equally likely to be up or down and $\mu = 0$. The central limit theorem tells us that the particle will be within a distance of order $\sqrt{n}$ from its starting point after n jumps. This does not exclude the possibility of large excursions from the origin and in (iii) the singular fact will be shown that the particle is certain to return to the origin but is also certain to make arbitrarily large excursions.

Example 2.5. A numerical example. We examine some of the above results in particular numerical instances. Suppose we take the cases

(a) $p = 0.6, \quad q = 0.2, \quad 1-p-q = 0.2;$

(b) $p = 0.4, \quad q = 0.3, \quad 1-p-q = 0.3;$

(c) $p = 0.4, \quad q = 0.4, \quad 1-p-q = 0.2.$

In (a) the particle takes on average three times as many positive as negative steps so that there is a strong positive drift. We use the normal approximation (9) to examine the relation between n, j and $\text{prob}(X_n \geqslant j)$. In this case (9) becomes

$$\text{prob}(X_n \geqslant j) \simeq 1 - \Phi\left(\frac{j - \frac{1}{2} - n\mu}{\sigma\sqrt{n}}\right),$$

where $\mu = p - q = 0.4$ and $\sigma^2 = p + q - (p-q)^2 = 0.64$.

Thus for example after exactly 100 steps the probability is about 0.99 that the particle is 24 or more units up from the origin, i.e.

$$\text{prob}(X_{100} \geqslant 24) \simeq 0.99.$$

Similarly

$$\text{prob}(X_{10,000} \geqslant 3834) \simeq 0.99.$$

In case (b) the drift is positive but weaker than in case (a). We find that with probability 0.99 the particle is above position -8 for $n = 100$ and above position 807 for $n = 10,000$. In case (c) there is no drift and we find that with probability 0.99 the particle is within ± 24 units of the origin for $n = 100$ and within ± 231 units for $n = 10,000$.

(ii) TWO ABSORBING BARRIERS

Suppose that the particle starts at the origin and moves in the presence of absorbing barriers at the points $-b$ and a $(a, b > 0)$ so that the motion ceases when the particle enters either of the states $-b$ or a. When the particle enters state a we say that absorption occurs at a, and similarly for state $-b$. Of the three possible ultimate outcomes to the motion, only two, namely absorption at $-b$ and absorption at a, have positive probability in general; we now show that the third possibility, that the particle continues wandering indefinitely between the two barriers, has zero probability. For the probability that the particle is still in motion at time n. i.e. that it occupies one of the non-absorbing states $-b+1$,

$-b+2$, ..., $a-1$, cannot exceed the probability that an unrestricted particle occupies one of these states at time n, for in computing the former probability we must exclude all possible journeys via states outside $(-b+1, a-1)$. Now it can be deduced from (9) that the latter probability tends to zero as $n \to \infty$. Thus the probability that the particle is not yet absorbed at time n also tends to zero as $n \to \infty$, and we shall show later that this vanishing is geometrically fast, i.e.

$$\text{prob}(-b < X_n < a) = O(\rho^n)$$

for some ρ ($0 < \rho < 1$).

Having deduced that the probability of absorption is unity, let us determine the distribution of this probability between the two absorbing states and among the possible times at which absorption may occur. It is convenient to consider an arbitrary starting point $X_0 = j$ ($-b \leqslant j \leqslant a$) and let $f_{ja}^{(n)}$ ($n = 0, 1, ...$) be the distribution of the total probability of absorption at a among the possible times $n = 0$, 1, 2, ... at which this event may occur. Thus $f_{ja}^{(n)}$ is the probability that the particle is absorbed at exactly time n. It is instructive to observe that $f_{ja}^{(n)}$ also represents the probability that an unrestricted particle reaches position a for the first time at time n without position $-b$ being occupied at any of the times 1, 2, ..., $n-1$, all conditional on starting at j. More precisely

$$f_{ja}^{(n)} = \text{prob}(-b < X_1 < a, ..., -b < X_{n-1} < a, \ X_n = a | X_0 = j)$$
$$(n = 1, 2, ...),$$

while for $n = 0$ we have the initial conditions

$$f_{ja}^{(0)} = \begin{array}{ll} 1 & (j = a), \\ 0 & (j \neq a), \end{array} \tag{10}$$

since if the particle starts at a, absorption occurs at time 0 with probability 1 and if it starts at any point other than a absorption cannot occur at time 0.

Let A_n denote the event 'absorption at a at time n'. Then $f_{ja}^{(n)} = \text{prob}(A_n | \text{start at } j)$. In order to obtain an equation for $f_{ja}^{(n)}$ we make a decomposition of A_n based on the first step. This method, which we describe here in detail, is used frequently later on to obtain what are called backward equations. If the first step is $+1$ then the particle moves to $j+1$ and all subsequent steps are independent; for A_n now to occur we must have A_{n-1} occurring conditional on starting at $j+1$. The probability that the first step is $+1$ and that A_n occurs is therefore

$$p \, \text{prob}(A_{n-1} | \text{start at } j+1).$$

We apply similar arguments to the possibilities of a first step of 0 and -1, and since $+1$, 0 and -1 are mutually exclusive and exhaustive possibilities for the first step, we add the three probabilities and obtain

$$f_{ja}^{(n)} = p \operatorname{prob}(A_{n-1} | \text{start at } j+1) + (1-p-q) \operatorname{prob}(A_{n-1} | \text{start at } j)$$
$$+ q \operatorname{prob}(A_{n-1} | \text{start at } j-1)$$

or

$$f_{ja}^{(n)} = p f_{j+1,a}^{(n-1)} + (1-p-q) f_{ja}^{(n-1)} + q f_{j-1,a}^{(n-1)}$$
$$(j = -b+1, \ldots, a-1; \; n = 0, 1, \ldots). \tag{11}$$

Together with (11) we must have the initial conditions (10) and the boundary conditions

$$f_{aa}^{(n)} = 0, \qquad f_{-b,a}^{(n)} = 0 \qquad (n = 1, 2, \ldots). \tag{12}$$

In equation (11) the unknown f is a function of the two discrete variables n and j and we have a difference equation of the first order in n and of the second order in j. We therefore need one condition, namely (10), which prescribes all values of f when $n = 0$ and the two boundary conditions (12) which prescribe all values of f when $j = -b$ and $j = a$.

The form of equation (11) suggests the use of generating functions which in effect eliminates one of the variables. Accordingly we define the generating function over the time variable n,

$$F_{ja}(s) = \sum_{n=0}^{\infty} f_{ja}^{(n)} s^n = F_j(s), \tag{13}$$

say, where for simplicity we omit the suffix a. We now multiply (11) by s^n and sum over $n = 1, 2, \ldots$ obtaining

$$F_j(s) = s\{p F_{j+1}(s) + (1-p-q) F_j(s) + q F_{j-1}(s)\}. \tag{14}$$

This is a second-order difference equation for F_j, with boundary conditions, obtained from (10) and (12),

$$F_a(s) = 1, \qquad F_{-b}(s) = 0.$$

We now have only the one variable j in (14). A common method of solving linear difference equations like (14) is to substitute a trial solution $F_j(s) = \lambda^j$. We find that

$$\lambda^j = s\{p \lambda^{j+1} + (1-p-q) \lambda^j + q \lambda^{j-1}\}$$

or

$$ps\lambda^2 - \lambda\{1 - s(1-p-q)\} + qs = 0, \tag{15}$$

a quadratic equation in λ, the two solutions of which are

$$\lambda_1(s), \lambda_2(s) = \frac{1 - s(1-p-q) \pm [\{1 - s(1-p-q)\}^2 - 4pqs^2]^{\frac{1}{2}}}{2ps}. \tag{16}$$

We assume that s is real and positive and that the function inside the square root in (16) is positive, i.e.

$$\{1-s(1-p-q)\}^2 > 4pqs^2,$$

or

$$0 < s < \frac{1}{(1-p-q)+2\sqrt{(pq)}} = \frac{1}{1-(\sqrt{p}-\sqrt{q})^2}$$

Further, we take the positive square root in (16). The general solution of (14) is now given by

$$A\{\lambda_1(s)\}^j + B\{\lambda_2(s)\}^j$$

and A and B, possibly functions of s, must be determined from the boundary conditions to give the solution to our particular problem. We find that

$$F_j(s) = F_{ja}(s) = \frac{\{\lambda_1(s)\}^{j+b} - \{\lambda_2(s)\}^{j+b}}{\{\lambda_1(s)\}^{a+b} - \{\lambda_2(s)\}^{a+b}}. \tag{17}$$

We have thus obtained the solution to (11) in the form of a generating function. In our original problem the particle started at the origin and so by putting $j = 0$ in (17) we obtain

$$F_{0a}(s) = \frac{\{\lambda_1(s)\}^b - \{\lambda_2(s)\}^b}{\{\lambda_1(s)\}^{a+b} - \{\lambda_2(s)\}^{a+b}}. \tag{18}$$

Equation (26) below gives the corresponding expression for $F_{0,-b}(s)$, the generating function for the probability of absorption at $-b$.

In order to obtain the probabilities $f_{0a}^{(n)}$ we must expand (18) as a power series in s. Before embarking on this by no means trivial calculation we can quite easily obtain the probabilities of the two mutually exclusive and exhaustive outcomes of the process, namely absorption at a or absorption at $-b$. To set $s = 1$ in (18) is equivalent to the summation over n of the probabilities $f_{0a}^{(n)}$, thus obtaining the probability that the particle is absorbed at a, namely $F_{0a}(1)$. This requires initially the substitution $s = 1$ in the roots $\lambda_1(s)$ and $\lambda_2(s)$. Since we have chosen the positive square root in (16) we see that for $s \geqslant 0$, $\lambda_1(s) \geqslant \lambda_2(s)$. Thus, writing $\lambda_1(1) = \lambda_1$ and $\lambda_2(1) = \lambda_2$, we have

$$\lambda_1 = \frac{q}{p} > \lambda_2 = 1 \quad (p < q),$$

$$\lambda_1 = 1 > \lambda_2 = \frac{q}{p} \quad (p > q), \tag{19}$$

$$\lambda_1 = 1 = \lambda_2 \quad (p = q).$$

Hence, from (18)

$$\text{prob(absorption occurs at } a) = F_{0a}(1) = \begin{array}{l} p^a \dfrac{p^b - q^b}{p^{a+b} - q^{a+b}} \quad (p \neq q), \\[3mm] \dfrac{b}{a+b} \quad\quad (p = q), \end{array} \tag{20}$$

where for $p = q$ we obtain an indeterminate form from the first expression which may be evaluated by taking the limit as $p \to q$. Thus the function $F_{0a}(s)/F_{0a}(1)$ is the p.g.f. of the time to absorption conditional on absorption occurring at a, and if we denote by N the random variable representing the time to absorption, we may write

$$F_{0a}(s) = F_{0a}(1) E(s^N | X_N = a)$$
$$= \text{prob}(X_N = a) E(s^N | X_N = a).$$

Since absorption is certain, it immediately follows that

$$\text{prob}(X_N = -b) = \text{prob}(\text{absorption occurs at} -b) = 1 - F_{0a}(1)$$

$$= \begin{cases} q^b \dfrac{p^a - q^a}{p^{a+b} - q^{a+b}} & (p \neq q), \\[2ex] \dfrac{a}{a+b} & (p = q). \end{cases} \tag{21}$$

We now describe briefly how to obtain the individual probabilities $f_{0a}^{(n)}$ by expanding (18) as a power series in s. This is a special case of an important general method, but nevertheless it may be omitted on a first reading. The method depends on the fact, not immediately obvious, that (18) is a rational function of s, as can be seen by expanding both the numerator and denominator in powers of

$$\phi(s) = [\{1 - s(1-p-q)\}^2 - 4pqs^2]^{\frac{1}{2}}$$

and noting that only odd powers of $\phi(s)$ occur; thus $\phi(s)$ cancels, leaving a ratio of two polynomials in s, each of degree $a+b-1$ in general. (If $p+q = 1$, the degree of the numerator is reduced by 1 if b is even and similarly for the denominator if $a+b$ is even.) An expansion of $F_{0a}(s)$ in partial fractions is obtainable which may be expressed as

$$F_{0a}(s) = (2sp)^a \sum_{\nu=1}^{a+b-1} \frac{\alpha_\nu}{1 - s/s_\nu}, \tag{22}$$

where

$$s_\nu = \frac{1}{1 - p - q + 2(pq)^{\frac{1}{2}} \cos\left(\dfrac{\nu\pi}{a+b}\right)} \quad (\nu = 1, \ldots, a+b-1), \tag{23}$$

are the roots of the denominator (the root given by $\nu = 0$ is also a root of the numerator). It is useful to observe that

$$\lambda_1(s_\nu), \lambda_2(s_\nu) = \left(\frac{q}{p}\right)^{\frac{1}{2}} \exp\left(\pm \frac{\nu\pi i}{a+b}\right) \tag{24}$$

and that

$$\frac{d}{ds}\lambda_j(s) = \frac{\lambda_j(s)}{s^2[q\{\lambda_j(s)\}^{-1} - p\lambda_j(s)]} \quad (j = 1, 2),$$

the latter relation being found by differentiating implicitly the equation (15). The constants α_ν in the partial fraction expansion are given by

$$\alpha_\nu = \lim_{s \to s_\nu} \left(1 - \frac{s}{s_\nu}\right)(2ps)^{-a}F_{0a}(s)$$

$$= (2ps_\nu)^{-a} \frac{\{\lambda_1(s_\nu)\}^b - \{\lambda_2(s_\nu)\}^b}{-s_\nu \dfrac{d}{ds}[\{\lambda_1(s)\}^{a+b} - \{\lambda_2(s)\}^{a+b}]_{s=s_\nu}}$$

$$= \frac{(-1)^{\nu+1} \sin\left(\dfrac{b\nu\pi}{a+b}\right) \sin\left(\dfrac{\nu\pi}{a+b}\right)}{2(a+b)(4pq)^{\frac12 a} s_\nu^{a-1}} \qquad (\nu = 1,\ldots,a+b-1).$$

Finally, from (22) the coefficient of s^n in $F_{0a}(s)$ is

$$f_{0a}^{(n)} = (2p)^a \sum_{\nu=1}^{a+b-1} \frac{\alpha_\nu}{s_\nu^{n-a}} \qquad (n = a, a+1, \ldots)$$

$$= \frac{\sqrt{(4pq)}}{(a+b)}\left(\frac{p}{q}\right)^{\frac12 a} \sum_{\nu=1}^{a+b-1} \frac{(-1)^{\nu+1}\sin\left(\dfrac{b\nu\pi}{a+b}\right)\sin\left(\dfrac{\nu\pi}{a+b}\right)}{s_\nu^{n-1}}$$

$$(n = a, a+1, \ldots). \qquad (25)$$

This is the solution which we set out to obtain from the difference equation (11).

It remains to determine the corresponding distribution over time of the probability that absorption occurs at $-b$. By an almost identical argument, in fact by solving the difference equation (14) with boundary conditions

$$F_a(s) = 0, \qquad F_{-b}(s) = 1,$$

we obtain

$$F_{0,-b}(s) = \frac{\{\lambda_2(s)\}^{-a} - \{\lambda_1(s)\}^{-a}}{\{\lambda_2(s)\}^{-a-b} - \{\lambda_1(s)\}^{-a-b}}. \qquad (26)$$

A similar application of partial fractions gives the result

$$f_{0,-b}^{(n)} = \frac{\sqrt{(4pq)}}{(a+b)}\left(\frac{q}{p}\right)^{\frac12 b} \sum_{\nu=1}^{a+b-1} \frac{(-1)^{\nu+1}\sin\left(\dfrac{a\nu\pi}{a+b}\right)\sin\left(\dfrac{\nu\pi}{a+b}\right)}{s_\nu^{n-1}}$$

$$(n = b, b+1, \ldots).$$

Alternatively this result may be obtained from (25) by interchanging a with b and p with q.

By examining the roots $s_1, s_2, \ldots$ more closely we can see that the root of smallest modulus is

$$s_1 = \frac{1}{1 - p - q + 2(pq)^{\frac12}\cos\left(\dfrac{\pi}{a+b}\right)}, \qquad (27)$$

and further we have the inequalities

$$s_1 > \frac{1}{1-p-q+2(pq)^{\frac{1}{2}}} = \frac{1}{1-(p^{\frac{1}{2}}-q^{\frac{1}{2}})^2} \geqslant 1,$$

so that s_1 always exceeds unity. Now s_1 is the singularity of $F_{0a}(s)$ nearest to the origin and is therefore the radius of convergence of the power series defining $F_{0a}(s)$. Therefore it follows that $f_{0a}^{(n)} = O(\rho^n)$ for some ρ ($0 < \rho < 1$). More precisely, we can obtain the following inequality from the exact expression (25) for $f_{0a}^{(n)}$:

$$f_{0a}^{(n)} < \frac{1}{2(a+b)} \left(\frac{p}{q}\right)^{\frac{1}{2}a} \frac{a+b-1}{s_1^{n-1}} \leqslant \left(\frac{p}{q}\right)^{\frac{1}{2}a} \frac{1}{s_1^{n-1}}, \qquad (28)$$

since $s_1 \geqslant |s_\nu|$ ($\nu = 2, \ldots, a+b-1$), and similarly

$$f_{0,-b}^{(n)} \leqslant \left(\frac{q}{p}\right)^{\frac{1}{2}b} \frac{1}{s_1^{n-1}}. \qquad (29)$$

Further, we have

$$\mathrm{prob}(N = n) = f_{0a}^{(n)} + f_{0,-b}^{(n)},$$

so that we have determined completely the probability distribution of N. Its p.g.f. is clearly

$$E(s^N) = F_{0a}(s) + F_{0,-b}(s).$$

For results on the expected value of N, see Exercise 12. Now the probability that the particle is still in motion at time n is precisely the probability that absorption occurs after time n, i.e.

$$\mathrm{prob}(-b < X_n < a) = \mathrm{prob}(N > n)$$

$$= \sum_{r=n+1}^{\infty} \{f_{0a}^{(r)} + f_{0,-b}^{(r)}\} \qquad (30)$$

$$= O(s_1^{-n})$$

in virtue of the inequalities (28) and (29), and so $\mathrm{prob}(-b < X_n < a)$ tends to zero geometrically in n.

In order to complete the description of the process in probability terms it is necessary to determine the probability $p_k^{(n)}$ that the particle is in a state k at a time n before absorption occurs. For if this is known then the state probabilities of the particle are known for all states, both absorbing and non-absorbing, and for all times. If we define the generating function

$$P_k(s) = \sum_{n=0}^{\infty} p_k^{(n)} s^n \quad \text{with} \quad p_k^{(0)} = \delta_{0k},$$

then we find, by relating the position of the particle at time n with its possible positions at time $n-1$, that $P_k(s)$ satisfies the difference equation

$$P_k(s) - 0k = s\{pP_{k-1}(s) + (1-p-q)P_k(s) + qP_{k+1}(s)\}, \tag{31}$$

an inhomogeneous version of (14) with p and q interchanged. The appropriate boundary conditions are

$$P_a(s) = P_{-b}(s) = 0.$$

However, we shall not proceed with the solution of (31) but rather obtain the result by a more general method in Section 2.3 (iv).

(iii) ONE ABSORBING BARRIER

Let us now suppose that the particle starts in the state $X_0 = 0$ and that an absorbing barrier is placed at the point $a > 0$, so that the particle is free to move among the states $x < a$ if and until it reaches the state a which, when once entered, holds the particle permanently. The time to absorption, or the duration of the walk, is clearly also the first passage time from the state 0 to the state a in the unrestricted random walk.

It is of interest to examine the probability that the particle will ever reach an absorbing state at a. Let $f_a^{(n)}$ denote the probability that absorption occurs at a at time n, i.e.

$$f_a^{(n)} = \text{prob}\{X_m < a \quad (m = 1, \ldots, n-1), \quad X_n = a\} \tag{32}$$

and define the generating function

$$F_a(s) = \sum_{n=1}^{\infty} f_a^{(n)} s^n.$$

By letting $b \to \infty$ in the formulae for the two-barrier case we can obtain our results for the present problem. Thus

$$F_a(s) = \lim_{b \to \infty} \frac{\{\lambda_1(s)\}^b - \{\lambda_2(s)\}^b}{\{\lambda_1(s)\}^{a+b} - \{\lambda_2(s)\}^{a+b}}$$
$$= \{\lambda_1(s)\}^{-a},$$

since $\lambda_1(s) > \lambda_2(s)$. Further, since $\lambda_1(s)\lambda_2(s) = q/p$, we also have

$$F_a(s) = \left(\frac{p}{q}\right)^a \{\lambda_2(s)\}^a. \tag{33}$$

There is, however, an alternative and instructive method of obtaining this result. Let us suppose that $p > q$ so that there is a drift upwards. In order to reach the state a the particle must at some intermediate times occupy each of the intervening states $1, 2, \ldots, a-1$, since at each jump the particle can move at most unit distance upwards. The probability of ultimately effecting the passage from state 0 to state 1 is unity, for otherwise, there is a positive probability of forever remaining in the states $0, -1, -2, \ldots$, i.e.

$$\lim_{n \to \infty} \operatorname{prob}(X_1 \leqslant 0, X_2 \leqslant 0, \ldots, X_n \leqslant 0) \geqslant C > 0.$$

But, by a similar argument to that used at the beginning of (ii) of this section, $\operatorname{prob}(X_1 \leqslant 0, \ldots, X_n \leqslant 0)$ cannot exceed the probability that an unrestricted particle lies below the origin at time n, which tends to zero by the central limit theorem. It follows that the passage from state 0 to state 1 is a certain event. Thus the first passage time N_1 from state 0 to state 1 is a random variable taking positive integral values. Let us denote its p.g.f. by $F_1(s)$. Since each jump is independent of the time and the state from which the jump is made, it follows that the first passage time from any given state k to the state $k+1$ is a random variable having the same distribution as N_1 and that the successive first passage times from state 0 to state 1, from state 1 to state 2, and so on, are independent random variables. Thus the first passage time from state 0 to state a is the sum of a independent random variables each with the distribution of N_1. Hence

$$F_a(s) = \{F_1(s)\}^a. \tag{34}$$

We can determine $F_1(s)$ by considering the three possible positions of the particle after making the first jump. Firstly, with probability p the first passage to state 1 is effected at the first step. Secondly, with probability $1-p-q$ the particle remains at 0 after the first step and takes a further time N_1' to reach state 1, where N_1' is a random variable with the distribution of N_1 and independent of the first step. Thirdly, with probability q the particle jumps to -1 at the first stage and then takes time $N_1'' + N_1'''$ to reach 1, where N_1'' and N_1''' are independent random variables each with the distribution of N_1. These three statements may be expressed as follows

$$\operatorname{prob}(X_1 = 1)\, E(s^{N_1}|X_1 = 1) = ps,$$
$$\operatorname{prob}(X_1 = 0)\, E(s^{N_1}|X_1 = 0) = (1-p-q)\, sF_1(s),$$
$$\operatorname{prob}(X_1 = -1)\, E(s^{N_1}|X_1 = -1) = qs\{F_1(s)\}^2,$$

and, on summing these three expressions, we obtain $E(s^{N_1})$ unconditionally,

$$E(s^{N_1}) = F_1(s) = ps + (1-p-q)\, sF_1(s) + qs\{F_1(s)\}^2. \tag{35}$$

This is a quadratic equation for $F_1(s)$ and if we take the solution satisfying $F_1(1) = 1$ we find that

$$F_1(s) = \frac{p}{q}\lambda_2(s) = \{\lambda_1(s)\}^{-1},$$

and the result (33), giving the p.g.f. $F_a(s)$, follows from (34).

The above argument can be adapted to cover the cases where $p \leqslant q$. Here we must interpret $F_1(s)$ as the generating function of a probability distribution which does not necessarily add up to 1. More precisely, if A denotes the event that the particle ever reaches state 1 when it starts from state 0 then

$$\text{prob}(A) = \sum_{n=1}^{\infty} f_1^{(n)},$$

where $f_1^{(n)}$ is defined in (32), and

$$F_1(s) = \sum_{n=1}^{\infty} f_1^{(n)} s^n \qquad (36)$$

$$= \text{prob}(A) E(s^{N_1}|A).$$

The random variable N_1 now has a distribution conditional on A. We obtain the same equation (35) for $F_1(s)$ and again we take the root

$$F_1(s) = \frac{p}{q} \lambda_2(s);$$

the other root, $(p/q)\lambda_1(s)$, is disqualified from being a function of the form (36) by the fact that $\lambda_1(s) \to \infty$ as $s \to 0$.

Thus in all cases we have the result

$$F_a(s) = \left(\frac{p}{q}\right)^a \{\lambda_2(s)\}^a = \{F_1(s)\}^a. \qquad (37)$$

Letting N_a denote the time to absorption at a, or equivalently the first passage time to state a from state 0, we may write

$$F_a(s) = \sum_{n=1}^{\infty} s^n \text{prob}(N_a = n).$$

For $s = 1$,

$$F_a(1) = \text{prob}(N_a < \infty),$$

which is the probability that absorption occurs at all. Thus

$$\text{prob}(N_a < \infty) = \begin{array}{ll} \left(\dfrac{p}{q}\right)^a & (p < q), \\ 1 & (p \geqslant q). \end{array} \qquad (38)$$

Now N_a also denotes the duration of the random walk and so, when the drift is towards the barrier or when there is no drift, N_a is finite with probability one. When the drift is away from the barrier N_a is finite with probability $(p/q)^a$ while with probability $1-(p/q)^a$ the walk continues indefinitely, the particle remaining forever in states below a.

Example 2.6. *A numerical example.* If $p = 0.3$, $q = 0.4$, then when the barrier is 10 units from the origin ($a = 10$) the probability (38) that the particle ever reaches the barrier is $0 \cdot 0563$, while if $a = 100$ it is about $10^{-12 \cdot 5}$. If $p = 0 \cdot 2$, $q = 0 \cdot 6$ then the corresponding probabilities are about $1/50{,}000$ and $10^{-47 \cdot 7}$ respectively.

If absorption is certain then $F_a(s) = \{F_1(s)\}^a$ is the p.g.f. of the number N_a of steps to absorption. In this case we have

$$E(N_a) = aF_1'(1) = \begin{cases} \dfrac{a}{p-q} & (p > q), \\[2mm] \infty & (p = q). \end{cases} \tag{39}$$

Thus although absorption is certain when $p = q$ the distribution of time to absorption has infinite mean (and therefore infinite moments of all orders). When $p > q$ we have for the variance

$$V(N_a) = aV(N_1) = a[F_1''(1) + F_1'(1) - \{F_1'(1)\}^2]$$

$$= \frac{a\{p + q - (p-q)^2\}}{(p-q)^3} \quad (p > q). \tag{40}$$

Further N_a is the sum of a independent random variables with finite variance and by the central limit theorem will be approximately normally distributed for large a. In terms of the mean $\mu = p - q$ and variance $\sigma^2 = p + q - (p-q)^2$ of a single jump we have

$$E(N_a) = \frac{a}{\mu}, \qquad V(N_a) = a\frac{\sigma^2}{\mu^3}.$$

Example 2.7. A numerical example. Taking the numerical values $p = 0 \cdot 4$, $q = 0 \cdot 3$, we find that when $a = 10$, N_a has mean 100 and standard deviation 83 indicating a distribution with a wide spread. When $a = 100$ the mean and standard deviation of N_a are 1000 and 263 respectively; using the normal approximation in this case we see that N_a will lie between 500 and 1500 with probability about $0 \cdot 95$. If $p = 0 \cdot 6$, $q = 0 \cdot 2$ the drift is stronger. For $a = 10$ we find that N_a has mean 33 and standard deviation 31 while for $a = 100$ the corresponding figures are 333 and 97.

When $p = q$ the behaviour of the particle is somewhat singular. It follows from our results that starting from state 0, the particle reaches any other given state with probability one, but that the mean time to achieve this passage is infinite. Having reached the given state it will return to state 0 with probability one, again with infinite mean passage time. Thus an unrestricted particle, if allowed sufficient time, is certain to make indefinitely large excursions from its starting point and is also certain to return to its starting point.

Example 2.8. *Simulation of the symmetrical random walk.* It is simple and interesting to simulate a symmetrical random walk with $p = q = \frac{1}{2}$ and with $a = 1$. Thus we toss an unbiased coin and score $+1$ for heads, -1 for tails and when the cumulative score is $+1$ we stop. Alternatively a table of random digits may be used. In 12 independent realizations it was found that the numbers of steps to absorption were 5, 1, 30, 1, 3, 1, 1, 1, 3, 1, 3, 412. These are observations from a distribution with infinite mean and variance.

(*iv*) FURTHER ASPECTS OF THE UNRESTRICTED RANDOM WALK

We have seen that the imposition of an absorbing barrier is a useful method of studying certain aspects of the unrestricted random walk, in particular the passage times. There is another interesting aspect revealed by the use of absorbing barriers, namely the maximum distance from its starting point reached by the particle in either direction.

More precisely let

$$U_n = \max_{0 \leqslant r \leqslant n} X_r, \qquad L_n = \min_{0 \leqslant r \leqslant n} X_r,$$

so that for a given realization of the process up to time n, U_n and $|L_n|$ are the maximum distances upwards and downwards respectively reached by the particle in the time interval $[0, n]$. We can find the joint probability distribution of U_n and L_n by imposing absorbing barriers at j and $-k$ $(j, k > 0)$ and observing that $\mathrm{prob}(U_n < j, L_n > -k)$ is precisely the probability (30) for $a = j$, $b = k$, i.e., the probability that absorption is later than time n. Hence, using the results for the case of two absorbing barriers, it is possible to calculate explicitly the joint distribution of U_n and L_n, while from the single barrier results the marginal distribution of each may be obtained.

Interesting results are obtained for U_n say, when we let $n \to \infty$, for here there is the possibility of a limiting distribution. Consider the behaviour of U_n for a particular realization. When the mean jump is positive $(p > q)$ then the particle will tend to drift to $+\infty$ and therefore U_n will tend to become indefinitely large as n increases; this will also happen when the mean jump is zero since the particle will make indefinitely large excursions. However, when the mean jump is negative the particle will drift to $-\infty$ after reaching a maximum distance U in the positive direction. We call U the *supremum* of the process and it is denoted by

$$U = \sup_n(X_n);$$

is finite with probability one only when $p < q$. When $p < q$ we have seen that for an absorbing barrier at $j > 0$ the probability is $1 - (p/q)^j$ that the particle never reaches the barrier. Hence it follows that

$$\text{prob}(U < j) = 1 - \left(\frac{p}{q}\right)^j \quad (j = 1, 2, \ldots)$$

and therefore that

$$\text{prob}(U = j) = \left(\frac{p}{q}\right)^j \left(1 - \frac{p}{q}\right) \quad (j = 0, 1, \ldots), \tag{41}$$

a geometric distribution. For $j = 0$, we obtain the probability that the particle never enters the non-negative states, i.e. that the particle remains forever below its starting point. Further, the mean and variance of the supremum are

$$E(U) = \frac{p}{q-p}, \qquad V(U) = \frac{pq}{(q-p)^2}.$$

Similarly we define the *infimum* of the process, $L = \inf_n(X_n)$.

The extrema of the process, i.e. the supremum and infimum, provide a useful illustration of the concept of the sample function of a stochastic process. A sample function, in the case of the simple random walk, is obtained by considering a full realization x_0, x_1, x_2, ..., and regarding position x as a function of the real variable time. We obtain a step function of a positive real variable with at most unit jump discontinuities at the integers. Figure 1.1 illustrates portions of two sample functions. It is possible to regard the collection of all such step functions, for fixed x_0, as a non-countable sample space which, together with an appropriate probability measure, may be regarded as a definition of the stochastic process. It is on this sample space that the random variables supremum and infimum are defined, for corresponding to each element, i.e. sample function, there is a unique supremum and infimum (not necessarily finite).

Another aspect of the unrestricted walk on which results for the absorbing barrier give information is the phenomenon of *return to the origin*. Conditional on starting at j, let $f_{jj}^{(n)}$ be the probability that the *first* return to j occurs at time n, and let $f_{jk}^{(n)}$ be the probability that the particle first enters position k ($\neq j$) at time n. Since the steps of the random walk are independent we clearly have

$$f_{jk}^{(n)} = f_{0, k-j}^{(n)}.$$

By a familiar decomposition of the first step we find that

$$f_{00}^{(n)} = p f_{10}^{(n-1)} + q f_{-1,0}^{(n-1)} = p f_{10}^{(n-1)} + q f_{01}^{(n-1)} \quad (n = 2, 3, \ldots)$$

and by definition $f_{00}^{(1)} = 1 - p - q$. On taking generating functions we obtain

$$\sum_{n=1}^{\infty} f_{00}^{(n)} s^n = s(1 - p - q) + sp \sum_{n=1}^{\infty} f_{10}^{(n)} s^n + sq \sum_{n=1}^{\infty} f_{01}^{(n)} s^n.$$

The second series on the right-hand side is the generating function $F_1(s)$ of absorption probabilities given by equation (36), and the first series is a similar generating function $F_1^*(s)$ for a random walk in which p and q are interchanged. Denoting the series on the left-hand side by $F_{00}(s)$ we obtain

$$F_{00}(s) = s(1-p-q) + spF_1^*(s) + sqF_1(s).$$

On interchanging p and q, $\lambda_2(s)$ is replaced by $\lambda_2^*(s) = (p/q)\lambda_2(s)$ and $F_1(s)$ of equation (37) is replaced by $(q/p)\lambda_2^*(s) = \lambda_2(s) = F_1^*(s)$. Hence

$$F_{00}(s) = s(1-p-q) + sp\lambda_2(s) + sp\lambda_2(s)$$

$$= s\{1-p-q+2p\lambda_2(s)\}.$$

Now $F_{00}(1)$ is the probability of ever returning to the origin and, using equations (19), we have

$$\text{prob(return to the origin)} = F_{00}(1) = \begin{array}{ll} 1 & (p = q), \\ 1+p-q < 1 & (p < q), \\ 1+q-p < 1 & (p > q). \end{array}$$

Thus return to the origin is certain only if the walk has zero drift ($p = q$) and uncertain if the drift is positive or negative ($p \neq q$). In the case of zero drift $F_{00}(s)$ is the p.g.f. of the time of first return to the origin, and the mean of this distribution, $F'_{00}(1)$, is infinite since $\lambda_2'(1) = \infty$ when $p = q$.

The starting point was chosen to be the origin but clearly the above remarks apply equally to any starting point j. Thus return to j is certain only if the walk has zero drift and is uncertain otherwise. In the case of certain return the mean time to return is infinite.

(v) TWO REFLECTING BARRIERS

In Example 2.2 we saw how reflecting barriers arise naturally in a probability model of a dam. We now discuss in more detail the simple random walk in the presence of such barriers.

For a simple random walk we may define a reflecting barrier as follows. Suppose a is a point above the initial position. If the particle reaches a then at the next step it either remains at a or returns to the neighbouring interior state $a-1$ with specified probabilities. In most of what follows we take these probabilities to be $1-q$ and q respectively, so that in effect steps taken when the particle is at a are truncated in the positive direction. However, in particular problems these probabilities may be different from $1-q$ and q as is the case in Example 2.9. Similar remarks apply to a reflecting barrier below the initial position.

Suppose that the particle is initially in the state j and that the states 0

and a ($a > 0$) are reflecting barriers. If X_n is the position of the particle·
immediately after the nth jump Z_n, then we have $X_0 = j$, and

$$
X_n = \begin{cases} X_{n-1} + Z_n & (0 \leqslant X_{n-1} + Z_n \leqslant a), \\ a & (X_{n-1} + Z_n > a), \\ 0 & (X_{n-1} + Z_n < 0). \end{cases} \tag{42}
$$

Thus the particle remains forever among the states $0, 1, \ldots, a$. On reaching
one of the barrier states the particle remains there until a jump of the
appropriate sign returns it to the neighbouring interior state.

Since there is now no possibility of the motion ceasing at any stage we
have a different type of long-term behaviour from the absorbing barrier
case. We shall show in Chapter 3 that after a long time the motion of the
particle settles down to a condition of statistical equilibrium in which
the occupation probabilities of the various states depend only on the
relative position of the barriers but neither on the initial position of the
particle nor on the time.

Let $p_{jk}^{(n)}$ be the probability that the particle occupies the state k at
time n having started in the state j. Since the jumps are independent
the position of the particle at time n depends only on two independent
quantities, its position at time $n-1$ and the nth jump. If k is an internal
state then during the time interval $(n-1, n)$ it can be reached by one of
three mutually exclusive ways, namely by a jump of $+1$, 0 or -1 from
states $k-1$, k or $k+1$ respectively. Since the state occupation probabilities at time $n-1$ are independent of the nth jump it follows that the
probabilities of these three ways are $pp_{j,k-1}^{(n-1)}$, $(1-p-q)p_{jk}^{(n-1)}$, and $qp_{j,k+1}^{(n-1)}$
respectively. Hence

$$
p_{jk}^{(n)} = pp_{j,k-1}^{(n-1)} + (1-p-q)p_{jk}^{(n-1)} + qp_{j,k+1}^{(n-1)} \quad (0 < k < a). \tag{43}
$$

Since motion above state a and below state 0 is not permitted, similar
reasoning will show that at the barrier states we must have

$$
p_{ja}^{(n)} = pp_{j,a-1}^{(n-1)} + (1-q)p_{ja}^{(n-1)}, \qquad p_{j0}^{(n)} = (1-p)p_{j0}^{(n-1)} + qp_{j1}^{(n-1)}. \tag{44}
$$

If there is a limiting equilibrium distribution of the state occupation
probabilities then we have as $n \to \infty$

$$
p_{jk}^{(n)} \to \pi_k \quad (k = 0, 1, \ldots, a)
$$

and from equations (43) and (44) the π_k must satisfy

$$
\begin{aligned} \pi_k &= p\pi_{k-1} + (1-p-q)\pi_k + q\pi_{k+1} \quad (k = 1, \ldots, a-1), \\ \pi_0 &= (1-p)\pi_0 + q\pi_1, \\ \pi_a &= p\pi_{a-1} + (1-q)\pi_a. \end{aligned} \tag{45}
$$

We may solve for π_1 in terms of π_0, obtaining

$$\pi_1 = \left(\frac{p}{q}\right)\pi_0.$$

Solving (45) recursively, we have that

$$\pi_2 = \{(p+q)\pi_1 - p\pi_0\}/q$$
$$= \left(\frac{p}{q}\right)^2 \pi_0$$

and in general

$$\pi_k = \left(\frac{p}{q}\right)^k \pi_0 \quad (k = 0, \ldots, a). \tag{46}$$

We require the solution to be a probability distribution, i.e. $\sum \pi_k = 1$, and this enables us to find π_0. Hence we obtain the truncated geometric distribution

$$\pi_k = \frac{1 - \frac{p}{q}}{1 - \left(\frac{p}{q}\right)^{a+1}} \left(\frac{p}{q}\right)^k \quad (k = 0, \ldots, a) \tag{47}$$

as the equilibrium set of state occupation probabilities. If $p > q$ then π_k decreases geometrically away from the upper barrier whereas if $p < q$, π_k decreases geometrically away from the lower barrier. If $p = q$ then from (46) we see that $\pi_k = \pi_0 = 1/(a+1)$ for all k, so that in the equilibrium situation all states are equally likely to be occupied by the particle.

In the following example we describe a model of a finite queue in discrete time and we shall see that it has a representation like a random walk with reflecting barriers. It will serve to illustrate both the method of solving these problems and the distinctive property of independent increments possessed by the random walk.

Example 2.9. A queueing system in discrete time. Consider the following model of a queueing system. A single server operates a service. Customers may arrive at the service point only at discrete time instants $n = 0, 1, 2, \ldots$ and form a queue if the server is occupied. The statistical laws of arrivals and service are such that at any discrete time instant there is a probability α that a customer arrives and independently a probability β that the customer already being served, if any, completes his service. Suppose further that the queue is restricted in size to a customers including the one being served. Thus a customer arriving to find the queue full is turned away.

Let X_n denote the number of customers in the queue, including the one being served, immediately after the nth time instant. Let Z_n denote the number of new arrivals (either 0 or 1) minus the number of customers completing service (also either 0 or 1) at the nth time instant. Then it follows that

$$X_n = X_{n-1} + Z_n \quad (0 \leqslant X_{n-1} + Z_n \leqslant a). \tag{48}$$

If $X_{n-1} = a$, i.e. if the queue is full at time $n-1$, and if at time n there is a new arrival, but no completion of service, i.e. if $Z_n = 1$, then this new arrival is turned away and the queue size remains a at time n. Thus

$$X_n = a \quad (X_{n-1} + Z_n > a). \tag{49}$$

If $X_{n-1} = 0$, i.e. the queue is empty at time $n-1$, then some care is needed in describing the situation. For if there is no arrival at time n, then $X_n = 0$ and if there is an arrival at time n then $X_n = 1$. Since the queue is empty at time $n-1$ there is no possibility of a customer completing service. To describe the distribution of Z_n we have to distinguish two cases. Firstly if X_{n-1} has one of the values 1, 2, ..., a then, since arrivals and service completions are independent, we have

$$\text{prob}(Z_n = 1) = \text{prob}(\text{new arrival and no service completion})$$
$$= \alpha(1-\beta),$$
$$\text{prob}(Z_n = 0) = \text{prob}(\text{new arrival and a service completion})$$
$$\qquad + \text{prob}(\text{no new arrival and no service completion})$$
$$= \alpha\beta + (1-\alpha)(1-\beta),$$
$$\text{prob}(Z_n = -1) = \text{prob}(\text{no new arrival and a service completion})$$
$$= (1-\alpha)\beta.$$

Secondly if $X_{n-1} = 0$ then

$$\text{prob}(Z_n = 1) = \alpha, \qquad \text{prob}(Z_n = 0) = 1-\alpha.$$

Equations (48) and (49) are the same as the first two equations in (42) but now $\{Z_n\}$ is no longer a sequence of identically distributed random variables, the distribution of Z_n when $X_n = 0$ being different from that when $X_n \neq 0$. This means that the reflecting conditions at the lower barrier are different from those of the process described by equation (42). In all other respects, however, the two processes are the same and we can use the same methods for dealing with them. Thus we may set up equations such as (43) and (44) and these will give rise to the following equations for the equilibrium probability distribution π_0, ..., π_a of queue size.

$$\pi_k = \alpha(1-\beta)\pi_{k-1} + \{(1-\alpha)(1-\beta) + \alpha\beta\}\pi_k +$$
$$\qquad + (1-\alpha)\beta\pi_{k+1} \quad (k = 2, \ldots, a-1),$$
$$\pi_1 = \alpha\pi_0 + \{(1-\alpha)(1-\beta) + \alpha\beta\}\pi_1 + (1-\alpha)\beta\pi_2,$$
$$\pi_0 = (1-\alpha)\pi_0 + (1-\alpha)\beta\pi_1,$$
$$\pi_a = \alpha(1-\beta)\pi_{a-1} + \{1 - \beta(1-\alpha)\}\pi_a. \tag{50}$$

These equations may be solved in a similar way to (45) and, on writing $\rho = \{\alpha(1-\beta)\}/\{\beta(1-\alpha)\}$, we obtain the solution

$$\pi_k = \frac{\rho^k}{1-\beta}\pi_0 \quad (k = 1, \ldots, a), \qquad \pi_0 = \frac{\beta-\alpha}{\beta-\alpha\rho^a}.$$

(vi) ONE REFLECTING BARRIER

Let us now consider a simple random walk over the positive integers starting in the state $j > 0$ with the zero state as a reflecting barrier. If the drift is negative ($q > p$) then our studies of the absorbing barrier case have shown that the barrier state is certain to be reached. Once in the barrier state the particle remains there for a time T, say, which has a geometric distribution

$$\text{prob}(T = r) = (1-p)^{r-1}p. \tag{51}$$

When the particle leaves the barrier state it is again certain to return and so the particle continues to move on and off the reflecting barrier. Hence after a long time we may expect a condition of statistical equilibrium to be established in which the initial state j plays no part. In fact we can obtain the equilibrium distribution of the position of the particle by letting $a \to \infty$ in equation (47), remembering that $p < q$. Hence we get the geometric distribution

$$\pi_k = \left(1-\frac{p}{q}\right)\left(\frac{p}{q}\right)^k \quad (k = 0, 1, \ldots). \tag{52}$$

Since the steps of a random walk are independent, it follows that successive times spent on and away from the barrier are independent random variables. The periods spent on the barrier have the geometric distribution (51). The periods spent away from the barrier (excluding the initial one) clearly have the same distribution as the first passage time from state 1 to state 0, the p.g.f. of which may be obtained from equation (26) by setting $b = 1$ and letting $a \to \infty$. Thus if U denotes a period spent away from the barrier its p.g.f. is

$$E(s^U) = \lambda_2(s) = \frac{1-s(1-p-q)-[\{1-s(1-p-q)\}^2-4pqs^2]^{\frac{1}{2}}}{2ps}. \tag{53}$$

If $N_1, M_1, N_2, M_2, \ldots$ denote the successive times at which the particle enters and leaves the reflecting barrier state then $(M_1 - N_1)$, $(M_2 - N_2)$, $\ldots$ are independent random variables each with the distribution (51) while $(N_2 - M_1)$, $(N_3 - M_2)$, $\ldots$ is a sequence of independent random variables each having the p.g.f. (53), and the two sequences are mutually independent. The sequence of time points $N_1, M_1, N_2, M_2, \ldots$ form what is called an *alternating renewal process* and such processes will be dealt with in continuous time in Chapter 9.

In the queue of Example 2.9 periods spent on and off the barrier have a definite interpretation. The former correspond to the server's idle periods and the latter to his busy periods.

Let us now consider what happens when the drift is away from the barrier, i.e. $p > q$. We saw in the case of the unrestricted random walk that the particle ultimately drifts off to $+\infty$. It follows that in the present case this will happen *a fortiori*, since corresponding to each sample function of the unrestricted walk there is a sample function of the random walk with the reflecting barrier and the latter never lies below the former. Hence there is no equilibrium distribution in this case.

When $p = q$, i.e. zero drift, there is again no limiting distribution. This can best be seen by examining the solution to the case of two reflecting barriers in which all states are equally likely in the limit, each having probability $1/(a+1)$. Letting $a \to \infty$ we see that each state occupation probability tends to zero. In one respect the particle behaves similarly to one with negative drift; it is always certain to return to the barrier. Periods spent on the barrier still have the geometric distribution (51) while periods spent away from the barrier have the p.g.f. (53) with $p = q$, corresponding to a distribution with infinite mean. We shall see in the theory of the Markov chain (Chapter 3) that when an equilibrium distribution exists the limiting occupation probability for any particular state is inversely proportional to the mean recurrence time, i.e. the mean time between successive visits to that state. In the case of zero drift all recurrence times have infinite mean and so, roughly speaking, visits to a particular state are too infrequent to allow an equilibrium situation to establish itself.

(*vii*) OTHER TYPES OF BARRIER

There are numerous other types of behaviour possible at a given barrier. For example, on reaching a barrier a particle may be immediately returned to the state occupied just before reaching the barrier. This is another type of reflection. Or on reaching a barrier a particle may return to its initial position. Also, in two barrier problems the barriers may be of different types; for example, we may have an absorbing barrier at 0 and a reflecting barrier at a. We shall not go into details here. See, however, Exercise 5.

2.3. The general one-dimensional random walk in discrete time

It is clear that if we are to make any progress in examining the random walk representations of the insurance company and the dam discussed in Examples 2.1 and 2.2 then we must allow the steps $Z_1, Z_2, \ldots$ to be more general than those of a simple random walk. We consider in this section the general one-dimensional random walk defined by equations (1) or (2) in which the steps, which may be either discrete or continuous, have a given distribution function $F(x)$. We shall find that the mathematical analysis of this process presents more difficult problems.

The simple random walk does however have important uses in the examination of the general random walk, firstly in providing analogy and suggestion and secondly in providing an actual approximation by means of replacing the distribution function $F(x)$ by a discrete distribution over the points $-l$, 0, l having the same mean and variance as $F(x)$ (see Exercise 1).

(i) UNRESTRICTED

We suppose the particle starts at the origin. Then, as in the case of the simple random walk, its position X_n after n steps is a sum of n mutually independent random variables each with the distribution function $F(x)$. Suppose that μ and σ^2 are respectively the mean and variance of $F(x)$, assumed finite. Then if n is large, X_n is approximately normally distributed with mean $n\mu$ and variance $n\sigma^2$. We may use similar arguments to those of Section 2.2 (i) to show that when n is large

$$X_n = n\mu + O(\sqrt{n}) \tag{54}$$

with high probability. It follows again that when $\mu > 0$ the particle ultimately drifts off to $+\infty$ and correspondingly to $-\infty$ when $\mu < 0$. When $\mu = 0$ the behaviour is of the same singular type as in the simple random walk, but this fact is somewhat more difficult to demonstrate in the general case. If we wish to compute the probability that after a large number of steps the particle lies in a given interval (y_1, y_2) then we can use the normal approximation

$$\text{prob}(y_1 \leqslant X_n \leqslant y_2) \simeq (2\pi\sigma^2 n)^{-\frac{1}{2}} \int_{y_1}^{y_2} \exp\left\{-\frac{(x-n\mu)^2}{2n\sigma^2}\right\} dx. \tag{55}$$

Statements like (54) and (55) are based on the central limit theorem whose truth depends essentially on the assumption that σ^2 is finite. However, assuming only that the first moment is finite, we may demonstrate, for example, that the particle drifts off to $+\infty$ when $\mu > 0$, by using the strong law of large numbers. If c is an arbitrarily large positive number then we can assert that $\text{prob}(X_n > c, X_{n+1} > c, \ldots)$ can be made as close to 1 as we please by taking n sufficiently large. For according to the strong law of large numbers, if $\epsilon < \mu$ is fixed and η is arbitrarily small then for n sufficiently large

$$\text{prob}\left(\left|\frac{X_n}{n} - \mu\right| < \epsilon, \left|\frac{X_{n+1}}{n+1} - \mu\right| < \epsilon, \ldots\right) > 1 - \eta,$$

i.e. $\text{prob}\{X_n > n(\mu - \epsilon), X_{n+1} > (n+1)(\mu - \epsilon), \ldots\} > 1 - \eta.$

By choosing $n(\mu - \epsilon) > c$ we have the result. Thus if the mean step is positive, then after a sufficiently large number of steps the particle will remain beyond any arbitrarily chosen point in the positive direction with probability as close to unity as we please. When $\mu = 0$, the behaviour is of the same singular type as in the simple random walk, but this fact is more difficult to prove (Chung and Fuchs, 1951); see also Section 2.4(ii).

(ii) SOME PROPERTIES OF MOMENT GENERATING FUNCTIONS

In the subsequent discussion of the random walk we require some properties of moment generating functions which we now establish.

Suppose X is a random variable with p.d.f. $f(x)$ such that X is strictly two sided, i.e. X takes both positive and negative values with positive probability. Mathematically we may define this condition by saying that there exists a number $\delta > 0$ such that

$$\text{prob}(X < -\delta) > 0 \quad \text{and} \quad \text{prob}(X > \delta) > 0. \tag{56}$$

The moment generating function (m.g.f.) of X is defined as the two-sided Laplace transform of the p.d.f. of X, namely

$$f^*(\theta) = \int_{-\infty}^{\infty} e^{-\theta x} f(x)\, dx, \tag{57}$$

and we shall suppose that $f(x)$ tends to zero sufficiently fast as $x \to \pm \infty$ to ensure that the integral in (57) converges for all real values of θ. For example, this is true of a normal density function or of a density function which vanishes outside a finite interval. We know that

$$\mu = E(X) = \{-(d/d\theta) f^*(\theta)\}_{\theta=0}$$

and hence if we plot a curve of $f^*(\theta)$ against θ, the slope of the curve at $\theta = 0$ is $-\mu$. Further $f^*(\theta)$ is a convex (downwards) function of θ. This can be seen by considering

$$\frac{d^2 f^*(\theta)}{d\theta^2} = \int_{-\infty}^{\infty} x^2 e^{-\theta x} f(x)\, dx,$$

which is positive for all real values of θ. Also $f^*(\theta) \to \infty$ as $\theta \to \pm \infty$, for, taking $\theta < 0$ as an example, we have that

$$f^*(\theta) > \int_{\delta}^{\infty} e^{-\theta x} f(x)\, dx$$

$$> e^{-\theta \delta} \int_{\delta}^{\infty} f(x)\, dx$$

$$= e^{-\theta \delta} \text{prob}(X > \delta) \to \infty$$

as $\theta \to -\infty$ in virtue of (56). By a similar argument it can be seen that

$f*(\theta) \to \infty$ as $\theta \to \infty$. Thus if we plot a graph of $f*(\theta)$ against θ we obtain a shape illustrated in Fig. 2.1.

Two further results follow from the convexity property of $f*(\theta)$. Firstly there is a unique value of θ, say θ_1, at which $f*(\theta)$ attains its minimum value; θ_1 has the same sign as μ and $\theta_1 = 0$ if $\mu = 0$. Also

$$f*(\theta_1) \begin{cases} < 1 & (\mu \neq 0), \\ = 1 & (\mu = 0). \end{cases} \tag{58}$$

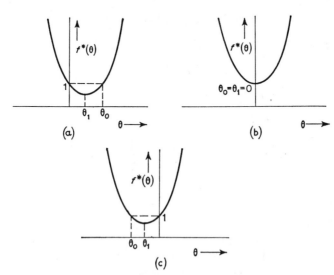

Fig. 2.1. A moment generating function $f*(\theta)$ of a random variable X for (a) $\mu > 0$, (b) $\mu = 0$, (c) $\mu < 0$, where $\mu = E(X) = \{-(d/d\theta) f*(\theta)\}_{\theta=0}$.

Secondly let us consider the roots of the equation

$$f*(\theta) = 1. \tag{59}$$

Clearly $\theta = 0$ is one root. If $\mu \neq 0$ there is a unique second real root $\theta = \theta_0 \neq 0$ also having the same sign as μ (see Fig. 2.1). If $\mu = 0$ then $\theta = 0$ is a double root of (59).

We have assumed in the above discussion that the p.d.f. $f(x)$ of the random variable X exists and, moreover, that the integral (57) defining $f*(\theta)$ converges for all real θ. For a general distribution function $F(x)$, we define the m.g.f. as the two-sided Laplace–Stieltjes transform

$$\int_{-\infty}^{\infty} e^{-\theta x} \, dF(x),$$

provided that the integral converges in a real θ-interval of which $\theta = 0$ is an interior point. The convexity property still holds in this interval and for a fairly general class of distributions we can still assert the existence of the numbers θ_0 and θ_1 (see Exercises 6 and 7).

For a random variable discrete over the integers, instead of the m.g.f. we usually use the p.g.f., for which similar results may be obtained (see Exercise 8).

Example 2.10. *The normal distribution.* Consider the normal distribution with mean μ and variance σ^2. The moment generating function is given by

$$f^*(\theta) = \exp\left(-\mu\theta + \tfrac{1}{2}\sigma^2\theta^2\right).$$

The real roots of $f^*(\theta) = 1$ are given by

$$-\mu\theta + \tfrac{1}{2}\sigma^2\theta^2 = 0,$$

i.e. are $\theta = 0$ and $\theta = 2\mu/\sigma^2$. Thus we have

$$\theta_0 = 2\mu/\sigma^2.$$

The unique real root θ_1, of $(d/d\theta)f^*(\theta) = 0$ is given by

$$-\mu + \sigma^2\theta = 0,$$

i.e. $\theta_1 = \mu/\sigma^2$.

(*iii*) ABSORBING BARRIERS

In the simple random walk it was sufficient to define two absorbing states $-b$ and a without paying any attention to possible states beyond these points. In the more general random walk, however, the particle may jump over or land on a barrier and we shall define absorption to include both these possibilities. Accordingly we suppose that all states in the open interval $(-b, a)$ $(a, b > 0)$ are non-absorbing, whereas all states in the two semi-infinite intervals $(-\infty, -b]$ and $[a, \infty)$ are absorbing states. If the particle starts at the origin then the number N of steps to absorption is the time of first exit or first passage from the interval (a, b), i.e. the time at which the particle first crosses one or other of the barriers. When absorption occurs the process ceases.

The first thorough investigation of the general random walk with absorbing barriers was made by Wald (1947) in the theory of a statistical technique known as sequential analysis. We shall give a somewhat more general version of Wald's results.

We shall confine our discussion for the present to continuous random variables although the method we use applies also to the discrete case. Accordingly we suppose that the random variables Z_1, Z_2, ... are mutually independent each with the same p.d.f. $f(x)$ and m.g.f. $f^*(\theta)$.

Let $f_n(x)\,dx$ denote the probability that the particle remains inside the interval $(-b, a)$ for the first $n-1$ steps and that at the nth step it lies in the interval $(x, x+dx)$, where x may be either an absorbing or a non-absorbing state, i.e.

$$f_n(x)\,dx = \text{prob}(-b < X_1, X_2, \ldots, X_{n-1} < a, x < X_n < x+dx)$$
$$(n = 1, 2, \ldots; \ -\infty < x < \infty). \quad (60)$$

We define

$$f_0(x) = \delta(x),$$

$\delta(x)$ representing the p.d.f. of a probability distribution located entirely at the point $x = 0$, i.e. $\delta(x)$ is the so-called Dirac delta function. It should be noted that $f_n(x)$ depends on a and b, but to avoid a cumbersome notation such as $f_n(x; a, b)$ we write simply $f_n(x)$. In the single barrier case, e.g. $b = \infty$, we would define

$$f_n(x)\,dx = \text{prob}(X_1, X_2, \ldots, X_{n-1} < a, x < X_n < x+dx)$$
$$(n = 1, 2, \ldots; \ -\infty < x < \infty). \quad (61)$$

The function $f_n(x)$ is an important one in the present discussion and so we shall examine it a little more closely. It represents the probability distribution of the particle's position at time n but it does not in general integrate to unity because the particle may be absorbed before time n. The sample functions of a general unrestricted random walk are step functions with discontinuities at the integers. On the graph of a sample function the barriers are represented by horizontal lines at a distance a above and b below the time axis. If we terminate these lines at time $n-1$ thus forming a 'pipe' then

$$\int_{-\infty}^{\infty} f_n(x)\,dx$$

is the probability measure of all sample functions emanating from the origin and passing within the 'pipe'.

Further we note that

$$\text{prob}(N > n) = \int_{-b}^{a} f_n(x)\,dx, \quad (62)$$

for the right-hand side is the probability that the particle is still between the barriers at time n, i.e. that absorption occurs after time n. Let p_A denote the probability that absorption ultimately occurs. We shall use (62) to show that $p_A = 1$ if a and b are both finite. We have

$$p_A = \text{prob}(N < \infty) = \lim_{n \to \infty} \text{prob}(N < n)$$

$$= 1 - \lim_{n \to \infty} \int_{-b}^{a} f_n(x)\,dx.$$

Now by an argument we used for the simple random walk the right-hand side of (62) cannot exceed the probability that an unrestricted particle lies in the interval (a, b) at time n, which by the normal approximation (55), tends to zero as $n \to \infty$, provided that $\sigma^2 < \infty$. For a more general argument, not depending on the assumption of finite variance, see Exercise 10, which shows in addition that the probability of the particle being between the barriers at time n tends to zero geometrically fast in n.

By using the moment generating function we can obtain an upper bound for (62) when $\mu \neq 0$. Let $g_n(x)$ be the p.d.f. of an unrestricted particle at time n, so that

$$\int_{-\infty}^{\infty} e^{-\theta x} g_n(x)\, dx = \{f^*(\theta)\}^n.$$

Now for $\theta > 0$ we have

$$\int_{-b}^{a} f_n(x)\, dx \leqslant \int_{-b}^{a} g_n(x)\, dx \leqslant \int_{-b}^{a} e^{-\theta(x-a)} g_n(x)\, dx$$

$$\leqslant e^{\theta a} \int_{-\infty}^{\infty} e^{-\theta x} g_n(x)\, dx$$

$$= e^{\theta a} \{f^*(\theta)\}^n.$$

This holds for all $\theta > 0$ and the left-hand side is independent of θ. When $\mu > 0$ then $\theta_1 > 0$, so that

$$\int_{-b}^{a} f_n(x)\, dx \leqslant e^{\theta_1 a} \{f^*(\theta_1)\}^n, \tag{63}$$

again showing that (62) tends to zero geometrically in n, since $f^*(\theta_1) < 1$. The above argument does not depend on b being finite, so that (63) also holds for the case of a single barrier at a with a drift towards the barrier. The corresponding inequality for $\mu < 0$ (see Exercise 9) is

$$\int_{-b}^{a} f_n(x)\, dx \leqslant e^{-\theta_1 b} \{f^*(\theta_1)\}^n. \tag{64}$$

A further property of $f_n(x)$ is obtained by observing that

$$f_n(x)\, dx = \operatorname{prob}(N = n,\ x < X_N < x + dx) \quad (x < -b \text{ or } x > a,$$
$$n = 1, 2, \ldots),$$

i.e. $f_n(x)\, dx$ $(x < -b$ or $x > a)$ is the joint probability that the time N to absorption is n and that the position reached when absorption occurs is between x and $x + dx$. Hence if we take a moment generating function with respect to n and with respect to x over absorbing states, we have

$$E(e^{-\theta X_N} s^N) = \sum_{n=1}^{\infty} s^n \left\{ \int_{-\infty}^{-b} e^{-\theta x} f_n(x) \, dx + \int_{a}^{\infty} e^{-\theta x} f_n(x) \, dx \right\} \quad (65)$$

Now define the generating function $K(\theta, s)$ of $f_n(x)$ taken with respect to x over the non-absorbing states and with respect to n over $0, 1, 2, \ldots$, i.e.

$$K(\theta, s) = \sum_{n=0}^{\infty} s^n \int_{-b}^{a} e^{-\theta x} f_n(x) \, dx. \quad (66)$$

We shall establish the following identity:

$$E(e^{-\theta X_N} s^N) = 1 - \{1 - sf^*(\theta)\} K(\theta, s). \quad (67)$$

First we obtain a recurrence relation between $f_n(x)$ and $f_{n-1}(x)$. If at time $n-1$ the particle is in the position y $(-b < y < a)$ then to reach the position x at time n the jump Z_n must take the value $x - y$, the distribution of Z_n being independent of y. Hence

$$\text{prob}(x < X_n < x + dx | X_{n-1} = y) = \text{prob}(x - y < Z_n < x - y + dx)$$
$$= f(x - y) \, dx$$

and

$$f_n(x) \, dx = \int_{-b}^{a} \text{prob}(x < X_n < x + dx | X_{n-1} = y) f_{n-1}(y) \, dy$$

$$= \left\{ \int_{-b}^{a} f(x - y) f_{n-1}(y) \, dy \right\} dx.$$

Thus we have the recurrence relation

$$f_n(x) = \int_{-b}^{a} f_{n-1}(y) f(x - y) \, dy \quad (n = 1, 2, \ldots). \quad (68)$$

Taking the Laplace transform of (68), we have that

$$\int_{-\infty}^{\infty} e^{-\theta x} f_n(x) \, dx = \int_{-\infty}^{\infty} e^{-\theta x} \left\{ \int_{-b}^{a} f_{n-1}(y) f(x - y) \, dy \right\} dx$$

$$= \int_{-b}^{a} e^{-\theta y} f_{n-1}(y) \left\{ \int_{-\infty}^{\infty} e^{-\theta(x-y)} f(x - y) \, dx \right\} dy$$

$$= f^*(\theta) \int_{-b}^{a} e^{-\theta y} f_{n-1}(y) \, dy. \quad (69)$$

The product of generating functions on the right-hand side of (69) arises because the position of the particle at time n is given by the addition of an independent random variable Z_n to its position at time $n-1$, provided that it is still between the barriers at time $n-1$.

Second, the expression in parentheses on the right-hand side of (65) may be written

$$\int_{-\infty}^{\infty} e^{-\theta x} f_n(x)\, dx - \int_{-b}^{a} e^{-\theta x} f_n(x)\, dx. \tag{70}$$

Finally combining (65), (66), (69) and (70), we obtain

$$E(e^{-\theta X_N} s^N) = sf^*(\theta) K(\theta, s) - \{K(\theta, s) - 1\}$$
$$= 1 - \{1 - sf^*(\theta)\} K(\theta, s),$$

which is the identity (67).

This identity provides a somewhat general expression for $E(e^{-\theta X_N} s^N)$. The barriers do not enter explicitly and (67) still holds if one of the barriers is placed at ∞, thus producing a single absorbing barrier. In such a case, however, the expectation may be with respect to a probability distribution whose total mass is less than unity.

If the random walk is discrete over the integers, i.e. if the steps $Z_1, Z_2, \ldots$ take on only integral values with p.g.f. $G(z)$, then (67) takes the form

$$E(z^{X_N} s^N) = 1 - \{1 - sG(z)\} L(z, s), \tag{71}$$

where $L(z, s)$ is the discrete analogue of $K(\theta, s)$.

(iv) APPLICATION TO THE SIMPLE RANDOM WALK

The identity (71) is particularly useful for the simple random walk because absorption occurs exactly on a barrier and X_N can take only the values $-b$ and a. If we use the notation of Section 2.2, (71) takes the form

$$z^a F_{0a}(s) + z^{-b} F_{0,-b}(s) = 1 - \{1 - s(pz + 1 - p - q + qz^{-1})\} L(z, s), \tag{72}$$

where

$$L(z, s) = \sum_{k=-b+1}^{a-1} z^k \sum_{n=0}^{\infty} s^n p_k^{(n)}$$
$$= \sum_{k=-b+1}^{a-1} z^k P_k(s), \tag{73}$$

where $P_k(s)$ satisfies equation (31). Now we have already determined $F_{0a}(s)$ and $F_{0,-b}(s)$ in (18) and (26). Thus from (72) we can find $L(z, s)$ and hence $P_k(s)$ by picking out the coefficient of z^k.

We could alternatively use the identity (71) to determine explicitly $E(z^{X_N} s^N)$. For we know that it must be of the form given in the left-hand side of (72). Now $\lambda_1(s)$ and $\lambda_2(s)$ are the roots of

$$1 - sG(z) = 0.$$

If we set $z = \lambda_1(s)$ and $z = \lambda_2(s)$ in turn in (72) the second term on the right-hand side vanishes; we obtain the two equations

$$\{\lambda_1(s)\}^a F_{0a}(s) + \{\lambda_1(s)\}^{-b} F_{0,-b}(s) = 1,$$

$$\{\lambda_2(s)\}^a F_{0a}(s) + \{\lambda_2(s)\}^{-b} F_{0,-b}(s) = 1,$$

which may be solved for $F_{0a}(s)$ and $F_{0,-b}(s)$. Hence in the simple random walk the identity (71) gives us all the generating functions of interest and hence all the results of Section 2.2(ii).

(v) WALD'S IDENTITY

Suppose in (67) that we set $s = \{f^*(\theta)\}^{-1}$. Then the second term on the right-hand side vanishes and we obtain

$$E[e^{-\theta X_N} \{f^*(\theta)\}^{-N}] = 1. \tag{74}$$

This is Wald's fundamental identity of sequential analysis and is a general result for random walks with absorbing barriers having some useful applications. If there is only a single absorbing barrier and if the drift is away from the barrier, then we saw for the simple random walk that ultimate absorption is not a certain event. Hence in such a case the total probability mass in the joint distribution of N and X_N is less than unity and we have to interpret the expectation sign in (74) as being taken over such a distribution. This is quite in order since in all cases the function $f_n(x)$ is well defined and so therefore is the generating function on the right-hand side of (67). If absorption is uncertain, say with a single barrier at a, then

$$\sum_{n=1}^{\infty} \int_a^{\infty} f_n(x)\,dx < 1.$$

Alternatively, if we denote the probability of absorption by p_A and let E_A denote expectation conditional upon the occurrence of absorption, then we may write (74) quite generally in the form

$$p_A E_A[e^{-\theta X_N} \{f^*(\theta)\}^{-N}] = 1. \tag{75}$$

The setting of $s = \{f^*(\theta)\}^{-1}$ in (67) in order to obtain Wald's identity requires justification, for if $K(\theta,s)$ becomes infinite when s takes this value the substitution would not be valid.

For fixed $\theta > 0$ the function $K(\theta,s)$ satisfies

$$K(\theta,s) < \sum_{n=0}^{\infty} s^n e^{\theta b} \int_{-b}^{a} f_n(x)\,dx$$

and in virtue of the inequality (63), the series on the right-hand side converges, provided that

$$sf^*(\theta_1) < 1, \quad \text{i.e. } s < \{f^*(\theta_1)\}^{-1}. \tag{76}$$

The same is easily shown to be true for $\theta < 0$, and so $K(\theta, s)$ is finite for any real θ and for s satisfying (76). It follows that, provided a and b are finite, we may safely make the substitution $s = \{f^*(\theta)\}^{-1}$; for then (76) is automatically satisfied for all real s, since θ_1 is the point at which $f^*(\theta)$ attains its unique minimum. It follows that for finite a and b, (74) is valid for all values of θ for which $|f^*(\theta)| > f^*(\theta_1)$. When either a or b is infinite, the single barrier case, a little more care is needed in order to justify (74), but in practice such cases can be examined by proceeding to the limit.

We have seen that $p_A = 1$ when a and b are both finite. Let p_a denote the probability that the particle is absorbed at the upper barrier and p_{-b} the corresponding probability for the lower barrier. We shall use Wald's identity to obtain an approximation to these quantities. We may write (74) as

$$p_a E[e^{-\theta X_N} \{f^*(\theta)\}^{-N} | X_N \geqslant a] + p_{-b} E[e^{-\theta X_N} \{f^*(\theta)\}^{-N} | X_N \leqslant -b] = 1,$$

(77)

since $X_N \geqslant a$ and $X_N \leqslant -b$ are mutually exclusive and exhaustive events. Now when $\mu \neq 0$ then $\theta_0 \neq 0$ and $f^*(\theta_0) = 1$, so that

$$p_a E[e^{-\theta_0 X_N} | X_N \geqslant a] + p_{-b} E[e^{-\theta_0 X_N} | X_N \leqslant -b] = 1.$$

If we neglect the excess over the barrier and write

$$X_N \simeq a \text{ when } X_N \geqslant a \quad \text{and} \quad X_N \simeq -b \text{ when } X_N \leqslant -b, \quad (78)$$

then we have

$$p_a e^{-\theta_0 a} + p_{-b} e^{\theta_0 b} \simeq 1.$$

Finally, using the fact that

$$p_a + p_{-b} = 1,$$

we obtain the approximations for the absorption probabilities when $\mu \neq 0$

$$p_a \simeq \frac{1 - e^{\theta_0 b}}{e^{-\theta_0 a} - e^{\theta_0 b}}, \qquad p_{-b} \simeq \frac{e^{-\theta_0 a} - 1}{e^{-\theta_0 a} - e^{\theta_0 b}}. \quad (79)$$

These ought to be good approximations for large a and b, for with reasonable distributions any excess over the barrier will then be small compared with a and b. When $\mu = 0$ then $\theta_0 = 0$ and the expressions in (79) become indeterminate. However, if we let $\theta_0 \to 0$ we obtain when $\mu = 0$

$$p_a \simeq \frac{b}{a+b}, \qquad p_{-b} \simeq \frac{a}{a+b}. \quad (80)$$

The approximations (80) are independent of the particular distribution $f(x)$ involved.

Using the same method, we can obtain approximately the p.g.f. of N, the number of steps to absorption. Let us reconsider the substitution

$$s = \{f^*(\theta)\}^{-1} \quad \text{or} \quad 1/s = f^*(\theta).$$

By the convexity property of $f^*(\theta)$ this equation will have two real roots in θ, provided that $1/s > f^*(\theta_1)$. Let us denote the roots by $\lambda_1(s)$ and $\lambda_2(s)$, where $\lambda_1(1) = 0$ and $\lambda_2(1) = \theta_0$. We now obtain Wald's identity in another form, expressed by the following two equations, by setting $\theta = \lambda_1(s)$ and $\theta = \lambda_2(s)$ in turn in the identity (67). Thus

$$E[\exp\{-\lambda_i(s) X_N\} s^N] = 1 \quad (i = 1, 2). \tag{81}$$

Again, using the approximation (78), we have

$$p_a e^{-\lambda_i(s) a} E_a(s^N) + p_{-b} e^{\lambda_i(s) b} E_{-b}(s^N) \simeq 1 \quad (i = 1, 2), \tag{82}$$

where E_a denotes expectation conditional on absorption at the upper barrier a, etc. Using the expressions (79) and (80) for p_a and p_{-b}, the two equations (82) may be solved to obtain approximately the p.g.f. of N, namely

$$E(s^N) = p_a E_a(s^N) + p_{-b} E_{-b}(s^N).$$

Further interesting results can be obtained by observing that the expression on the left of Wald's identity is a function of θ, whereas the right-hand side is a constant. We may therefore expand the expression inside the expectation sign in powers of θ and equate the expected value of the coefficients to zero. This is equivalent to differentiating inside the expectation sign and then setting $\theta = 0$. Remembering that the cumulant generating function $\log f^*(\theta) = -\mu\theta + \frac{1}{2}\sigma^2\theta^2 - \ldots$, we may write (74) as

$$E[\exp\{-(X_N - N\mu)\theta - \tfrac{1}{2}N\sigma^2\theta^2 + \ldots\}] = 1.$$

Hence

$$E(X_N - N\mu) = 0$$

and

$$E\{(X_N - N\mu)^2 - N\sigma^2\} = 0,$$

from the coefficients of θ and θ^2 respectively. Hence

$$E(X_N) = \mu E(N) \tag{83}$$

and

$$E\{(X_N - N\mu)^2\} = \sigma^2 E(N). \tag{84}$$

Thus

$$E(N) = \begin{cases} \dfrac{1}{\mu} E(X_N) & (\mu \neq 0), \\[2mm] \dfrac{1}{\sigma^2} E(X_N^2) & (\mu = 0). \end{cases} \tag{85}$$

Now $E(X_N) \simeq ap_a - bp_{-b}$ and $E(X_N^2) \simeq a^2 p_a + b^2 p_{-b}$, so that from (85) we obtain the following approximation for the expected number of steps to absorption:

$$E(N) \simeq \begin{cases} \dfrac{(a+b) - a\,e^{\theta_0 b} - b\,e^{-\theta_0 a}}{e^{-\theta_0 a} - e^{\theta_0 b}} & (\mu \neq 0), \\[2ex] \dfrac{ab}{\mu \sigma^2} & (\mu = 0). \end{cases} \tag{86}$$

These formulae have important statistical applications in sequential analysis.

Example 2.11. *Random walk with normally distributed steps.* Let us apply some of these results to a random walk in which the steps have a normal distribution with mean μ and variance σ^2. Then

$$f^*(\theta) = \exp\left(-\mu\theta + \tfrac{1}{2}\sigma^2\theta^2\right)$$

and we saw in Example 2.10 that $\theta_0 = 2\mu/\sigma^2$. Then applying the results (79) we find that the probability of absorption at the upper barrier is given by

$$p_a \simeq \frac{1 - \exp\left(2\mu b/\sigma^2\right)}{\exp\left(-2\mu a/\sigma^2\right) - \exp\left(2\mu b/\sigma^2\right)} \quad (\mu \neq 0)$$

and that at the lower barrier by

$$p_{-b} \simeq \frac{\exp\left(-2\mu a/\sigma^2\right) - 1}{\exp\left(-2\mu a/\sigma^2\right) - \exp\left(2\mu b/\sigma^2\right)} \quad (\mu \neq 0).$$

For the expected number of steps to absorption we have

$$E(N) \simeq \frac{(a+b) - a\exp\left(2\mu b/\sigma^2\right) - b\exp\left(-2\mu a/\sigma^2\right)}{\exp\left(-2\mu a/\sigma^2\right) - \exp\left(2\mu b/\sigma^2\right)} \quad (\mu \neq 0).$$

When $\mu = 0$, p_a and p_{-b} are approximately independent of σ^2 and are given by (80), whereas $E(N)$ is given by (86).

Example 2.12. *Mixture of two exponential distributions.* We consider a random walk in which the distribution of each step is a mixture of two exponential distributions, one on $(0, \infty)$ and the other on $(-\infty, 0)$. We choose this somewhat unrealistic example since it illustrates a case where Wald's identity can be used to obtain the exact probabilities of absorption and the exact p.g.f. of N. The simplification in this case arises from the fact, which we shall demonstrate below, that X_N and N are independently distributed. We define for $0 < \alpha < 1$

$$f(x) = \begin{cases} \alpha\nu\,e^{-\nu x} & (x > 0), \\[1ex] (1-\alpha)\lambda\,e^{x\lambda} & (x < 0). \end{cases}$$

Thus

$$f^*(\theta) = \alpha\left(\frac{\nu}{\nu+\theta}\right) + (1-\alpha)\left(\frac{\lambda}{\lambda-\theta}\right).$$

In particular, if $\alpha = \lambda/(\lambda+\nu)$ then

$$f^*(\theta) = \frac{\nu\lambda}{(\nu+\theta)(\lambda-\theta)}, \tag{87}$$

and each step is then the sum of two independent components, one having an exponential p.d.f. $\nu e^{-\nu x}$ on $(0, \infty)$ and the other having an exponential p.d.f. $\lambda e^{\lambda x}$ on $(-\infty, 0)$. Alternatively, each step is the difference of two independent random variables, each with an exponential distribution.

If absorption occurs at, say, the upper barrier, then the step which carries the particle over the barrier must arise from the positive component of the mixture. The excess over the barrier, namely $X_N - a$, is the excess of the exponentially distributed random variable Z_N over the quantity $a - X_{N-1}$ conditional on this excess being positive and, by a well-known property of the exponential distribution, this excess has the same exponential distribution. Hence

$$E(e^{-\theta(X_N-a)}|X_N \geqslant a) = \frac{\nu}{\nu+\theta}$$

and

$$E(e^{-\theta X_N}|X_N \geqslant a) = \frac{\nu e^{-\theta a}}{\nu+\theta}$$

independently of N. Similar remarks apply to the lower barrier and Wald's identity takes the form

$$p_a\left(\frac{\nu e^{-\theta a}}{\nu+\theta}\right)E[\{f^*(\theta)\}^{-N}|X_N \geqslant a]$$
$$+ p_{-b}\left(\frac{\lambda e^{\theta b}}{\lambda-\theta}\right)E[\{f^*(\theta)\}^{-N}|X_N \leqslant -b] = 1. \tag{88}$$

When $f^*(\theta)$ is given by (87), then we find that $\theta_0 = \lambda-\nu$ and on setting $\theta = \lambda-\nu$ in (88) we obtain, remembering that $f^*(\theta_0) = 1$,

$$\left(\frac{\nu}{\lambda}\right)p_a e^{-a(\lambda-\nu)} + \left(\frac{\lambda}{\nu}\right)p_{-b} e^{b(\lambda-\nu)} = 1.$$

Because $p_a + p_{-b} = 1$, we have

$$p_a = \begin{cases} \dfrac{1-(\lambda/\nu)\,e^{b(\lambda-\nu)}}{(\nu/\lambda)\,e^{-a(\lambda-\nu)} - (\lambda/\nu)\,e^{b(\lambda-\nu)}} & (\lambda \neq \nu), \\[4mm] \dfrac{b+\lambda^{-1}}{a+b+2\lambda^{-1}} & (\lambda = \nu), \end{cases} \tag{89}$$

where the second expression is obtained from the first by letting $\nu \to \lambda$. This exact expression for the probability of absorption at the upper barrier may be compared with the approximation derived from (79), namely

$$p_a \simeq \begin{cases} \dfrac{1 - e^{b(\lambda-\nu)}}{e^{-a(\lambda-\nu)} - e^{b(\lambda-\nu)}} & (\lambda \neq \nu), \\[3mm] \dfrac{b}{a+b} & (\lambda = \nu). \end{cases}$$

For further details, see Exercise 14.

(vi) ONE ABSORBING BARRIER

Suppose that the random walk starts at the origin and that there is an absorbing barrier at a. We may examine this case by letting $b \to \infty$ in the two-barrier situation discussed above. Before doing this we can infer from the properties of the unrestricted random walk discussed in (i) that the barrier is certain to be crossed if $\mu = E(Z_i) \geqslant 0$. Hence, letting p_a denote the probability of absorption, i.e. the probability that the barrier is crossed at all, we have, for $\mu \geqslant 0$,

$$p_a = 1.$$

If $\mu < 0$, then we can obtain an approximation by letting $b \to \infty$ in (79). Thus, for $\mu < 0$,

$$p_a \simeq e^{\theta_0 a}, \tag{90}$$

where $\theta_0 < 0$, and a is assumed to be large.

When absorption is certain we can use (82) to obtain approximately the p.g.f. of N, the number of steps to absorption. Setting $p_a = 1$ and $p_{-b} = 0$ in (82), we obtain

$$E(s^N) = E_a(s^N) \simeq e^{a\lambda_i(s)} \quad (i = 1, 2).$$

We choose the root $\lambda_1(s)$ since $\lambda_1(1) = 0$, thus making $E(s^N)$ a proper p.g.f. Hence

$$E(s^N) \simeq e^{a\lambda_1(s)}.$$

Again the approximation involved is that of neglecting the excess over the barrier, i.e. writing $X_N = a$ whenever $X_N \geqslant a$. By differentiating the above expression for $E(s^N)$, we obtain for the first two moments of N when $\mu > 0$

$$E(N) \simeq \frac{a}{\mu}, \qquad V(N) \simeq \frac{\sigma^2 a}{\mu^3}.$$

When $\mu = 0$ these moments become infinite. Since in fact $E(X_N) \geqslant a$, the approximation for $E(N)$ is really an inequality $E(N) \geqslant a/\mu$.

Example 2.13. *Insurance risk.* Consider the insurance company described in Example 2.1. We saw that its capital X_n at time n can be represented as a random walk starting at X_0 (the company's initial capital) with an absorbing barrier at 0. Absorption corresponds to the exhaustion of the capital which implies the ruin of the company.

Let us assume that the net income in each period is constant. Hence let

$$Y_r = \gamma \quad (r = 1, 2, \ldots).$$

The claim totals $W_1, W_2, \ldots$ may each be regarded as the sum of a large number of independent claims and it is therefore reasonable to assume that W_r is normally distributed with, say, positive mean μ and variance σ^2, although in practice this assumption is likely to be vitiated by occasional very large claims. Hence $Z_r = Y_r - W_r$, the change in capital in period r, is normally distributed with mean $\gamma - \mu$ and variance σ^2. Now if $\gamma - \mu \leqslant 0$, then ruin is certain. Therefore the company must arrange its liabilities so that $\gamma > \mu$, i.e. its policies must be arranged so that the net income per period exceeds the average claim. Let $\pi(x_0)$ denote the probability of ruin for an initial capital x_0. Then from (90) it follows that

$$\pi(x_0) \simeq e^{-\theta_0 x_0},$$

where $\theta_0 > 0$ since the barrier is now on the negative side of the starting point. From Example 2.10

$$\theta_0 = \frac{2(\gamma - \mu)}{\sigma^2}$$

and therefore

$$\pi(x_0) \simeq \exp\left\{-\frac{2(\gamma - \mu)x_0}{\sigma^2}\right\}. \tag{91}$$

Using this formula we could find, for example, what initial capital a company would have to borrow to achieve a sufficiently small probability of ruin for given γ, μ and σ^2 (see Exercise 15).

The results for the single-barrier case have been derived by letting one of the barriers in the two-barrier case approach infinity. However, Wald's identity in the general form (75) holds for a single absorbing barrier although the values of θ for which it is valid are not quite so general as in the case of two barriers. See Exercise 11. All the results for the single-barrier case may be obtained directly from Wald's identity.

(vii) REFLECTING BARRIERS

Suppose now that the barriers at $-b$ and a are reflecting barriers. By this we mean that once the particle crosses a barrier, it instantaneously returns to that barrier and remains there until a step of the appropriate sign allows it to move into the region between the barriers. We assume $X_0 = 0$. The equations defining such a process are

$$X_n = \begin{array}{ll} X_{n-1}+Z_n & (-b < X_{n-1}+Z_n < a), \\ a & (X_{n-1}+Z_n \geqslant a), \\ -b & (X_{n-1}+Z_n \leqslant -b), \end{array}$$

where Z_1, Z_2, ..., the steps of the particle, are independent random variables. If the steps are continuously distributed, then after n steps the particle will have a probability distribution consisting of discrete probabilities at $-b$ and a, representing the probabilities that the particle will be located on the respective barriers, and a continuous distribution of probability in the interval $(-b, a)$. In view of this fact it is more convenient to work with the distribution function rather than the density function. Thus let

$$H_n(x) = \operatorname{prob}(X_n \leqslant x).$$

Then $H_n(x)$ is non-decreasing in $-b < x < a$, continuous on the right and

$$H_n(x) = 0 \quad (x < -b), \qquad H_n(-b) = \operatorname{prob}(X_n = -b),$$
$$1 - H_n(a-0) = \operatorname{prob}(X_n = a), \qquad H_n(x) = 1 \quad (x \geqslant a). \tag{92}$$

The jump discontinuities of $H_n(x)$ at $-b$ and a represent the discrete probabilities of locating the particle on the respective barriers. To obtain a recurrence relation for $H_n(x)$ we note that if it is given that the particle is at the position y ($-b \leqslant y \leqslant a$) at time $n-1$ then

$$\operatorname{prob}(X_n \leqslant x | X_{n-1} = y) = \begin{array}{ll} 0 & (x < -b), \\ F(x-y) & (-b \leqslant x < a), \\ 1 & (x \geqslant a), \end{array}$$

where $F(x)$ is the distribution function of Z_n, the nth step. Hence $H_n(x) = 0$ $(x < -b)$, $H_n(x) = 1$ $(x \geqslant a)$, whereas for $-b \leqslant x \leqslant a$

$$\operatorname{prob}(X_n \leqslant x) = H_n(x) = \int_{-b-0}^{a+0} F(x-y) \, dH_{n-1}(y).$$

Integrating by parts on the right-hand side, we obtain, assuming $F'(x) = f(x)$,

$$H_n(x) = F(x-a) H_{n-1}(a+0) - F(x+b) H_{n-1}(-b-0)$$

$$+ \int_{-b}^{a} H_{n-1}(y) f(x-y) \, dy$$

$$= F(x-a) + \int_{-b}^{a} H_{n-1}(y) f(x-y) \, dy \quad (-b \leqslant x < a), \tag{93}$$

in virtue of (92). As in the case of the simple random walk we expect an equilibrium situation to become established after a long time has elapsed, i.e. as $n \to \infty$, $H_n(x) \to H(x)$, the equilibrium distribution function. Letting $n \to \infty$ in (93), we see that $H(x)$ must satisfy the integral equation

$$H(x) = F(x-a) + \int_{-b}^{a} H(y) f(x-y) \, dy \quad (-b \leqslant x < a). \qquad (94)$$

If the steps are not continuously distributed then (94) holds in the form

$$H(x) = F(x-a-0) + \int_{-b}^{a} H(y) \, dF(x-y) \quad (-b \leqslant x < a). \qquad (95)$$

In particular, if the steps Z_n have a discrete distribution $\{f_j\}$ over the integers, then there will be an equilibrium probability distribution $\{h_k\}$ over the integers $k = -b, -b+1, \ldots, a$. Such a process is a Markov chain with discrete states in discrete time and is more appropriately discussed in the context of the next chapter.

The equilibrium distribution has the properties (a) that it is independent of the point at which the random walk starts, since the known quantities in the equation (94) are a, b and $f(x)$ and the only unknown is $H(x)$, and (b) that if the initial position of the particle is random with the probability distribution $H(x)$, then at each subsequent stage its distribution is $H(x)$, i.e. if we set $n = 1$ in (93) and $H_0(x) = H(x)$ then we see that $H_1(x) = H(x)$, and similarly $H_n(x) = H(x)$ for all $n \geqslant 1$.

It is not known in general how to obtain a solution in reasonably explicit terms to the equation (94) and thus it is not possible in general to determine explicitly the equilibrium distribution. However, if one of the barriers is moved to infinity, e.g. if we set $b = 0$, $a = \infty$ then (94) gives that $H(x) = 0$ $(x < 0)$ and, for $x \geqslant 0$,

$$H(x) = \int_{0}^{\infty} H(y) f(x-y) \, dy, \qquad (96)$$

an equation of the *Wiener–Hopf* type which can be solved explicitly in certain special cases; see, for example, Titchmarsh (1948, p. 339) and Noble (1958). It has been shown by Lindley (1952) that there is a unique probability distribution $H(x)$ satisfying (96) if $E(Z_n) < 0$ (negative drift) and no solution if $E(Z_n) \geqslant 0$ (positive or zero drift). This is plausible by the same sort of general argument which was used for the simple random walk in Section 2.2(vi). Spitzer (1957) has shown that the equilibrium distribution $H(x)$ exists under somewhat more general conditions on the density function $f(x)$ and has given a rather complicated explicit expression for the characteristic function of $H(x)$.

Example 2.14. *Waiting time in the single-server queue.* We give here an example of a process which, on close examination, is seen to have the structure of a random walk although initially it appears to be quite unlike one. The process in question is the waiting time process for the queue with a single server and the formulation is due to Lindley (1952).

Suppose that the queue is initially empty and that the server serves the customers in order of arrival. Customers arrive at times T_0, $T_0 + T_1$, $T_0 + T_1 + T_2$, ... Thus T_n is the time interval between the arrival of the nth and $(n+1)$th customer. We have here a process in continuous time but we use the method of imbedding mentioned in Example 1.4 and examine the process only at the time instants when successive customers commence their service. Essentially, our discrete time variable now represents the serial number of customers in order of arrival. We assume that the inter-arrival times T_n form a sequence of mutually independent, identically distributed random variables. Let S_n denote the service-time of the nth customer and we assume that these also form a sequence of mutually independent, identically distributed random variables, independent of the T_n.

Let W_n denote the waiting time of the nth customer so that W_n is the time that elapses between the arrival of the nth customer and the commencement of his service. If the queue is empty when the nth customer arrives, then $W_n = 0$. Finally let $Z_n = S_n - T_n$. Then we shall show that

$$W_{n+1} = \begin{array}{ll} W_n + Z_n & (W_n + Z_n > 0), \\ 0 & (W_n + Z_n \leqslant 0). \end{array} \qquad (97)$$

This is the equation of a random walk on the positive half-axis with a reflecting barrier at 0.

To prove (97), consider the nth customer. He arrives at time $Y_n = T_0 + T_1 + \ldots + T_{n-1}$, waits for a time W_n and is served for a time S_n. Thus he completes his service at time $Y_n + W_n + S_n$. If by this time the next customer has already arrived, i.e. if $Y_n + T_n < Y_n + W_n + S_n$, the nth customer now commences his service and

$$Y_n + T_n + W_{n+1} = Y_n + W_n + S_n \quad (Y_n + T_n < Y_n + W_n + S_n),$$

i.e. $\quad W_{n+1} = W_n + (S_n - T_n) \qquad \{W_n + (S_n - T_n) > 0\}. \qquad (98)$

On the other hand if the $(n+1)$th customer has not arrived by the time that the nth customer completes his service, i.e. if $Y_n + T_n \geqslant Y_n + W_n + S_n$, then the queue will be empty when the $(n+1)$th customer does arrive and he will not have to wait. Thus

$$W_{n+1} = 0 \quad (Y_n + T_n \geqslant Y_n + W_n + S_n),$$

i.e. $\quad W_{n+1} = 0 \quad \{W_n + (S_n - T_n) \leqslant 0\}. \qquad (99)$

Equations (98) and (99) are the same as (97).

Thus the structure of the waiting time process is that of a random walk with a reflecting barrier. This illustrates the usefulness of trying to identify the process under consideration, or some aspect of it, with a common or more basic process. It follows from the results on the random walk that W_n will have an equilibrium distribution if $E(S_n - T_n) < 0$, i.e. if $E(S_n) < E(T_n)$ or, in words, the mean service-time is less than the mean arrival interval. This means that the queue will be stable if the average rate at which customers are served is greater than the average rate at which they arrive. On the other hand if $E(S_n) \geqslant E(T_n)$ then W_n will not have an equilibrium distribution as $n \to \infty$; as more customers arrive their waiting times will tend to become longer and not to settle down to a state of statistical equilibrium.

(viii) EQUIVALENCE BETWEEN ABSORBING AND REFLECTING BARRIER PROBLEMS

Hitherto we have considered separately the cases of absorbing and reflecting barriers. We shall show now that these two types of random walk are in a sense mathematically equivalent. In particular, we shall show that if we have determined the absorption probability at one of the barriers (and hence at the other) for an arbitrary starting point then we can immediately write down the equilibrium distribution for the corresponding reflecting barrier situation. More specifically, suppose we have absorbing barriers at $-a$ and a (we can without loss of generality choose the origin midway between the barriers) and let $Q(x)$ denote the probability that absorption occurs at $-a$ when the walk starts at x $(-a \leqslant x \leqslant a)$. Let $H(x)$ denote the equilibrium distribution function when the barriers are reflecting. Then we shall show that

$$H(x) = Q(-x), \tag{100}$$

a result due to Lindley (1959).

Consider first the case where the steps are continuously distributed with p.d.f. $f(x)$. Suppose that the barriers are reflecting. Let $H_n(x)$ $(-a \leqslant x \leqslant a)$ be the distribution function of the particle's position after n steps when initially it is at the upper barrier. Then

$$H_0(x) = \begin{matrix} 1 & (x \geqslant a), \\ 0 & (x < a), \end{matrix} \tag{101}$$

and $H_n(x)$ satisfies equation (93), namely

$$H_n(x) = F(x-a) + \int_{-a}^{a} H_{n-1}(y) f(x-y) \, dy \quad (n \geqslant 1, \, -a \leqslant x < a). \tag{102}$$

Next suppose that the barriers are absorbing and for a walk starting at x let $Q_n(x)$ be the probability that absorption occurs at the barrier $-a$

at or before the nth step. Then in order to find a recurrence relation for $Q_n(x)$ we note that absorption can occur in two mutually exclusive ways: either it occurs at the first step with probability $F(-x-a)$ or the particle moves to y ($-a < y < a$) at the first step and absorption occurs at one of the steps $2, \ldots, n$. Thus

$$Q_n(x) = F(-x-a) + \int_{-a}^{a} f(y-x) Q_{n-1}(y) \, dy \quad (n \geq 1, \, -a < x < a),$$

and

$$Q_0(x) = \begin{array}{ll} 1 & (x = -a), \\ 0 & (x > -a). \end{array}$$

Rewriting in terms of $-x$, we have

$$Q_0(-x) = \begin{array}{ll} 1 & (x = a), \\ 0 & (x < a), \end{array} \tag{103}$$

and

$$Q_n(-x) = F(x-a) + \int_{-a}^{a} f(x-y) Q_{n-1}(-y) \, dy \quad (n \geq 1, \, -a < x < a).$$

From (101), (102) and (103) we see that $H_n(x)$ and $Q_n(-x)$ satisfy the same initial conditions and the same recurrence relation. It follows that for all n

$$Q_n(-x) = H_n(x) \tag{104}$$

and the result (100) follows on taking the limit as $n \to \infty$.

Some care is needed when interpreting the results (100) and (104) for $x = a$ and $x = -a$ since $H(x)$ has discontinuities at these points. We can overcome the difficulty by interpreting (100) and (104) as holding for $-a < x < a$. The discontinuities of $H_n(x)$ at $-a$ and a will then be given by

$$H_n(-a+0) = \lim_{x \to a-0} Q_n(x),$$
$$1 - H_n(a-0) = 1 - \lim_{x \to -a+0} Q_n(x). \tag{105}$$

When the steps of the random walk are discrete over the integers then we must define $Q_n(x)$ in terms of absorbing barriers at $a+1$ and $-a$ and $H_n(x)$ (right-continuous) in terms of reflecting barriers at a and $-a$. Then the results (100) and (104) hold for $x = -a, \ldots, a$.

Example 2.15. *Simple random walk.* In the simple random walk with reflecting barriers at $-a$ and a the equilibrium probability distribution is by equation (47)

$$\pi_k = \frac{1-\left(\dfrac{p}{q}\right)}{1-\left(\dfrac{p}{q}\right)^{2a+1}}\left(\frac{p}{q}\right)^{a+k} \qquad (k = -a, \ldots, a; p \neq q).$$

Now define

$$H(x) = \sum_{k=-a}^{x} \pi_k$$

$$= \frac{1-\left(\dfrac{p}{q}\right)^{a+1+x}}{1-\left(\dfrac{p}{q}\right)^{2a+1}} \qquad (x = -a, \ldots, a).$$

Now consider the simple random walk with absorbing barriers at $-a$ and $a+1$ and let $Q(x)$ denote the probability that absorption occurs at $-a$. Then from the result (20) we have

$$Q(x) = \frac{1-\left(\dfrac{p}{q}\right)^{a+1-x}}{1-\left(\dfrac{p}{q}\right)^{2a+1}} \qquad (x = -a, \ldots, a)$$

and hence $Q(-x) = H(x)$, illustrating the general result (100) in this instance.

Example 2.16. *Mixture of two exponential distributions.* Consider the random walk discussed in Example 2.12 where each step is the sum of independent positive and negative exponential variates. Adapting equation (89) to absorbing barriers at $-a$ and a and initial position x we have

$$Q(x) = \begin{cases} \dfrac{1-\left(\dfrac{\lambda}{\nu}\right)e^{(a-x)(\lambda-\nu)}}{1-\left(\dfrac{\lambda}{\nu}\right)^{2}e^{2a(\lambda-\nu)}} & (\lambda \neq \nu), \\[4mm] \dfrac{a+\lambda^{-1}-x}{2(a+\lambda^{-1})} & (\lambda = \nu). \end{cases} \qquad (106)$$

This is obtained from (89) by replacing a and b by $a-x$ and $b+x$ respectively and then putting $Q(x) = 1-p_a$. The general result (100) now enables us to obtain the equilibrium distribution $H(x)$ when the barriers are reflecting by simply writing $H(x) = Q(-x)$. The discrete probabilities of locating the particle on the lower and upper barriers in the equilibrium situation are given respectively by $H(-a)$ and $1 - H(a)$. The continuous distribution between the barriers is thus a truncated exponential one if $\lambda \neq \nu$ and uniform if $\lambda = \nu$.

It might seem that the approximation afforded by Wald's identity in the case of absorbing barriers ought to give an approximation to the equilibrium distribution function in the case of reflecting barriers. However, Wald's approximation holds when the barriers are far from the starting point and if we use it in (100) we can only expect to approximate to the equilibrium distribution function $H(x)$ when x is roughly midway between the barriers. But we get no approximation to the discrete probabilities on the barriers and these are important in the reflecting barrier case.

For a single barrier there is also a duality between the absorbing and reflecting barrier cases. Consider a random walk on the non-negative half-line with identically distributed steps $Z_1, Z_2, \ldots$ and suppose that zero is a *reflecting* barrier. If $E(Z_n) < 0$ then the equilibrium distribution $H(x)$ exists and satisfies equation (96) where $f(x)$ is the p.d.f. of the steps Z_n. Now consider a random walk with the same $f(x)$ on the non-positive half-line starting at $-x$ $(x > 0)$ with zero as an *absorbing* barrier. Let $Q(x)$ be the probability that absorption at zero ultimately occurs. Then

$$H(x) = 1 - Q(x) \quad (x > 0), \tag{107}$$

and

$$H(0+) = \lim_{x \to 0+} \{1 - Q(x)\}$$

is the discrete probability at zero in the reflecting barrier case. The proof of (107), which is similar to that of (100), is left as an exercise for the reader.

Equation (107) holds quite generally, even without assumptions concerning the continuity of the steps, provided we define $H(x)$ as being left-continuous, i.e. define $H(x) = \text{prob}(X_\infty < x)$, not $\text{prob}(X_\infty \leq x)$.

Example 2.17. *Waiting time in the single-server queue.* We can use the result (107) to examine the equilibrium distribution of the waiting time in a single-server queue, Example 2.14. We shall show that when the service-times are exponentially distributed then for any distribution of inter-arrival times the equilibrium distribution of the waiting time is also an exponential distribution with a discrete probability at zero, provided of course that the queue is stable, i.e. the mean service-time is less than the mean inter-arrival time. This result is due to Smith (1953).

In the notation of Example 2.14, let

$$E(e^{-\theta S_n}) = \frac{\beta}{\beta+\theta}$$

$$E(e^{-\theta T_n}) = g^*(\theta)$$

where we assume that

$$\frac{1}{\beta} < -g^{*\prime}(0),$$

i.e. that $E(S_n) < E(T_n)$. Let $f^*(\theta)$ be the m.g.f. of the distribution of $Z_n = S_n - T_n$, i.e.

$$f^*(\theta) = \frac{\beta}{\beta+\theta} g^*(-\theta), \tag{108}$$

under the assumptions of Example 2.14 concerning the independence of the S_n and T_n. If we consider a random walk on the non-negative half-line whose steps have a distribution with m.g.f. (108), and with a reflecting barrier at zero, then we showed in Example 2.14 that the equilibrium distribution of this random walk is the equilibrium waiting time distribution of the queue; denote this distribution by $H(x)$.

In order to find $H(x)$ we consider a dual absorbing barrier problem, i.e. we consider a random walk starting at $-x$ with an absorbing barrier at zero and suppose the distribution of the steps has m.g.f. (108). Equivalently we may assume that the starting point is 0 and the absorbing barrier is at x. Let $Q(x)$ be the probability that absorption ultimately occurs. Then using Wald's identity in the form (75), we have

$$Q(x) E_A[e^{-\theta X_N}\{f^*(\theta)\}^{-N}] = 1. \tag{109}$$

By an argument similar to that used in Example 2.12 it will be seen that $X_N - x$, the excess over the barrier, must have the same exponential distribution as the positive component of the Z_n, conditional on crossing the barrier at all, i.e.

$$E_A\{e^{-\theta(X_N-x)}\} = \frac{\beta}{\beta+\theta},$$

independently of N. Thus

$$E_A\{e^{-\theta X_N}\} = \frac{\beta e^{-\theta x}}{\beta+\theta}$$

and (109) becomes

$$Q(x)\frac{\beta e^{-\theta x}}{\beta+\theta} E_A[\{f^*(\theta)\}^{-N}] = 1. \tag{110}$$

Now let θ_0 be the real non-zero root of $f^*(\theta) = 1$, i.e. of the equation

$$g^*(-\theta) = 1 + \frac{1}{\beta}\theta. \tag{111}$$

This root always exists since, for real θ, $g^*(-\theta)$ is a convex function satisfying $g^*(0) = 1$ and $g^{*'}(0) > 1/\beta$. Thus the curve $g^*(-\theta)$ and the line $1 + \theta/\beta$ must intersect at some value $\theta_0 < 0$. Setting $\theta = \theta_0$ in (110) we obtain

$$Q(x) = \frac{\beta + \theta_0}{\beta} e^{\theta_0 x} \quad (\theta_0 < 0).$$

Hence by the result (107), the equilibrium distribution of the waiting time is

$$H(x) = 1 - Q(x) = 1 - \frac{\beta + \theta_0}{\beta} e^{\theta_0 x} \quad (x > 0),$$

an exponential distribution with the discrete probability $H(0) = -\theta_0/\beta$ at $x = 0$.

2.4. Further topics

We now consider some further topics on the subject of the random walk. These are more advanced and may be omitted in a first reading. In any case, much of this section is of a fairly general and discursive nature rather than detailed and mathematical.

(i) THE MULTIDIMENSIONAL RANDOM WALK

So far in this chapter we have been concerned only with the random walk on a line, i.e. the one-dimensional random walk. From the point of view of the motion of a particle it is of course more natural to consider a random walk on a plane or in three-dimensional Euclidean space or even in non-Euclidean spaces, for example on a sphere. We shall confine our brief remarks to Euclidean space and for the sake of generality to the case of m dimensions.

When we leave one dimension the possibilities become much richer and the mathematical problems much more difficult, especially so when there are boundaries present. Consider for example a random walk in the plane. A particle starts at the origin and undergoes steps $\mathbf{Z}_1$, $\mathbf{Z}_2$, ... where the $\mathbf{Z}_i$ are independent two-dimensional random vectors with a given bivariate distribution. After n steps the position of the particle is given by the two-dimensional vector

$$\mathbf{X}_n = \mathbf{Z}_1 + \ldots + \mathbf{Z}_n. \tag{112}$$

If we try to generalize from one dimension the idea of absorbing barriers we find that any region Ω containing the origin may be defined as a non-absorbing region, while the remainder $\bar{\Omega}$ of the plane is an absorbing region. The possible kinds of boundaries are much richer. Even the simplest boundaries, e.g. a square or rectangle, produce intractable mathematical problems if the steps are other than of the simplest kind. Some boundaries reduce the problem to the one-dimensional case. For example, for continuous steps in the plane with a single straight line absorbing boundary we may take axes perpendicular and parallel to the boundary. From the point of view of the time to absorption, only motion along the axis perpendicular to the boundary is relevant and this is a one-dimensional problem. This is not so, however, if the random walk takes place on a lattice unless the boundary is parallel to one of the axes.

In m dimensions we can generalize directly the results given by the central limit theorem and the laws of large numbers. The central limit theorem gives an approximation to the distribution of the position vector after a large number of steps. If the individual steps are independent, identically distributed m-dimensional vectors $\mathbf{Z}_1$, $\mathbf{Z}_2$, ... with finite second moments then (112) again gives the position vector $\mathbf{X}_n$ after n steps. By the central limit theorem $\mathbf{X}_n$ has asymptotically the multivariate normal distribution with mean vector $n\mathbf{\mu}$ and dispersion matrix $n\mathbf{\Sigma}$ where $\mathbf{\mu}$ is the mean and $\mathbf{\Sigma}$ the dispersion matrix of the individual steps. If the $\mathbf{Z}_i$ have finite mean $\mathbf{\mu}$ but not necessarily finite second moments, then according to the strong law of large numbers we have, with probability as close to unity as we please and n_0 sufficiently large,

$$\mathbf{X}_n = n\mathbf{\mu} + o(n_0)$$

for all $n \geqslant n_0$, where $o(n_0)$ depends only on n_0 and not on n. More specifically if $\mathbf{X}_n = (X_n^{(1)}, \ldots, X_n^{(m)})$, $\mathbf{\mu} = (\mu^{(1)}, \ldots, \mu^{(m)})$ then

$$\text{prob}\left(\left|\frac{X_n^{(1)}}{n} - \mu_1\right| > \epsilon_1, \left|\frac{X_n^{(2)}}{n} - \mu_2\right| > \epsilon_2, \ldots, \left|\frac{X_n^{(m)}}{n} - \mu_m\right| > \epsilon_m \right.$$
$$\left. \text{for all } n \geqslant n_0\right) < \delta$$

for any positive ϵ_1, ..., ϵ_m, δ and n_0 sufficiently large.

We may note also that the identity (67), developed for the case of a one-dimensional random walk with absorbing barriers, extends to the random walk in several dimensions with an absorbing region. Suppose that we consider the random walk

$$\mathbf{X}_n = \mathbf{X}_0 + \mathbf{Z}_1 + \ldots + \mathbf{Z}_n; \tag{113}$$

where $\mathbf{X}_0$ is a given point in m-dimensional space (the initial position) and the steps $\mathbf{Z}_i$ are mutually independent, identically distributed m-dimensional random variables. We denote the non-absorbing region of m-dimensional space by Ω, where we suppose that $\mathbf{X}_0$ is a point in Ω, while the complementary region $\bar{\Omega}$ is the absorbing region. Let N denote the time at which the walk first enters $\bar{\Omega}$. Then if $F_n(dx)$ denotes the probability

$$F_n(dx) = \text{prob}(\mathbf{X}_1 \in \Omega, \ldots, \mathbf{X}_{n-1} \in \Omega, \mathbf{X}_n \in dx) \tag{114}$$

we have

$$E(e^{-\theta' X_N} s^N) = 1 - \{1 - sF^*(\theta)\}\left\{ \sum_{n=0}^{\infty} s^n \int_{\Omega} e^{-\theta' x} F_n(dx)\right\}, \quad (115)$$

where $F^*(\theta)$ is the m.g.f. of the Z_i. The result may be proved in a similar manner to the one-dimensional identity (67).

(*ii*) THE RESULTS OF CHUNG AND FUCHS

We have seen in Section 2.2(iii) that a particle undergoing a simple symmetrical random walk in one dimension is certain to reach any given position. In the next chapter on Markov chains we describe such behaviour by saying that the process is *recurrent*. Chung and Fuchs (1951) investigated the corresponding problem for general random variables in one, two and three dimensions. If

$$X_n = Z_1 + \ldots + Z_n$$

is a random walk in one, two or three dimensions then the value b is said to be *possible* if, for any $\epsilon > 0$, the region $|X - b| < \epsilon$ can be reached with non-zero probability. Here $|Y|$ denotes the maximum of the absolute values of the components of Y. The value b is said to be *recurrent* if the region $|X - b| < \epsilon$ is certain to be reached. Chung and Fuchs showed in the one-dimensional case that if

$$E(|Z_i|) < \infty, \qquad E(Z_i) = 0,$$

then every possible value is recurrent and in the two-dimensional case that if

$$E(Z_i^2) < \infty, \qquad E(Z_i) = 0,$$

where Z_i^2 is the square length of the vector Z_i, then again every possible value is recurrent. However in the three-dimensional random walk no value is recurrent.

(*iii*) SPITZER'S IDENTITY

Returning to the one-dimensional random walk, we consider the distribution of the maximum distance reached by a particle, say in the positive direction, during n steps. Accordingly let

$$U_n = \max(0, X_1, \ldots, X_n).$$

In Section 2.2(iv) we considered the limiting distribution of U_n for the simple random walk. Using purely combinatorial methods, Spitzer (1956) established a relation between the sequence of distributions of U_n and the sequence of distributions of the unrestricted sums X_n. Let

$$V_n = \max(0, X_n).$$

Then Spitzer proved, for an arbitrary distribution of the steps, the identity

$$\sum_{n=0}^{\infty} s^n E(e^{-\theta U_n}) = \exp\left\{\sum_{n=1}^{\infty} \frac{s^n}{n} E(e^{-\theta V_n})\right\}, \tag{116}$$

and various results concerning the limiting distribution of U_n and the number of positive X_n's follow from (116). For details the reader is referred to Spitzer's paper. Later authors, for example Wendel (1958), have proved (116) using purely analytical methods.

Bibliographic Notes

The insurance risk problem, Example 2.1, and more general forms of it, have been extensively studied, especially by Scandinavian mathematicians; for a discussion, historical note and further references, see Cramér (1954). Stochastic models for the content of storage systems, such as Example 2.2 are discussed in the monograph of Moran (1959) and the review papers of Gani (1957) and Prabhu (1964). The simple random walk, Section 2.2, is discussed extensively by Feller (1957, Chapter 14) for the case $p+q = 1$. The discussion of the general random walk with absorbing barriers, Section 2.3(iii), follows Miller (1961a). Further discussion of the general random walk and, in particular, the connexion with the Wiener–Hopf equation (96), may be found in the monograph of Kemperman (1961). In recent years, research has been directed towards exploring the connexion between the random walk and potential theory; for a discussion and references see Spitzer (1964).

Exercises

1. Let $F(x)$ be a distribution function with moments $\mu_1, \mu_2', \mu_3', \ldots$ about the origin. Examine how far it is possible to approximate to $F(x)$ by a discrete distribution over the three points $-l, 0, l$ by means of equating moments.

2. Show in Example 2.9 that the time intervals between the arrivals of successive customers and the service times of successive customers each have a geometric distribution.

3. If $a \to \infty$ in Example 2.9 show that there is an equilibrium distribution of queue size if $\alpha < \beta$ and find this distribution.

4. Use the methods of Section 2.2(iii) in Example 2.9 to find the p.g.f. of the server's busy period when $a = \infty$.

5. Consider a simple random walk with $p+q = 1$. The walk starts at j

$(0 < j \leqslant a)$ where 0 is an absorbing barrier and a a reflecting barrier. Let

$$F_j(s) = \sum_{n=0}^{\infty} s^n f_j^{(n)}$$

be the p.g.f. of N, the time to absorption and let

$$\lambda_1(s), \lambda_2(s) = \frac{1}{2ps}\{1 \pm (1 - 4pqs^2)^{\frac{1}{2}}\}.$$

Show that

(i)
$$F_j(s) = \left(\frac{q}{p}\right)^j \frac{\lambda_1^{a-j+1} - \lambda_2^{a-j+1} - \lambda_1^{a-j} + \lambda_2^{a-j}}{\lambda_1^{a+1} - \lambda_2^{a+1} - \lambda_1^{a} + \lambda_2^{a}}$$

(ii)
$$E(N) = \begin{cases} \dfrac{j}{q-p} + \dfrac{p^{a+1}}{q^a(q-p)^2}\left\{1 - \left(\dfrac{q}{p}\right)^j\right\} & (p \neq q), \\[2ex] j + j(2a - j) & (p = q = \tfrac{1}{2}) \end{cases}$$

(Weesakul, 1961).

6. Suppose X is a two-sided random variable (i.e. satisfying the condition (56)) with p.d.f. $f(x)$ and d.f. $F(x)$, where $F(x)$ satisfies

$$1 - F(x) = O(e^{-\mu x}) \quad (x \to \infty)$$
$$F(x) = O(e^{\lambda x}) \quad (x \to -\infty)$$

for some $\lambda > 0$, $\mu > 0$ (roughly speaking the distribution of X has exponentially small 'tails'). Show that the integral (57) defining $f^*(\theta)$ converges at least in the interval $-\mu < \theta < \lambda$.

7. Let $X = Y + Z$ where Y is a positive random variable with p.d.f. $\alpha e^{-\alpha y}$ and Z a negative random variable with p.d.f. $\beta e^{\beta z}$. Find the m.g.f. $f^*(\theta)$ of X and show that the interval of convergence of (57) is $-\alpha < \theta < \beta$. Find the roots θ_0 and θ_1 and evaluate $f^*(\theta_1)$, verifying (58). Sketch $f^*(\theta)$.

8. Translate the results of Section 2.3(ii) concerning m.g.f.'s into results for discrete random variables concerning p.g.f.'s by making the substitution $z = e^{-\theta}$, so that, for example, $\theta = 0$ corresponds to $z = 1$.

9. Obtain the inequality (64).

10. Let N denote the time to absorption in a general random walk between two absorbing barriers a and $-b$. Without assuming any conditions on the moments of the steps $Z_1, Z_2, \ldots$, use (62) and (68) to prove that $\text{prob}(N > n) = O(\rho^n)$ for some ρ satisfying $0 < \rho < 1$.

11. Using inequalities similar to (63) and (64) show that the substitution $s = \{f^*(\theta)\}^{-1}$ in (67), and hence Wald's identity (75), is valid in the case of a single absorbing barrier at a provided that θ satisfies

$$|f^*(\theta)| > f^*(\theta_1) \quad \text{and} \quad \mathscr{R}(\theta) < \theta_1.$$

12. In the simple random walk starting at 0 with absorbing barriers at $-b$ and a use (20) and (21) to show that the moments of X_N are given by

$$E(X_N^k) = \begin{cases} \dfrac{a^k p^a(p^b - q^b) + (-b)^k q^b(p^a - q^a)}{p^{a+b} - q^{a+b}} & (p \neq q), \\[3mm] \dfrac{ba^k + a(-b)^k}{a+b} & (p = q). \end{cases}$$

Hence, using (85), prove that

$$E(N) = \begin{cases} \dfrac{ap^a(p^b - q^b) - bq^b(p^a - q^a)}{(p-q)(p^{a+b} - q^{a+b})} & (p \neq q), \\[3mm] \dfrac{ab}{p+q} & (p = q). \end{cases}$$

13. Show how Wald's identity may be used to find the exact p.g.f. of N in Example 2.12.

14. Suppose we let $b \to \infty$ in Example 2.12 so that we have a single barrier at a. Show that absorption is certain if $\nu \leqslant \lambda$ and that in this case the p.g.f. of N is

$$E(s^N) = \frac{[\lambda + \nu - \{(\lambda+\nu)^2 - 4\lambda\nu s\}^{\frac{1}{2}}]}{2\nu} \exp\left[\frac{\lambda - \nu - \{(\lambda+\nu)^2 - 4\lambda\nu s\}^{\frac{1}{2}}}{2}\right].$$

Find the probability of absorption when $\nu > \lambda$.

15. A man wishing to start a certain type of insurance business assesses that he has m prospective policy holders. From past statistical data on policies of this type the yearly claim per policy can be assumed to have mean γ and standard deviation σ. Running expenses are δ per policy per annum and interest charges on borrowed capital are 100α per cent per annum. If the man is willing to accept a probability β of ruin then by using (91), show that in order to minimize the average annual premium per policy he must borrow the initial sum

$$x_0 = \sigma\left\{\frac{m}{2\alpha}\log\left(\frac{1}{\beta}\right)\right\}^{\frac{1}{2}}.$$

Find the minimum average annual premium.

CHAPTER THREE

Markov Chains

3.1. Introduction

In this chapter we consider the class of Markov processes in discrete time with a discrete state space. We call such processes Markov chains. An example of a Markov chain was given in Example 1.2 and a formal definition of a Markov process in Section 1.3. Thus we may define a Markov chain as a sequence $X_0, X_1, \ldots$ of discrete random variables with the property that the conditional distribution of X_{n+1} given $X_0, X_1, \ldots, X_n$ depends only on the value of X_n but not further on $X_0, X_1, \ldots, X_{n-1}$; i.e. for any set of values $h, j, \ldots, k$ belonging to the discrete state space,

$$\text{prob}(X_{n+1} = k | X_0 = h, \ldots, X_n = j) = \text{prob}(X_{n+1} = k | X_n = j).$$

As in Section 1.3, it is not difficult to see that an equivalent definition of a Markov chain is one which states that for any finite set of time points $l < \ldots < m < n$

$$\text{prob}(X_n = k | X_l = h, \ldots, X_m = j) = \text{prob}(X_n = k | X_m = j).$$

Thus for the conditional distribution of X_n, given values assumed by the process at a set of times before time n, only the value assumed at the latest time in that set is relevant.

A sequence of independent discrete random variables is a trivial example of a Markov chain. Also the simple random walk discussed in the previous chapter is an example of a Markov chain; for if we are given the position of the particle at time $n-1$, we can, knowing the distribution of the steps, calculate the distribution of its position at time n, and indeed at any later time m $(m = n, n+1, n+2, \ldots)$. Further, this calculation is not affected by information about the position before time $n-1$. Examples 1.7 and 1.8 illustrate processes with discrete states in discrete time that are not Markov chains.

From the theoretical point of view the subject of Markov chains is an elegant one and lends itself to a logical and rigorous treatment at a fairly elementary mathematical level. For this reason the present chapter is rather more formal than the remainder of the book.

The present chapter is nearly self-contained. Some of the examples, however, are drawn from Chapter 2. Section 3.8, which deals with the systems of linear equations associated with irreducible chains, is more advanced and may be omitted at a first reading. Sections 3.10 and 3.11 which deal with the application of the theory of non-negative square matrices to finite Markov chains are rather specialized and may also be omitted on a first reading, although these sections are necessary for a complete treatment of the finite Markov chain.

In describing the simple random walk in Chapter 2 we often referred to the position of the particle at a given time as the state occupied by that particle. In the present chapter we shall frequently use this terminology. Since we are dealing with a discrete state space we shall usually denote the possible states of the system by the integers, usually non-negative, and the state of the system at time n by X_n. Thus $X_n = k$ means that the system is in state k at time n. The somewhat vague term system means the physical phenomenon which we are trying to represent. The random variable X_n need not be a genuine numerical quantity. Thus, in Example 1.2, the state space of the stochastic process representing the day-to-day inspection of a component is the set of three integers $(0, 1, 2)$ denoting satisfactory, unsatisfactory and failed, respectively. The statement $X_n = 1$ means that the component is found to be unsatisfactory on day n. On the other hand in the case of the simple random walk, X_n is a genuine numerical quantity representing the distance of the particle from the origin on a given axis at time n.

Example 3.1. *The mover–stayer model.* In a study of the mobility of labour in American industry Blumen, Kogan and McCarthy (1955) developed what they called a mover–stayer model, of which the following is a very brief description. Industry is divided into a number, say c, of different categories. By taking a sample of workers and observing the records of their past histories, the movement of each member of the sample among the categories and the durations of stay in each category can be recorded. To explain the data the authors of the study proposed that the population of workers be divided into two classes – the movers and the stayers. A stayer remains in the same industrial category throughout his working life. A mover changes categories in such a way that if at the end of period $n-1$ he is in category j, then at the end of period n he will be in category k with probability p_{jk}, where $p_{j1} + p_{j2} + \ldots + p_{jc} = 1$. Over a long period of time, changes in economic conditions will make the p_{jk} vary with n, but for the sake of simplicity we assume that p_{jk} does not change with time. Thus as far as the movers are concerned, the process of an individual's movement among the c categories is a Markov chain: if a mover is in category j at the end of period n then we can calculate the probability that he will be in category k at the end of m further periods. The realism of the assumption that a mover's category at the end of any given period depends only upon where he was at the end of one earlier period and not on more than one earlier periods can of course be questioned but we shall not do so here. In this model the states of the system are the industrial categories $1, \ldots, c$ and X_n denotes the category of a randomly selected mover at the end of period n. Each individual mover's history is a realization of the process. From the statistical point of view the problem is to estimate from the data the quantities p_{jk} and the proportion of movers in each category and then to test the adequacy of the model.

For further details, see the original study and also Goodman (1961).

3.2. A two-state Markov chain

Let us consider in some detail a Markov chain with two states. This is the simplest non-trivial state space. We may regard one of the states as 'success', denoted by 1 and the other as 'failure', denoted by 0. We thus have an example of dependent Bernoulli trials in which the probability of success or failure at each trial depends on the outcome of the previous trial.

Suppose that if the nth trial results in failure then the probability of failure at the $(n+1)$th trial is $1-\alpha$ and the probability of success at the $(n+1)$th trial is α. Similarly if the nth trial results in success then there are probabilities $1-\beta$ and β of success and failure respectively at the $(n+1)$th trial. Alternatively we may say that if the system is in state 0 at time n then there is a probability $1-\alpha$ of being in state 0 at time $n+1$ and a probability α of being in state 1 at time $n+1$. Similarly if the system is in state 1 at time n then the probabilities of being in state 1, state 0 at time $n+1$ are $1-\beta$, β respectively. These probabilities are called transition probabilities and we may write them in a matrix array, as in Example 1.2, equation (1.5),

$$\mathbf{P} = \begin{array}{c} \\ 0 \\ 1 \end{array} \begin{array}{cc} 0 & 1 \\ \begin{bmatrix} 1-\alpha & \alpha \\ \beta & 1-\beta \end{bmatrix} \end{array}.$$

The matrix element in position (j, k) denotes the conditional probability of a transition to state k at time $n+1$ given that the system is in state j at time n. Note that we are making the assumption that the transition probabilities are independent of time. Also we exclude the somewhat trivial cases (i) $\alpha + \beta = 0$, i.e. $\alpha = 0$, $\beta = 0$; in this case the system remains forever in its initial state; (ii) $\alpha + \beta = 2$, i.e. $\alpha = 1$, $\beta = 1$; in this case the system alternates deterministically between the two states, and if the initial state is given the behaviour of the system is non-random.

Example 3.2. *Rainfall in Tel Aviv.* In a study of rainfall in Tel Aviv, Gabriel and Neumann (1962) found that a two-state Markov chain gave a good description of the occurrence of wet and dry days during the rainy period December, January, February. If we take dry as state 0 and wet as state 1 then Gabriel and Neumann estimated, using relative frequencies from data over 27 years, that during this season the probability α of a wet day following a dry day is 0·250 and the probability β of a dry day following a wet day is 0·338. The data, comprising a total of 2437 days, were

		Dry	Wet	Total
		Actual day		
	Dry	1049	350	1399
Preceding day				
	Wet	351	687	1038

from which the estimated transition probabilities are

$$\mathbf{P} = \begin{bmatrix} 0\cdot750 & 0\cdot250 \\ 0\cdot338 & 0\cdot662 \end{bmatrix}.$$

In the sequel we shall use this example to illustrate some further results for the two-state chain.

Let the row vector $\mathbf{p}^{(n)} = (p_0^{(n)}, p_1^{(n)})$ denote the probabilities of finding the system in states 0 or 1 at time n when the initial probabilities of the two states are given by $\mathbf{p}^{(0)} = (p_0^{(0)}, p_1^{(0)})$. Consider the event of being in state 0 at time n. This event can occur in two mutually exclusive ways; either state 0 was occupied at time $n-1$ and no transition out of state 0 occurred at time n; this has probability $p_0^{(n-1)}(1-\alpha)$. Alternatively state 1 was occupied at time $n-1$ and a transition from state 1 to state 0 occurred at time n; this has probability $p_1^{(n-1)}\beta$. Considerations such as these lead to the following recurrence relations,

$$p_0^{(n)} = p_0^{(n-1)}(1-\alpha) + p_1^{(n-1)}\beta,$$
$$p_1^{(n)} = p_0^{(n-1)}\alpha + p_1^{(n-1)}(1-\beta), \tag{1}$$

which in matrix notation may be compactly written

$$\mathbf{p}^{(n)} = \mathbf{p}^{(n-1)}\mathbf{P}, \tag{2}$$

and on iteration

$$\mathbf{p}^{(n)} = \mathbf{p}^{(n-2)}\mathbf{P}^2 = \ldots = \mathbf{p}^{(0)}\mathbf{P}^n. \tag{3}$$

Thus, given the initial probabilities $\mathbf{p}^{(0)}$ and the matrix of transition probabilities $\mathbf{P}$, we can find the state occupation probabilities at any time n using the relation (3). Denote the element (j, k) of $\mathbf{P}^n$ by $p_{jk}^{(n)}$. If the system is initially in state 0, then $\mathbf{p}^{(0)} = (1, 0)$ and $\mathbf{p}^{(n)} = (p_{00}^{(n)}, p_{01}^{(n)})$. If the system is initially in state 1 then $\mathbf{p}^{(0)} = (0, 1)$ and $\mathbf{p}^{(n)} = (p_{10}^{(n)}, p_{11}^{(n)})$. Thus

$$p_{jk}^{(n)} = \text{prob(state } k \text{ at time } n|\text{state } j \text{ at time } 0).$$

The quantities $p_{jk}^{(n)}$ are called n-step transition probabilities.

Example 3.3. *Rainfall in Tel Aviv (continued)*. Let us evaluate $\mathbf{P}^5$ using the matrix $\mathbf{P}$ of Example 3.2. This can be done by computing successively $\mathbf{P}^2$, $\mathbf{P}^3 = \mathbf{P}^2\mathbf{P}$, $\mathbf{P}^5 = \mathbf{P}^3\mathbf{P}^2$. We find that to three places of decimals,

$$\mathbf{P}^5 = \begin{bmatrix} 0\cdot580 & 0\cdot420 \\ 0\cdot568 & 0\cdot432 \end{bmatrix}.$$

Thus, for example, given that January 1 is a dry day the probability that January 6 is a dry day is $0\cdot580$, while if January 1 is a wet day the probability that January 6 is a dry day is $0\cdot568$.

One question that arises is whether after a sufficiently long period of time the system settles down to a condition of statistical equilibrium in which the state occupation probabilities are independent of the initial conditions. If this is so then there is an equilibrium probability distribution $\boldsymbol{\pi} = (\pi_0, \pi_1)$ and, on letting $n \to \infty$ in (2), $\boldsymbol{\pi}$ will clearly satisfy

$$\boldsymbol{\pi} = \boldsymbol{\pi}\mathbf{P},$$

or
$$\boldsymbol{\pi}(\mathbf{I} - \mathbf{P}) = 0. \qquad (4)$$

Thus
$$\pi_0\alpha - \pi_1\beta = 0, \qquad -\pi_0\alpha + \pi_1\beta = 0.$$

This is a homogeneous system of equations and will only have a non-zero solution if the determinant $|\mathbf{I} - \mathbf{P}|$ vanishes. Clearly $|\mathbf{I} - \mathbf{P}|$ does vanish and we can make the solution unique by noting that we need the condition

$$\pi_0 + \pi_1 = 1$$

for a probability distribution. Thus

$$\pi_0 = \frac{\beta}{\alpha + \beta}, \qquad \pi_1 = \frac{\alpha}{\alpha + \beta}. \qquad (5)$$

Note that if the initial probability distribution is $\boldsymbol{\pi}$, then

$$\mathbf{p}^{(1)} = \boldsymbol{\pi}\mathbf{P} = \boldsymbol{\pi}, \qquad \mathbf{p}^{(2)} = \mathbf{p}^{(1)}\mathbf{P} = \boldsymbol{\pi}\mathbf{P} = \boldsymbol{\pi},$$

and
$$\mathbf{p}^{(n)} = \boldsymbol{\pi} \quad (n = 1, 2, \ldots).$$

Thus the distribution $\mathbf{p}^{(n)}$ is stationary if $\mathbf{p}^{(0)} = \boldsymbol{\pi}$, i.e. it does not change with time.

Example 3.4. *Rainfall in Tel Aviv (continued)*. Using the matrix $\mathbf{P}$ from Example 3.2 we find that the equilibrium probabilities are

$$\pi_0 = 0\cdot575, \qquad \pi_1 = 0\cdot425.$$

In fact if we compute $\mathbf{P}^{10}$, then we find that to three decimal places

$$\mathbf{P}^{10} = \begin{bmatrix} 0\cdot575 & 0\cdot425 \\ 0\cdot575 & 0\cdot425 \end{bmatrix},$$

so that after only 10 days the equilibrium condition has for all practical purposes been reached. Thus, for example, if we call December 31 day 0 and January 10 day 10, then whatever distribution $\mathbf{p}^{(0)}$ we take for day 0 we find from equation (3) that

$$\mathbf{p}^{(10)} = (0\cdot575, 0\cdot425).$$

In order to find the time-dependent probabilities $\mathbf{p}^{(n)}$ for given initial probabilities $\mathbf{p}^{(0)}$ it is necessary to use equation (3) and therefore to evaluate $\mathbf{P}^n$. In the present case when $\mathbf{P}$ is a 2×2 matrix this can be done quite conveniently by simply computing directly the powers of $\mathbf{P}$ as we have done in Example 3.3. For higher order matrices, which become necessary when the number of states, though finite, is more than two, the direct computation is extremely laborious without the aid of an electronic computer. However, a useful mathematical representation for $\mathbf{P}^n$ is obtained by using the diagonal or spectral representation of $\mathbf{P}$. We now proceed to discuss this.

Suppose that $\mathbf{P}$ has distinct eigenvalues (also called characteristic roots or latent roots) λ_1, λ_2. Then it is a standard result of matrix theory (Bellman, 1960, p. 187) that we can find a 2×2 matrix $\mathbf{Q}$ such that

$$\mathbf{P} = \mathbf{Q} \begin{bmatrix} \lambda_1 & 0 \\ 0 & \lambda_2 \end{bmatrix} \mathbf{Q}^{-1}, \tag{6}$$

where the columns $\mathbf{q}_1, \mathbf{q}_2$ of $\mathbf{Q}$ are solutions of the equations

$$\mathbf{P}\mathbf{q}_i = \lambda_i \mathbf{q}_i \quad (i = 1, 2).$$

Hence we have

$$\mathbf{P}^n = \mathbf{Q} \begin{bmatrix} \lambda_1^n & 0 \\ 0 & \lambda_2^n \end{bmatrix} \mathbf{Q}^{-1}.$$

Let us therefore attempt to carry out this representation. The eigenvalues of $\mathbf{P}$ are the solutions of the determinantal equation

$$|\mathbf{P} - \lambda\mathbf{I}| = 0, \tag{7}$$

i.e.
$$(1 - \alpha - \lambda)(1 - \beta - \lambda) - \alpha\beta = 0. \tag{8}$$

Hence $\lambda_1 = 1$ and $\lambda_2 = 1 - \alpha - \beta$ and $\lambda_1 \neq \lambda_2$ provided $\alpha + \beta \neq 0$. We find that we may take $\mathbf{Q}$ as

$$\mathbf{Q} = \begin{bmatrix} 1 & \alpha \\ 1 & -\beta \end{bmatrix}.$$

The multiplication of the columns of $\mathbf{Q}$ by constants does not affect the result. Then

$$\mathbf{Q}^{-1} = \frac{1}{\alpha+\beta}\begin{bmatrix} \beta & \alpha \\ 1 & -1 \end{bmatrix}. \tag{9}$$

Hence

$$\mathbf{P} = \mathbf{Q}\begin{bmatrix} 1 & 0 \\ 0 & 1-\alpha-\beta \end{bmatrix}\mathbf{Q}^{-1}. \tag{10}$$

Note also that the eigenvalue $\lambda_2 = 1-\alpha-\beta$ is less than unity in modulus unless $\alpha+\beta = 0$ or $\alpha+\beta = 2$.

We have

$$\mathbf{P}^n = \frac{1}{\alpha+\beta}\begin{bmatrix} 1 & \alpha \\ 1 & -\beta \end{bmatrix}\begin{bmatrix} 1 & 0 \\ 0 & (1-\alpha-\beta)^n \end{bmatrix}\begin{bmatrix} \beta & \alpha \\ 1 & -1 \end{bmatrix}$$

$$= \frac{1}{\alpha+\beta}\begin{bmatrix} \beta & \alpha \\ \beta & \alpha \end{bmatrix} + \frac{(1-\alpha-\beta)^n}{\alpha+\beta}\begin{bmatrix} \alpha & -\alpha \\ -\beta & \beta \end{bmatrix}, \tag{11}$$

and so for any initial probability vector $\mathbf{p}^{(0)}$ we can use (3) and (11) to find $\mathbf{p}^{(n)}$. The first term in (11) is constant and is seen to be

$$\begin{bmatrix} \pi_0 & \pi_1 \\ \pi_0 & \pi_1 \end{bmatrix},$$

while the second term is a transient term and tends to zero rapidly as n increases, since $|1-\alpha-\beta| < 1$. Thus as $n \to \infty$,

$$\mathbf{P}^n \to \begin{bmatrix} \pi_0 & \pi_1 \\ \pi_0 & \pi_1 \end{bmatrix}$$

and from (3)

$$\mathbf{p}^{(n)} \to \mathbf{p}^{(0)}\begin{bmatrix} \pi_0 & \pi_1 \\ \pi_0 & \pi_1 \end{bmatrix} = (\pi_0, \pi_1) = \boldsymbol{\pi}.$$

Thus the limiting state occupation probabilities π_0, π_1 exist and are independent of the initial conditions represented by $\mathbf{p}^{(0)}$.

Example 3.5. *Rainfall in Tel Aviv (continued)*. If we carry out the representation (11) for the matrix $\mathbf{P}$ of Example 3.2 we find that

$$\mathbf{P}^n = \begin{bmatrix} 0.575 & 0.425 \\ 0.575 & 0.425 \end{bmatrix} + (0.412)^n\begin{bmatrix} 0.425 & -0.425 \\ -0.575 & 0.575 \end{bmatrix}.$$

Clearly the factor $(0.412)^n$ makes the second or transient term tend rapidly to zero as n increases.

Further quantities of interest are the proportions of a long period of time spent in each of the two states. Let X_n ($= 0$ or 1) denote the state of the system at time n. Then $X_1 + \ldots + X_n$ is the number of times out of n that state 1 is occupied. Since $\text{prob}(X_n = 1) = p_{j1}^{(n)}$ where j is the initial state it follows that

$$E(X_1 + \ldots + X_n | X_0 = j) = p_{j1}^{(1)} + \ldots + p_{j1}^{(n)}.$$

The expected proportion of time spent in state 1 is therefore

$$\frac{1}{n}\{p_{j1}^{(1)} + \ldots + p_{j1}^{(n)}\}.$$

It is well known that if a sequence $\{a_n\}$ has a limit a then the sequence

$$\left\{\frac{1}{n}(a_1 + \ldots + a_n)\right\}$$

also has the limit a. Since $p_{j1}^{(n)} \to \pi_1$ as $n \to \infty$ it follows that the expected proportion of time spent in state 1 tends to π_1 as $n \to \infty$, and similarly to π_0 for state 0. These limits are independent of the initial state. The probability distribution of $X_1 + \ldots + X_n$ is a more complicated matter to which we return in Example 3.22.

Suppose we are initially in state 0. Consider the random variable T denoting the time of first return to state 0. Thus $T = n$ if at times $1, 2, \ldots,$ $n-1$ state 1 is occupied, while at time n there is a return to state 0. The random variable T is called the recurrence time of state 0 and we denote its probability distribution by $\{f_{00}^{(n)}; n = 1, 2, \ldots\}$. (Clearly, in this case, $T-1$ is the length of a spell spent continuously in state 1.) Thus

$$f_{00}^{(n)} = \text{prob}(T = n) = \alpha\beta(1-\beta)^{n-2} \quad (n = 2, 3, \ldots)$$

and we also have that

$$f_{00}^{(1)} = 1 - \alpha.$$

This is a distribution of the geometric type. A little calculation will show that the mean recurrence time is

$$E(T) = \sum_{n=1}^{\infty} nf_{00}^{(n)} = \frac{\alpha + \beta}{\beta}. \tag{12}$$

Example 3.6. *Rainfall in Tel Aviv (continued).* The preceding results enable us to calculate the distributions of the lengths of wet and dry spells in the rainfall model of Example 3.2. A wet spell of length W is defined as W successive wet days followed by a dry day. Thus given that state 1 is occupied at time 1, W is the length of time up to (but not including) the next occurrence of state 0. Hence

$$\text{prob}(W = n) = \text{prob}(X_1 = \ldots = X_n = 1, X_{n+1} = 0 | X_1 = 1)$$

$$= (1-\beta)^{n-1}\beta = (0 \cdot 662)^{n-1}(0 \cdot 338),$$

a geometric distribution. We find

$$E(W) = \sum_{n=1}^{\infty} n(1-\beta)^{n-1}\beta = \beta^{-1} = 2\cdot96 \text{ days.}$$

Similarly, for the length D of a dry spell, we find

$$\text{prob}(D = n) = (1-\alpha)^{n-1}\alpha$$

and

$$E(D) = \alpha^{-1} = 4\cdot00 \text{ days.}$$

A weather cycle may be defined as a wet spell followed by a dry spell. The distribution of the length C of a cycle is therefore the convolution of two independent geometric distributions. Hence, for example,

$$E(C) = E(W) + E(D) = 6\cdot96 \text{ days,}$$

or approximately one week. For further details the reader is referred to Gabriel and Neumann (1957).

3.3. General definitions and some examples

We now turn to the general Markov chain in which the number of states may be finite or denumerably infinite. We therefore have a sequence of discrete random variables X_0, X_1, ... having the property that given the value of X_m for any time instant m, then for any later time instant $m+n$ the probability distribution of X_{m+n} is completely determined and the values of X_{m-1}, X_{m-2}, ... at times earlier than m are irrelevant to its determination. In other words, if the present state of the system is known we can determine the probability of any future state without reference to the past. Thus if $m_1 < m_2 < ... < m_r < m < m+n$

$$\text{prob}(X_{m+n} = k \,|\, X_{m_1}, ..., X_{m_r}, X_m) = \text{prob}(X_{m+n} = k \,|\, X_m). \quad (13)$$

The theory of Markov chains is most highly developed for homogeneous chains and we shall mostly be concerned with these. A Markov chain is said to be *homogeneous* or to possess a *stationary transition mechanism* when the probability in (13) depends on the time interval n but not on the time m. For such a chain we define the *n-step transition probabilities*

$$p_{jk}^{(n)} = \text{prob}(X_{m+n} = k \,|\, X_m = j) = \text{prob}(X_n = k \,|\, X_0 = j)$$
$$(m, n = 1, 2, ...). \quad (14)$$

Of particular importance are the *one-step transition probabilities* which we write simply as p_{jk}; thus

$$p_{jk} = p_{jk}^{(1)} = \text{prob}(X_{m+1} = k \,|\, X_m = j). \quad (15)$$

Thus for a homogeneous chain the probability of a transition in unit time from one given state to another depends only on the two states and not on the time.

Since the system must move to some state from any state j, we have for all j

$$\sum_{k=0}^{\infty} p_{jk} = 1. \tag{16}$$

The matrix $\mathbf{P}$ of transition probabilities

$$\mathbf{P} = (p_{jk}) = \begin{bmatrix} p_{00} & p_{01} & \cdots \\ p_{10} & p_{11} & \cdots \\ \cdot & \cdot & \\ \cdot & \cdot & \end{bmatrix} \tag{17}$$

is called a *stochastic matrix*, the defining property of a stochastic matrix being that its elements are non-negative and that all its row sums are unity. Thus any homogeneous Markov chain has a stochastic matrix of transition probabilities and any stochastic matrix defines a homogeneous Markov chain. Note that in common with most authors we adhere to the convention of Example 1.2 that p_{jk} denotes the probability of a transition from state j (row suffix) to state k (column suffix). Some authors, for example Bartlett (1955), use the reverse convention.

In the non-homogeneous case the transition probability

$$\text{prob}(X_s = k | X_r = j) \quad (s > r)$$

will depend on both r and s. In this case we write

$$p_{jk}(r, s) = \text{prob}(X_s = k | X_r = j).$$

In particular the one-step transition probabilities $p_{jk}(r, r+1)$ will depend on the time r and we will have a sequence of stochastic matrices corresponding to (17),

$$\mathbf{P}(r) = \{p_{jk}(r, r+1)\} = \begin{bmatrix} p_{00}(r, r+1) & p_{01}(r, r+1) & \cdots \\ p_{10}(r, r+1) & p_{11}(r, r+1) & \cdots \\ \vdots & \vdots & \end{bmatrix}$$

$$(r = 0, 1, 2, \ldots).$$

A *finite* Markov chain is one whose state space consists of a finite number of points and in the homogeneous case the matrix $\mathbf{P}$ will be a $K \times K$ square matrix where K is the number of states. In general the state space may be finite or denumerably infinite. If the state space is denumerably infinite then the matrix $\mathbf{P}$ has an infinite number of rows and columns. The two-state Markov chain discussed in Section 3.2 is an example of a finite chain and some of the random walks discussed in Chapter 2 are examples of chains with a denumerably infinite state space.

Example 3.7. *The simple random walk with a reflecting barrier at the origin.* We use the notation of Section 2.2. Here the states of the system are the possible positions 0, 1, 2, ... of the particle relative to the origin 0. It is clear that the system is a Markov chain since if we are given the position j of the particle at time n we can calculate, independently of the previous states, the probability of its moving to any position k at time $n+1$, for all values of j and k. Thus we can write down the transition probabilities as follows:

$$p_{j,j+1} = p \quad (j = 0, 1, \ldots),$$

$$p_{jj} = 1-p-q \quad (j = 1, 2, \ldots),$$

$$p_{00} = 1-p,$$

$$p_{j,j-1} = q \quad (j = 1, 2, \ldots),$$

and $p_{jk} = 0$ otherwise. The transition matrix, therefore, is

$$
\mathbf{P} =
\begin{array}{c}
\begin{array}{cccccc}
0 & 1 & 2 & 3 & 4 & \cdots
\end{array} \\
\begin{array}{c}
0 \\ 1 \\ 2 \\ 3 \\ 4 \\ \vdots
\end{array}
\left[
\begin{array}{cccccc}
1-p & p & 0 & 0 & 0 & \cdots \\
q & 1-p-q & p & 0 & 0 & \cdots \\
0 & q & 1-p-q & p & 0 & \cdots \\
0 & 0 & q & 1-p-q & p & \cdots \\
0 & 0 & 0 & q & 1-p-q & \cdots \\
\vdots & \vdots & \vdots & \vdots & \vdots &
\end{array}
\right]
\end{array}
\quad (18)
$$

Here we have a Markov chain with a denumerably infinite number of states.

Example 3.8. *The simple random walk between two absorbing barriers.* Suppose absorbing barriers are at 0 and a. Then these states cannot be left once they are entered, so that

$$p_{aa} = p_{00} = 1,$$

$$p_{ak} = 0 \quad (k = 0, 1, \ldots, a-1), \quad (19)$$

$$p_{0k} = 0 \quad (k = 1, \ldots, a).$$

This process is a finite Markov chain on the state space $(0, 1, \ldots, a)$. Apart from rows 0 and a the entries of the transition matrix $\mathbf{P}$ are the same as those of (18). Hence

$$
\mathbf{P} =
\begin{array}{c}
\\ 0 \\ 1 \\ 2 \\ \vdots \\ a-1 \\ a
\end{array}
\begin{array}{cccccccccc}
0 & 1 & 2 & . & . & & . & . & a \\
\left[\begin{array}{c} 1 \\ q \\ 0 \\ \vdots \\ 0 \\ 0 \end{array}\right. &
\begin{array}{c} 0 \\ 1-p-q \\ q \\ \vdots \\ 0 \\ 0 \end{array} &
\begin{array}{c} 0 \\ p \\ 1-p-q \\ \vdots \\ . \\ . \end{array} &
\begin{array}{c} 0 \\ 0 \\ p \\ \vdots \\ . \\ . \end{array} &
\begin{array}{c} 0 \\ 0 \\ 0 \\ \vdots \\ . \\ . \end{array} &
\begin{array}{c} \cdots \\ \cdots \\ \cdots \\ \\ \\ \end{array} &
\begin{array}{c} 0 \\ 0 \\ 0 \\ \vdots \\ q \\ 0 \end{array} &
\begin{array}{c} 0 \\ 0 \\ 0 \\ \vdots \\ 1-p-q \\ 0 \end{array} &
\left.\begin{array}{c} 0 \\ 0 \\ 0 \\ \vdots \\ p \\ 1 \end{array}\right]
\end{array}
\tag{20}
$$

Example 3.9. *The general discrete random walk.* Consider an unrestricted random walk, $X_n = Z_1 + \ldots + Z_n$, whose steps Z_n are independent and have a probability distribution $\{c_k\}$ $(k = \ldots -1, 0, 1, \ldots)$. The state space of such a process will be the set of all integers. Although the transition matrix can be written as a semi-infinite one such as (17) or (18), it is more meaningful to write it as doubly-infinite. We have

$$
\begin{aligned}
p_{jk} &= \operatorname{prob}(X_{n+1} = k | X_n = j) \\
&= \operatorname{prob}(X_{n+1} - X_n = k-j | X_n \doteq j) \\
&= \operatorname{prob}(Z_{n+1} = k-j) \\
&= c_{k-j} \quad (j, k = \ldots -1, 0, 1, \ldots)
\end{aligned}
$$

and thus

$$
\mathbf{P} =
\begin{array}{c}
\\ \\ -2 \\ -1 \\ 0 \\ 1 \\ 2 \\ \\
\end{array}
\begin{array}{ccccccc}
\cdots & -2 & -1 & 0 & 1 & 2 & \cdots \\
& \vdots & \vdots & \vdots & \vdots & \vdots & \\
\left[\begin{array}{c} \\ \cdots \\ \cdots \\ \cdots \\ \cdots \\ \cdots \\ \end{array}\right. &
\begin{array}{c} \vdots \\ c_0 \\ c_{-1} \\ c_{-2} \\ c_{-3} \\ c_{-4} \\ \vdots \end{array} &
\begin{array}{c} \vdots \\ c_1 \\ c_0 \\ c_{-1} \\ c_{-2} \\ c_{-3} \\ \vdots \end{array} &
\begin{array}{c} \vdots \\ c_2 \\ c_1 \\ c_0 \\ c_{-1} \\ c_{-2} \\ \vdots \end{array} &
\begin{array}{c} \vdots \\ c_3 \\ c_2 \\ c_1 \\ c_0 \\ c_{-1} \\ \vdots \end{array} &
\begin{array}{c} \vdots \\ c_4 \\ c_3 \\ c_2 \\ c_1 \\ c_0 \\ \vdots \end{array} &
\left.\begin{array}{c} \\ \cdots \\ \cdots \\ \cdots \\ \cdots \\ \cdots \\ \end{array}\right]
\end{array}
\tag{21}
$$

All elements on any one diagonal have the same value. This property is characteristic of transition matrices representing random walks. The imposition of boundary conditions, e.g. barriers of one kind or another, may change this property for boundary rows and columns but within the body of the matrix this property is retained, as in (18) and (19).

Example 3.10. *The imbedded Markov chain of a single-server queueing process.* We consider from a different point of view a queue similar to the one discussed in Example 2.14. We now assume that customers arrive in a Poisson process of rate λ and that their service-times are independently distributed with distribution function $B(x)$ $(0 \leqslant x < \infty)$. This is a continuous time process, but suppose we consider it only at the successive instants of service completion. Accordingly, let X_n denote the number of customers in the queue immediately after the nth customer has completed his service. Thus X_n includes the customer, if any, whose service is just commencing. Then we can write down the following equations:

$$X_{n+1} = \begin{array}{ll} X_n - 1 + Y_{n+1} & (X_n \geqslant 1), \\ Y_{n+1} & (X_n = 0), \end{array}$$

$$\hspace{6cm}(22)$$
$$\hspace{6cm}(23)$$

where Y_n is the number of customers arriving during the service-time of the nth customer. Equation (22) expresses the fact that if the nth customer does not leave an empty queue behind him $(X_n \geqslant 1)$ then during the service-time of the $(n+1)$th customer, Y_{n+1} new customers arrive and his own departure diminishes the queue size by 1. If the nth customer leaves an empty queue then the $(n+1)$th customer arrives (after an exponentially distributed time-interval) and departs after the completion of his service during which Y_{n+1} new customers arrive. This is expressed by (23).

The distribution of Y_{n+1} is that of the number of Poisson points in a random time-interval (the service-time). Since the service-times are independent the properties of the Poisson process ensure that $\{Y_n\}$ is a sequence of mutually independent random variables whose common distribution is given by

$$b_r = \mathrm{prob}(Y_n = r) = \int_0^\infty \frac{e^{-\lambda t}(\lambda t)^r}{r!}\, dB(t) \quad (r = 0, 1, \ldots). \quad (24)$$

Here we use the result of Example 1.4 and Section 4.1 that the number of arrivals in time t has a Poisson distribution of mean λt. From (22) and (23) the elements of the transition matrix for the process X_n are given by

$$p_{jk} = b_{k-j+1} \quad (k \geqslant j-1; j = 1, 2, \ldots),$$
$$p_{0k} = b_k \quad (k = 0, 1, \ldots),$$
$$p_{jk} = 0 \quad \text{otherwise.}$$

Hence

$$\mathbf{P} = \begin{bmatrix} b_0 & b_1 & b_2 & b_3 & \cdots \\ b_0 & b_1 & b_2 & b_3 & \cdots \\ 0 & b_0 & b_1 & b_2 & \cdots \\ 0 & 0 & b_0 & b_1 & \cdots \\ \vdots & \vdots & \vdots & \vdots & \end{bmatrix}. \quad (25)$$

This is again a transition matrix of the random walk type, since apart from the first row the elements on any one diagonal are the same. We have assumed that there is no restriction on the size of the queue and this leads to a denumerably infinite chain. If, however, the size of the queue is limited to, say, N customers (including the one being served), in such a way that arriving customers who find the queue full are turned away, then the resulting Markov chain is finite with N states. Immediately after a service completion there can be at most $N-1$ customers in the queue, so that the corresponding imbedded Markov chain has the state space $0, 1, \ldots, N-1$ and transition matrix

$$
\mathbf{P} = \begin{bmatrix}
b_0 & b_1 & b_2 & b_3 & \ldots & b_{N-2} & d_{N-1} \\
b_0 & b_1 & b_2 & b_3 & \ldots & b_{N-2} & d_{N-1} \\
0 & b_0 & b_1 & b_2 & \ldots & b_{N-3} & d_{N-2} \\
0 & 0 & b_0 & b_1 & \ldots & b_{N-4} & d_{N-3} \\
\vdots & & & & & & \\
0 & 0 & 0 & 0 & \ldots & b_1 & d_2 \\
0 & 0 & 0 & 0 & \ldots & b_0 & d_1
\end{bmatrix}, \tag{26}
$$

where $d_n = b_n + b_{n+1} + \ldots$ The system (25) is discussed further in Example 3.19.

For the corresponding representation of a queue with a general distribution of inter-arrival times and exponential service-times see Exercise 9.

Example 3.11. *A simple random walk with variable probabilities.* The random walks we have considered so far in this chapter have the property of spatial homogeneity, i.e. the probability of a step of any given magnitude and direction does not depend on the position from which the step is taken, with the possible exception of boundary conditions. Now consider a simple random walk in which each step may take the values $-1, 0, 1$ with probabilities ϕ_j, ψ_j, θ_j ($\phi_j + \psi_j + \theta_j = 1; j = 1, 2, \ldots$), where j denotes the position from which the step is made. For, say, a reflecting condition at the boundary $j = 0$ the transition probability elements are

$$
p_{j,j+1} = \theta_j, \quad p_{j,j} = \psi_j, \quad p_{j,j-1} = \phi_j \quad (j = 1, 2, \ldots),
$$

$$
p_{00} = \psi_0, \quad p_{01} = \theta_0 \quad (\psi_0 + \theta_0 = 1),
$$

$$
p_{jk} = 0 \quad \text{(otherwise)}.
$$

Hence we have

$$
\mathbf{P} = \begin{array}{c} \\ 0 \\ 1 \\ 2 \\ 3 \\ \\ \\ \end{array}
\begin{array}{ccccc}
0 & 1 & 2 & 3 & \cdot \; \cdot \\
\left[\begin{array}{ccccc}
\psi_0 & \theta_0 & 0 & 0 & \cdot \; \cdot \\
\phi_1 & \psi_1 & \theta_1 & 0 & \cdot \; \cdot \\
0 & \phi_2 & \psi_2 & \theta_2 & \cdot \; \cdot \\
0 & 0 & \phi_3 & \psi_3 & \cdot \; \cdot \\
\cdot & \cdot & \cdot & \cdot & \\
\cdot & \cdot & \cdot & \cdot &
\end{array}\right]
\end{array}.
\tag{27}
$$

Returning to the general case, we consider a homogeneous Markov chain with transition matrix $\mathbf{P}$. Let the row vector

$$
\mathbf{p}^{(0)} = (p_0^{(0)}, p_1^{(0)}, \ldots)
$$

be a given vector of initial state occupation probabilities and let

$$
\mathbf{p}^{(n)} = (p_0^{(n)}, p_1^{(n)}, \ldots)
$$

be the vector of state occupation probabilities at time n; thus the components of $\mathbf{p}^{(n)}$ $(n = 0, 1, \ldots)$ are defined by

$$
p_j^{(n)} = \text{prob}(X_n = j) \quad (n = 0, 1, \ldots; \; j = 0, 1, \ldots).
\tag{28}
$$

By arguments similar to those used for the two-state chain in deriving equations (1) we find that we have the recurrence relation

$$
p_k^{(n)} = \sum_{j=0}^{\infty} p_j^{(n-1)} p_{jk} \quad (n = 1, 2, \ldots).
\tag{29}
$$

There are no convergence difficulties since $\{p_j^{(n-1)}; j = 0, 1, \ldots\}$ is a probability distribution and p_{jk} is non-negative and bounded by 1 for all j, k; thus the series on the right of (29) is absolutely convergent. We may thus express (29) in matrix notation

$$
\mathbf{p}^{(n)} = \mathbf{p}^{(n-1)} \mathbf{P},
\tag{30}
$$

and on iteration we obtain

$$
\mathbf{p}^{(n)} = \mathbf{p}^{(0)} \mathbf{P}^n \quad (n = 1, 2, \ldots).
\tag{31}
$$

Thus the initial probability vector $\mathbf{p}^{(0)}$ and the set of transition probabilities (p_{jk}) suffice to determine the marginal distributions $\mathbf{p}^{(n)}$. If the system starts in any given state j, then $\mathbf{p}^{(0)}$ has zero components in all but the jth position and (31) tells us that for each k, the n-step transition probability $p_{jk}^{(n)}$ is simply the element in position (j, k) in the nth power of $\mathbf{P}$. Thus

$$
(p_{jk}^{(n)}) = \mathbf{P}^n.
\tag{32}
$$

From probabilistic reasoning, or simply from the fact that $\mathbf{P}^{m+n} = \mathbf{P}^m \mathbf{P}^n$, we have

$$p_{jk}^{(m+n)} = \sum_l p_{jl}^{(m)} p_{lk}^{(n)}. \tag{33}$$

This is known as the *Chapman–Kolmogorov relation* for the homogeneous Markov chain. All Markov chains must satisfy this relation but there are also non-Markovian processes which satisfy it; see Exercise 3.

In the non-homogeneous case, we have, corresponding to (32), the equation

$$\{p_{jk}(r,s)\} = \mathbf{P}(r)\,\mathbf{P}(r+1)\ldots\mathbf{P}(s-1) \quad (s > r)$$

and the Chapman–Kolmogorov equation (33) takes the form

$$p_{jk}(r,t) = \sum_l p_{jl}(r,s)\,p_{lk}(s,t) \quad (r < s < t).$$

The fact that all the row sums of a stochastic matrix are unity can be expressed in matrix notation by

$$\mathbf{P}\mathbf{1} = \mathbf{1}, \tag{34}$$

where $\mathbf{1}$ is a column vector of 1's. Pre-multiplying each side of (34) by $\mathbf{P}$, we have

$$\mathbf{P}^2 \mathbf{1} = \mathbf{P}\mathbf{1} = \mathbf{1}$$

and in general

$$\mathbf{P}^n \mathbf{1} = \mathbf{1}.$$

It follows that $\mathbf{P}^n$ is a stochastic matrix for each integral n.

We may add at this juncture that the method of diagonalizing the matrix $\mathbf{P}$ which we used in Section 3.2 may be extended with some modifications to finite chains of any order. But for denumerably infinite chains the corresponding method is not available for infinite matrices.

3.4. The classification of states and the limit theorem

The states of a Markov chain fall into distinct types according to their limiting behaviour. Suppose that the chain is initially in a given state j. If the ultimate return to this state is a certain event, the state is called *recurrent*; in this case the time of first return will be a random variable called the *recurrence time* and the state is called *positive-recurrent* or *null-recurrent* according as the mean recurrence time is finite or infinite. If the ultimate return to a state has probability less than unity the state is called *transient*.

Let us exclude from the discussion the following trivial type of state. A state j is called *ephemeral* if $p_{ij} = 0$ for every i. A chain can only be in an ephemeral state initially and pass out of it on the first transition. An ephemeral state can never be reached from any other state. The column of the transition matrix $\mathbf{P}$ corresponding to an ephemeral state is composed entirely of zeros.

Suppose that the chain is initially in the state j. Let $f_{jj}^{(n)}$ denote the probability that the next occurrence of state j is at time n, i.e. $f_{jj}^{(1)} = p_{jj}$ and for $n = 2, 3, \ldots$,

$$f_{jj}^{(n)} = \text{prob}(X_r \neq j, r = 1, \ldots, n-1; \ X_n = j | X_0 = j). \tag{35}$$

In other words we may say that conditional on state j being occupied initially, $f_{jj}^{(n)}$ is the probability that state j is avoided at times $1, 2, \ldots, n-1$ and entered at time n. We call $f_{jj}^{(n)}$ the first return probability for time n. In a similar way, we define the first passage probability from state j to state k for time n as being the conditional probability $f_{jk}^{(n)}$ that state k is avoided at times $1, \ldots, n-1$ and entered at time n, given that state j is occupied initially. Thus $f_{jk}^{(1)} = p_{jk}$ and for $n = 2, 3, \ldots$,

$$f_{jk}^{(n)} = \text{prob}(X_r \neq k, r = 1, \ldots, n-1; \ X_n = k | X_0 = j). \tag{36}$$

Given that the chain starts in state j the sum

$$f_j = \sum_{n=1}^{\infty} f_{jj}^{(n)} \tag{37}$$

is the probability that state j is eventually re-entered. If $f_j = 1$ then state j is recurrent while if $f_j < 1$ state j is transient. Thus, conditional on starting in a transient state j, there is a positive probability $1 - f_j$ that state j will never be re-entered, while for a recurrent state re-entrance is a certain event. For a recurrent state, therefore, $\{f_{jj}^{(n)}, n = 1, 2, \ldots\}$ is a probability distribution and the mean of this distribution

$$\mu_j = \sum_{n=1}^{\infty} n f_{jj}^{(n)} \tag{38}$$

is the mean recurrence time. If μ_j is infinite then state j is null-recurrent.

Similarly, given that the chain starts in state j the sum

$$f_{jk} = \sum_{n=1}^{\infty} f_{jk}^{(n)} \tag{39}$$

is the probability of ever entering state k. We may call f_{jk} the *first passage probability* from state j to state k. If $f_{jk} = 1$, then

$$\sum_{n=1}^{\infty} n f_{jk}^{(n)} \tag{40}$$

is the mean *first passage time* from state j to state k.

Suppose that when the chain starts in state j, subsequent occupations of state j can only occur at times $t, 2t, 3t, \ldots$ where t is an integer greater than 1; choose t to be the largest integer with this property. Then state j is called *periodic* with period t and $p_{jj}^{(n)}$ vanishes except when n is an integral multiple of t.

A state which is not periodic is called *aperiodic*. Essentially it has period 1. An aperiodic state which is positive-recurrent is called *ergodic*. The above definitions are summarized in Table 3.1.

Table 3.1. *Classification of states in a Markov chain*

Type of state	Definition of state (assuming, where applicable, state is occupied initially)
Periodic	Return to state possible only at times t, $2t$, $3t$, ..., where $t > 1$
Aperiodic	Not periodic
Recurrent	Eventual return to state certain
Transient	Eventual return to state uncertain
Positive-recurrent	Recurrent, finite mean recurrence time
Null-recurrent	Recurrent, infinite mean recurrence time
Ergodic	Aperiodic, positive-recurrent

Example 3.12. *The simple random walk.* To illustrate these definitions let us examine the simple random walk discussed in Section 2.2. We saw in Section 2.2(iv) that when there is zero drift and the process starts in state j, return to j is certain and the mean recurrence time is infinite. This means in Markov chain terminology that for a symmetric random walk (i.e. $p = q$) every state is null-recurrent. For an asymmetric random walk ($p \neq q$) starting at j, return to j is uncertain and so every state is transient.

Suppose now that $p + q = 1$, so that each step is either $+1$ or -1. If the walk starts at the origin then after one step it is either in state 1 or state -1. It may return to the origin at the second step and a simple enumeration will show that the origin can only be occupied at times 2, 4, 6, ... Thus the state 0 and, by a similar argument, all other states are periodic with period 2.

If we now consider a simple random walk with the origin as a reflecting barrier, then according to equation (2.53), $\lambda_2(s)$ is the generating function of recurrence times for the origin; although equation (2.53) was derived on the assumption that $p \leqslant q$, a little thought will show that it holds for $p > q$ as well, if we interpret $E(s^u)$ as $\text{prob}(U < \infty)E(s^u | U < \infty)$. Thus $\lambda_2(1)$ is the probability of returning to the origin given that the walk starts at the origin. It follows from (2.19) that the origin is a recurrent state if $p \leqslant q$ and is transient if $p > q$. Further, if $p = q$ then $\lambda_2'(1)$ is infinite. Thus the origin is null-recurrent if $p = q$. If $p < q$ the recurrence time is finite and the origin is a positive-recurrent state.

Let

$$F_{jj}(s) = \sum_{n=1}^{\infty} f_{jj}^{(n)} s^n \tag{41}$$

be the generating function of first return probabilities defined in equation (35), and let

$$P_{jj}(s) = \sum_{n=1}^{\infty} p_{jj}^{(n)} s^n \qquad (42)$$

be the generating function of the transition probabilities from state j to itself. We proceed to derive an important relation between these two generating functions and this will give us a classification criterion for the state j based on the transition probabilities $p_{jj}^{(n)}$.

If we start in state j and enter state j at time n (not necessarily for the first time) the probability of this event is $p_{jj}^{(n)}$. However the *first* return to state j must have occurred at one of the times $1, 2, \ldots, n$ and these are mutually exclusive possibilities. We may call this a decomposition based on first return times. Hence if we define $p_{jj}^{(0)} = 1$, then

$$p_{jj}^{(n)} = f_{jj}^{(1)} p_{jj}^{(n-1)} + f_{jj}^{(2)} p_{jj}^{(n-2)} + \cdots$$
$$+ f_{jj}^{(n-1)} p_{jj}^{(1)} + f_{jj}^{(n)} \quad (n = 1, 2, \ldots). \qquad (43)$$

Multiplying by s^n and summing over n, we have that

$$P_{jj}(s) = F_{jj}(s) + F_{jj}(s) P_{jj}(s). \qquad (44)$$

Rearranging, we obtain the following important relations:

$$F_{jj}(s) = \frac{P_{jj}(s)}{1 + P_{jj}(s)}, \qquad (45)$$

$$P_{jj}(s) = \frac{F_{jj}(s)}{1 - F_{jj}(s)}. \qquad (46)$$

For first passage probabilities define

$$F_{jk}(s) = \sum_{n=1}^{\infty} f_{jk}^{(n)} s^n \qquad (47)$$

and for transition probabilities define

$$P_{jk}(s) = \sum_{n=1}^{\infty} p_{jk}^{(n)} s^n. \qquad (48)$$

Then by considering a decomposition based on first passage times from state j to state k an argument similar to that preceding equation (44) leads to the relation

$$P_{jk}(s) = F_{jk}(s) + F_{jk}(s) P_{kk}(s) \qquad (49)$$

or, rearranging,

$$F_{jk}(s) = \frac{P_{jk}(s)}{1 + P_{kk}(s)}. \qquad (50)$$

A state j is said to *communicate* with a state k if it is possible to reach k from j in a finite number of transitions, i.e. if there is an integer r such that $p_{jk}^{(r)} > 0$; we define r_{jk} to be the smallest integer r for which this is true. If j communicates with k and k with j, then j and k are said to be *inter-communicating*.

We have classified the states of a Markov chain as recurrent, transient, etc., in terms of the properties of the first return probabilities $f_{jj}^{(n)}$. We shall now relate the classification to the limiting properties of the transition probabilities $p_{jj}^{(n)}$. We do this by applying the results on power series with non-negative coefficients given in Section 3.13 to the generating function relations (45), (46) and (50). For any generating function $P(s) = \sum p_n s^n$ with non-negative coefficients p_n, we write $P(1) = \lim P(s)$ as $s \to 1-$. From Theorem 1 of Section 3.13 we always have $P(1) = \sum p_n$, whether this is finite or infinite.

From (37), we see that $f_k = F_{kk}(1)$ is the probability that state k, once entered, is ever re-entered. From (38) we have that $\mu_k = F'_{kk}(1)$, whether μ_k is finite or infinite. Hence for a transient state $F_{kk}(1) < 1$ and for a recurrent state, $F_{kk}(1) = 1$. For a positive-recurrent state $F'_{kk}(1) < \infty$ and for a null-recurrent state $F'_{kk}(1) = \infty$.

For any states j and k we have $F_{kk}(1) \leqslant 1$ and $F_{jk}(1) \leqslant 1$. From (45) we see that $F_{kk}(1) < 1$ if and only if $P_{kk}(1) < \infty$; thus a necessary and sufficient condition for k to be a transient state is that the series $\sum_n p_{kk}^{(n)}$ is convergent, and in this case it follows from (50) that, for each j, $\sum_n p_{jk}^{(n)}$ is convergent.

Also from (45) we see that $F_{kk}(1) = 1$ if and only if $P_{kk}(1) = \infty$; thus a necessary and sufficient condition for k to be a recurrent state is that the series $\sum_n p_{kk}^{(n)}$ is divergent. If j communicates with k then $F_{jk}(1) > 0$ and from (50) we see that $P_{kk}(1) = \infty$ implies that $P_{jk}(1) = \infty$. Hence if k is recurrent the series $\sum_n p_{jk}^{(n)}$ is divergent for each state j which communicates with k.

Suppose k is a recurrent, aperiodic (i.e. ergodic) state. We now show that the limit as $n \to \infty$ of the transition probability $p_{kk}^{(n)}$ is the inverse of the mean recurrence time of state k. This result is known as the *ergodic theorem* for Markov chains. If we write (46) in the form

$$1 + P_{kk}(s) = \frac{1}{1 - F_{kk}(s)},$$

then appealing to Theorem 2 of Section 3.13 and identifying $1 + P_{kk}(s)$ with $P(s)$ and $F_{kk}(s)$ with $F(s)$ we see that

$$\lim_{n \to \infty} p_{kk}^{(n)} = \frac{1}{\mu_k},$$

the limit being zero if μ_k is infinite. For any other state j, we have, from (45) and (49), that

$$P_{jk}(s) = \frac{F_{jk}(s)}{1 - F_{kk}(s)}.$$

Hence, from Theorem 3 of Section 3.13, we have

$$\lim_{n \to \infty} p_{jk}^{(n)} = \frac{F_{jk}(1)}{\mu_k}.$$

If k is a recurrent periodic state with period t then a similar application of the corollary to Theorem 2 gives the limit

$$\lim_{n \to \infty} p_{kk}^{(nt)} = \frac{t}{\mu_k}.$$

If j communicates with k, then $p_{jk}^{(n)}$ and $f_{jk}^{(n)}$ vanish except when n is of the form $mt + r_{jk}$ $(m = 0, 1, \ldots)$, where r_{jk} is the smallest value of r for which $p_{jk}^{(r)} > 0$. Hence

$$P_{jk}(s) = s^{r_{jk}} \sum_m p_{jk}^{(mt+r_{jk})} s^{mt},$$

with a similar expression for $F_{jk}(s)$. Hence an application of Theorem 3 similar to the aperiodic case gives

$$\lim_{n \to \infty} p_{jk}^{(r_{jk}+nt)} = \frac{tF_{jk}(1)}{\mu_k}.$$

It is convenient at this stage to collect these results together.

The limit theorem for Markov chains:

A. *Let k be an arbitrary but fixed state. Then*

  (i) *k is transient if and only if the series $\sum_n p_{kk}^{(n)}$ is convergent (i.e. $P_{kk}(1) < \infty$) and in this case $\sum_n p_{jk}^{(n)}$ is convergent for each j.*
  (ii) *k is recurrent if and only if the series $\sum_n p_{kk}^{(n)}$ is divergent (i.e. $P_{kk}(1) = \infty$) and in this case $\sum_n p_{jk}^{(n)}$ is divergent for every j which communicates with k.*

B. *Let k be a recurrent state and let $\mu_k = \sum_n n f_{kk}^{(n)} = F_{kk}'(1)$ be its mean recurrence time. Define $1/\mu_k = 0$ if $\mu_k = \infty$.*

  (i) *If k is aperiodic then*

$$\lim_{n \to \infty} p_{kk}^{(n)} = 1/\mu_k, \tag{51}$$

  and

$$\lim_{n \to \infty} p_{jk}^{(n)} = F_{jk}(1)/\mu_k. \tag{52}$$

  (ii) *If k has period t then*

$$\lim_{n \to \infty} p_{kk}^{(nt)} = t/\mu_k, \tag{53}$$

  and for each state j which communicates with k

$$\lim_{n \to \infty} p_{jk}^{(r_{jk}+nt)} = tF_{jk}(1)/\mu_k. \tag{54}$$

We now discuss some implications of these results.

If for an aperiodic state k, we have $p_{kk}^{(n)} \to 0$, then k is either transient or null-recurrent depending on whether $\sum_n p_{kk}^{(n)}$ is convergent or divergent. On the other hand, if $\sum_n p_{kk}^{(n)}$ is divergent, then k is null-recurrent or ergodic according as $p_{kk}^{(n)}$ tends to zero or not. If k is ergodic then $p_{kk}^{(n)}$ tends to a non-zero limit and conversely; this limit is the reciprocal of the mean recurrence time for k. Similar remarks apply to periodic states except that $p_{kk}^{(n)}$ is zero except when n is a multiple of the period t and we allow n to tend to infinity through multiples of t.

From (52) we see that for an ergodic state k and any initial state j which communicates with k the limiting value of $p_{jk}^{(n)}$ is positive. It follows from this theorem that the limiting behaviour of the transition probabilities $p_{jk}^{(n)}$ for states j which communicate with k, or of $p_{kk}^{(n)}$, determines whether k is transient, null-recurrent or positive-recurrent.

The phenomenon of periodicity tends to complicate the description of the limiting behaviour of the transition probabilities $p_{jk}^{(n)}$. It is possible, however, to smooth out the periodicities by taking a suitable mean, such as the simple arithmetic average,

$$q_{jk}^{(n)} = \frac{1}{n}(p_{jk}^{(1)} + \ldots + p_{jk}^{(n)}),$$

the limit of which always exists.

By definition a state j is recurrent or transient according as $\sum_n f_{jj}^{(n)} = 1$ or < 1. An equivalent definition of the properties of recurrence and transience is that *a state is recurrent if and only if it is revisited infinitely often with probability one, and transient if and only if it is revisited finitely often with probability one.* Thus if $X_0 = j$, X_1, X_2, ... denotes a realization of the chain, the state j is recurrent or transient according as the integer j occurs with probability one an infinite or finite number of times in the sequence X_0, X_1, ... For let N denote the number of visits to j in a realization $X_0 = j$, X_1, X_2, ... Then using the method of first return times (see equation (43)) we have

$$\text{prob}(N > n) = \sum_{m=1}^{\infty} f_{jj}^{(m)} \text{prob}(N > n-1),$$

since, on account of homogeneity, the distribution of N is the same as the conditional distribution of the number of visits to j after time m, given $X_m = j$. Thus

$$\text{prob}(N > n) = F_{jj}(1)\,\text{prob}(N > n-1)$$
$$= [F_{jj}(1)]^2\,\text{prob}(N > n-2) = [F_{jj}(1)]^n.$$

Thus

$$\text{prob}(N = \infty) = \lim_{n \to \infty} [F_{jj}(1)]^n$$

and the right-hand side is one or zero according as j is recurrent or transient.

A further important property is that *for an ergodic state the limiting occupation probability is, with probability one, the proportion of time spent in that state in an indefinitely long realization of the process.* For let k be an arbitrary but fixed ergodic state which, we assume, is occupied initially. Let $N_1, N_2, \ldots$ denote the successive subsequent times at which state k is occupied. Then $N_1, (N_2 - N_1), (N_3 - N_2), \ldots$ is a sequence of mutually independent, identically distributed random variables with p.g.f. $F_{kk}(s)$ and finite mean μ_k. It follows from the strong law of large numbers that with probability one,

$$\lim_{r \to \infty} \frac{N_1 + (N_2 - N_1) + \ldots + (N_r - N_{r-1})}{r} = \lim_{r \to \infty} \frac{N_r}{r} = \mu_k.$$

In the time interval $[1, n]$ let $R(n)$ be the number of times state k is occupied. Then $R(n)/n$ is the proportion of the time in $[1, n]$ spent in state k. We have that

$$N_{R(n)} \leqslant n < N_{R(n)+1}$$

and as $n \to \infty$, $R(n) \to \infty$ with probability one, since an ergodic state is revisited infinitely often with probability one. Now

$$\frac{R(n)}{n} \leqslant \frac{R(n)}{N_{R(n)}}$$

and therefore, with probability one,

$$\limsup_{n \to \infty} \frac{R(n)}{n} \leqslant \lim_{n \to \infty} \frac{R(n)}{N_{R(n)}} = \frac{1}{\mu_k}.$$

Further,

$$\frac{R(n)+1}{n} \geqslant \frac{R(n)+1}{N_{R(n)+1}}$$

and so, with probability one,

$$\liminf_{n \to \infty} \frac{R(n)+1}{n} = \liminf_{n \to \infty} \frac{R(n)}{n} \geqslant \lim_{n \to \infty} \frac{R(n)+1}{N_{R(n)+1}} = \frac{1}{\mu_k}.$$

Hence it follows that with probability one,

$$\lim_{n \to \infty} \frac{R(n)}{n} = \frac{1}{\mu_k},$$

which, according to the limit theorem, is the limiting occupation probability for state k.

A slight modification of the argument will give the same result if some other state, from which k is certain to be reached, is occupied initially.

3.5. Closed sets of states

Having defined the basic types of states, we proceed to show that there is what has been called by Chung (1960, p. 11) a class solidarity among these types in the sense that a state of a given type can only inter-communicate with other states of the same type.

Let j and k be two inter-communicating states and let M and N be two integers such that

$$p_{jk}^{(M)} > 0, \qquad p_{kj}^{(N)} > 0. \tag{55}$$

From the Chapman–Kolmogorov equation (33) we have, for *any* three states j, k and l and any integers m and n, that

$$p_{jk}^{(m+n)} \geqslant p_{jl}^{(m)} p_{lk}^{(n)}. \tag{56}$$

Using (55) and a repeated application of the inequality (56), we obtain, for $n > M + N$, the inequalities

$$p_{kk}^{(n)} \geqslant p_{kj}^{(N)} p_{jj}^{(n-M-N)} p_{jk}^{(M)} \tag{57}$$

and

$$p_{jj}^{(n)} \geqslant p_{jk}^{(M)} p_{kk}^{(n-M-N)} p_{kj}^{(N)}. \tag{58}$$

Hence there are constants $A > 0$ and $B > 0$ and an integer R such that

$$p_{kk}^{(n)} \geqslant A p_{jj}^{(n-R)}, \qquad p_{jj}^{(n)} \geqslant B p_{kk}^{(n-R)} \quad (n > R). \tag{59}$$

Thus for inter-communicating states j and k the transition probabilities $p_{jj}^{(n)}$ and $p_{kk}^{(n)}$ have the same asymptotic behaviour, and further, if j is periodic, then so is k with the same period, and conversely. *Hence if j and k inter-communicate they must be of the same type, both transient or both null-recurrent or both positive-recurrent, and furthermore they must have the same period.*

Let j be a recurrent state and k another state with which j communicates; thus there is an integer M such that $p_{jk}^{(M)} > 0$. Since j is recurrent, k must communicate with j for otherwise there would be a positive probability of never returning to j. Hence j and k must inter-communicate and they are therefore of the same type and period.

A set C of states is called a *closed set* if each state in C communicates only with other states in C. The *closure* of a given set of states is the smallest closed set containing that set. A single state forming a closed set is an absorbing state. Once a closed set is entered it is never vacated. If j belongs to a closed set C then p_{jk} is zero for all k outside C. Hence if all rows and columns of **P** corresponding to states outside C are deleted, we are still left with a stochastic matrix.

We may summarize the above remarks in the following theorem.

Decomposition theorem. (a) *The states of an arbitrary Markov chain may be divided into two sets (one of which may be empty), one set being composed* of all the recurrent states, the other of all the transient states. (b) *The recurrent states may be decomposed uniquely into closed sets. Within each closed set all states inter-communicate and they are all of the same type and period. Between any two closed sets no communication is possible.*

Example 3.13. Decomposition of a finite chain. A finite Markov chain with two absorbing states, two closed sets of recurrent states and a set of transient states has a transition matrix of the following form

$$
\mathbf{P} = \begin{array}{c} \\ \\ C_1 \\ \\ C_2 \\ \\ C_3 \end{array}
\begin{array}{ccc} C_1 & C_2 & C_3 \\ \end{array}
\left[
\begin{array}{ccccc|ccc|ccc}
1 & 0 & 0 & \ldots & 0 & 0 & \ldots & 0 & 0 & \ldots & 0 \\
0 & 1 & 0 & \ldots & 0 & 0 & \ldots & 0 & 0 & \ldots & 0 \\
\hline
& \mathbf{0} & & \mathbf{P}_1 & & & \mathbf{0} & & & \mathbf{0} & \\
\hline
& \mathbf{0} & & & \mathbf{0} & & \mathbf{P}_2 & & & \mathbf{0} & \\
\hline
& \mathbf{A} & & & \mathbf{B} & & \mathbf{C} & & & \mathbf{D} &
\end{array}
\right].
\qquad (60)
$$

The first two rows and columns represent the absorbing states, C_1 and C_2 represent two closed sets of recurrent states and C_3 the transient states. The matrix $\mathbf{P}_1$ is a stochastic matrix of transition probabilities within the set C_1 and $\mathbf{P}_2$ that within C_2, The matrices $\mathbf{B}$ and $\mathbf{C}$ contain the transition probabilities from the transient states to the sets C_1 and C_2 respectively. The matrix $\mathbf{D}$ contains the probabilities of transition within the transient states and $\mathbf{A}$ contains the transition probabilities from the transient states to the absorbing states.

Example 3.14. The simple random walk (continued). In the simple random walk with a reflecting barrier at the origin (Example 3.7), having the transition matrix (18), it is clear that after a sufficient number of transitions any state can be reached from any other state. Hence the chain consists of a single closed set. In Example 3.12 we showed that the origin is a transient state if $p > q$, from which it follows that all states are transient if $p > q$. Similarly all states are null-recurrent if $p = q$ and ergodic if $p < q$ and $p+q < 1$. If $p+q = 1$ all states are periodic with period 2 and if in addition $p < q$ all states are recurrent but non-null.

For the random walk between two absorbing barriers with transition matrix (20), the barrier states 0 and a are absorbing states. The interior states $1, \ldots, a-1$ are all transient; they cannot be recurrent since each communicates with the absorbing states and each absorbing state is a closed set consisting of a single state.

Our results have important implications for finite chains. Firstly *a finite chain cannot consist only of transient states*; for if it did then all the transition probabilities $p_{jk}^{(n)}$ would tend to zero as $n \to \infty$. This is impossible, however, since the finite sum $\sum_k p_{jk}^{(n)}$ is unity. Secondly, *a finite chain cannot have any null-recurrent states*; for the one-step transition probabilities within a closed set of null-recurrent states form a finite stochastic matrix $\mathbf{P}$ such that $\mathbf{P}^n \to 0$ as $n \to \infty$ and this again is impossible.

3.6. Irreducible chains and equilibrium distributions

Perhaps the most important single class of Markov chains is the class of irreducible chains. An *irreducible chain* is one in which all states intercommunicate. Thus an irreducible chain forms a single closed set and all its states are of the same type. It is thus possible to speak of the *chain* or the *system* as being recurrent, periodic, etc. The random walk of Example 3.7 is an irreducible chain but that of Example 3.8 is not, since the latter has two absorbing states. It follows from the decomposition theorem that an arbitrary Markov chain is composed of transient states, if any, and a number of irreducible, recurrent sub-chains. If the system starts in a sub-chain or once it enters a sub-chain it never leaves that sub-chain. Hence the properties of a general Markov chain depend on those of irreducible chains and transient states.

The remarks on finite chains at the end of Section 3.5 show that an irreducible finite chain is necessarily positive-recurrent. If in addition it is aperiodic then it is ergodic and there is a unique row vector $\boldsymbol{\pi}$ of limiting occupation probabilities called the *equilibrium distribution* and formed by the inverses of the mean recurrence times. Hence in this case

$$\lim_{n \to \infty} \mathbf{P}^n = \begin{pmatrix} \boldsymbol{\pi} \\ \boldsymbol{\pi} \\ \vdots \\ \boldsymbol{\pi} \end{pmatrix} = \mathbf{1}\boldsymbol{\pi}, \tag{61}$$

where $\mathbf{1}$ is a column vector of 1's. For *any* initial probability distribution $\mathbf{p}^{(0)}$ we have from (31)

$$\lim_{n \to \infty} \mathbf{p}^{(n)} = \lim_{n \to \infty} \mathbf{p}^{(0)} \mathbf{P}^n = \mathbf{p}^{(0)} \mathbf{1}\boldsymbol{\pi} = \boldsymbol{\pi}. \tag{62}$$

Thus a finite ergodic system settles down in the long run to a condition of statistical equilibrium independent of the initial conditions. The distribution $\boldsymbol{\pi}$ is also a *stationary distribution* since if $\mathbf{p}^{(0)} = \boldsymbol{\pi}$, then $\mathbf{p}^{(n)} = \boldsymbol{\pi}$ for all n. To see this, we note that since $\lim \boldsymbol{\pi}\mathbf{P}^n = \boldsymbol{\pi}$ it follows that $\lim \boldsymbol{\pi}\mathbf{P}^{n-1}\mathbf{P} = \boldsymbol{\pi}$, i.e.

$$\boldsymbol{\pi}\mathbf{P} = \boldsymbol{\pi}. \tag{63}$$

Thus $\boldsymbol{\pi}\mathbf{P}^2 = \boldsymbol{\pi}\mathbf{P} = \boldsymbol{\pi}$ and by induction $\boldsymbol{\pi}\mathbf{P}^n = \boldsymbol{\pi}$, whence $\mathbf{p}^{(n)} = \boldsymbol{\pi}$ for

$n = 1, 2, \ldots$ Hence if the system starts with the distribution π over states, the distribution over states for all subsequent times is π. This is the defining property of a stationary distribution.

For denumerably infinite systems which are irreducible, we shall see in Section 3.8 that the limiting behaviour has richer possibilities. Such systems may be positive- or null-recurrent or transient. Ergodic systems retain the property of having unique equilibrium distributions and these are also unique stationary distributions; in this case too the equilibrium probabilities are the inverses of the mean recurrence times.

As we proved at the end of Section 3.4, the equilibrium probabilities have the property that in a long realization they are, with probability one, the proportions of time spent in the corresponding states. This is an important result because it enables us to examine the properties of the chain from a single realization rather than from a large number of independent realizations.

The states of an irreducible periodic chain of period t can be uniquely divided into t mutually exclusive *cyclic sub-sets* $C_0, C_1, \ldots, C_{t-1}$ such that if the chain initially occupies a state belonging to the sub-set C_r, at the next transition it will move to sub-set C_{r+1} and then to $C_{r+2}, \ldots, C_{t-1}$, $C_0, \ldots, C_{r-1}$, returning to C_r at the tth transition. The proof of this statement is left as an exercise for the reader.

Example 3.15. *A periodic chain.* An 8-state chain of period 4 with cyclic sub-sets $(0,1)$, (2), $(3,4,5)$, $(6,7)$ has a transition matrix of the following form, in which all non-zero entries are represented by x.

$$
\mathbf{P} = \begin{array}{c} \\ 0 \\ 1 \\ 2 \\ 3 \\ 4 \\ 5 \\ 6 \\ 7 \end{array}
\begin{array}{c} \begin{array}{cccccccc} 0 & 1 & 2 & 3 & 4 & 5 & 6 & 7 \end{array} \\
\left[\begin{array}{cccccccc}
0 & 0 & x & 0 & 0 & 0 & 0 & 0 \\
0 & 0 & x & 0 & 0 & 0 & 0 & 0 \\
0 & 0 & 0 & x & x & x & 0 & 0 \\
0 & 0 & 0 & 0 & 0 & 0 & x & x \\
0 & 0 & 0 & 0 & 0 & 0 & x & x \\
0 & 0 & 0 & 0 & 0 & 0 & x & x \\
x & x & 0 & 0 & 0 & 0 & 0 & 0 \\
x & x & 0 & 0 & 0 & 0 & 0 & 0
\end{array} \right]
\end{array}. \tag{64}
$$

3.7. Branching processes

We turn from the general discussion to consider a Markov chain arising in the theory of population growth. In a population of like individuals, e.g. males of a biological species or neutrons in a physical substance, suppose that each individual, independently of all other individuals, is capable of giving rise to a number of offspring, this number being a random variable taking values 0, 1, 2, ... with probabilities $g_0, g_1, g_2, \ldots$ Starting with one individual we examine the development of its descendants generation by generation. The initial individual is regarded as belonging to the zeroth generation, its offspring as belonging to the first generation, the total number of offspring of individuals in the first generation comprise the second generation and so on. If all the individuals of a generation fail to reproduce, the population becomes extinct. The process is called the simple discrete branching process.

A striking example of this process is that of family surnames. A family surname stemming from a given male survives until a generation fails to produce any male descendants. Thus the population under consideration in this case would be the male descendants of a given man. The man's sons comprise the first generation, the sons' sons the second generation, and so on.

We shall be interested in the problem of extinction and so we shall assume $g_0 > 0$, for if $g_0 = 0$ each individual must have at least one offspring and extinction is impossible. Further, we assume that $g_0 + g_1 < 1$; for if $g_0 + g_1 = 1$, then the process is trivial, since each individual can have at most one offspring and each generation can number at most one. Let

$$G(z) = \sum_{i=0}^{\infty} g_i z^i \tag{65}$$

be the p.g.f. of the distribution of the number of offspring per individual.

Let X_n denote the number of individuals in the nth generation. If $X_n = j$, say, then X_{n+1} will be the sum of j independent random variables $Z_1, \ldots, Z_j$ each having the distribution $\{g_i\}$. Hence

$$\text{prob}(X_{n+1} = k \,|\, X_n = j) = \text{prob}(Z_1 + \ldots + Z_j = k)$$
$$= \text{coefficient of } z^k \text{ in } \{G(z)\}^j. \tag{66}$$

In particular, if $X_n = 0$ then $X_{n+1} = 0$. It follows that $\{X_n\}$ is a denumerably infinite Markov chain, the state of the system at time n being the number of individuals in the nth generation. The state 0 is clearly an absorbing state, and the probability of absorption is the probability of ultimate extinction. The one-step transition probabilities p_{jk} are given by (66).

With the initial condition $X_0 = 1$ let us define

$$p_k^{(n)} = \text{prob}(X_n = k); \tag{67}$$

thus for given n, $\{p_k^{(n)}\}$ is the distribution of the number of nth generation descendants of a single individual. Since

$$p_k^{(n)} = \sum_{j=0}^{\infty} p_j^{(n-1)} p_{jk}, \tag{68}$$

it follows from (66) that

$$p_k^{(n)} = \text{coefficient of } z^k \text{ in } \sum_{j=0}^{\infty} p_j^{(n-1)}\{G(z)\}^j. \qquad (69)$$

Hence, defining the p.g.f. of the size of the nth generation as

$$F_n(z) = \sum_{k=0}^{\infty} p_k^{(n)} z^k,$$

we have

$$F_n(z) = \sum_{j=0}^{\infty} p_j^{(n-1)}\{G(z)\}^j \qquad (70)$$

$$= F_{n-1}(G(z)).$$

Thus $F_1(z) = G(z)$, $F_2(z) = F_1(G(z)) = G(G(z))$, ... and

$$F_n(z) = G(G(\ldots G(G(z))\ldots)). \qquad (71)$$

We can also write, therefore,

$$F_n(z) = G(F_{n-1}(z)). \qquad (72)$$

It is clear that an explicit expression for $F_n(z)$ will be difficult to obtain from the recurrence relation (72)

However, we can obtain the moments of the size of the nth generation. Let $z = e^{-\theta}$ and define the cumulant generating functions

$$K(\theta) = \log G(e^{-\theta}), \qquad K_n(\theta) = \log F_n(e^{-\theta})$$

Then (72) becomes

$$\log F_n(e^{-\theta}) = \log G(F_{n-1}(e^{-\theta}))$$

or

$$K_n(\theta) = K(-K_{n-1}(\theta)). \qquad (73)$$

Letting $\mu = -K'(0)$ and $\sigma^2 = K''(0)$, assumed finite, denote the mean and variance of the number of offspring per individual respectively and $\mu_n = -K_n'(0)$, $v_n = K_n''(0)$ the mean and variance of the size of the nth generation, we obtain from (73), on differentiation

$$\mu_n = \mu\mu_{n-1}, \qquad (74)$$

$$v_n = \sigma^2 \mu^{2n-2} + \mu v_{n-1} \qquad (75)$$

Hence the mean and variance of the size of the nth generation are given by

$$\mu_n = \mu^n, \qquad (76)$$

$$v_n = \sigma^2(\mu^{n-1} + \ldots + \mu^{2n-2})$$

$$= \sigma^2 \mu^{n-1} \frac{1-\mu^n}{1-\mu} \qquad (77)$$

Observe that when $\mu < 1$, both μ_n and v_n approach zero exponentially fast as $n \to \infty$. Since X_n takes only integral values, a straightforward application of Chebychev's inequality will show that

$$\text{prob}(X_n = 0) = p_0^{(n)} \to 1$$

and hence ultimate extinction is certain in this case. When $\mu = 1$, then $\mu_n = 1$ and $v_n = n\sigma^2$, whereas if $\mu > 1$, μ_n and v_n both tend to $+\infty$ exponentially fast.

It is possible to determine exactly the probability of extinction, which we already know to be unity when the mean number of offspring per individual is less than 1. We have that $p_0^{(n)} = F_n(0)$ is the probability that there are no individuals in the nth generation, i.e. that extinction occurs at or before the nth generation. Now $p_0^{(n)} = \text{prob}(X_n = 0 | X_0 = 1)$ and 0 is an aperiodic recurrent (in fact absorbing) state. Therefore it follows from (52) that $p_0^{(n)}$ tends to a limit ξ as $n \to \infty$, this limit being the probability of extinction. From (72) we have

$$F_n(0) = G(F_{n-1}(0)),$$

and letting $n \to \infty$ we see that ξ must satisfy the relation $\xi = G(\xi)$. We must therefore examine the roots of the equation

$$x = G(x) \tag{78}$$

and since ξ is to be a probability we must seek roots satisfying $0 \leqslant \xi \leqslant 1$. Graphically we can see that the roots of (78) are the values of x at which the curve $y = G(x)$ and the line $y = x$ intersect. It is easily verified from (65) that $G'(1) > 0$ and for $x > 0$, $G''(x) > 0$, so that $G(x)$ is a convex increasing function for $x > 0$. Thus the curve $y = G(x)$ can intersect $y = x$ in at most two points when $x > 0$. Now $G(0) = g_0 > 0$ and it is clear that $x = 1$ is one point of intersection. Let $\xi_1 \leqslant \xi_2$ denote the two positive roots of (78). Then graphically we see the following results, remembering that $\mu = G'(1)$,

$$\xi_1 < \xi_2 = 1, \text{ if } \mu > 1; \qquad \xi_1 = \xi_2 = 1, \text{ if } \mu = 1;$$
$$\xi_1 = 1 < \xi_2, \text{ if } \mu < 1. \tag{79}$$

Since $G(x)$ is increasing, we have for any positive root ξ_i of (78)

$$F_1(0) = G(0) < G(\xi_i) = \xi_i$$
$$F_2(0) = G(G(0)) < G(\xi_i) = \xi_i$$

and proceeding thus we have that

$$F_n(0) < \xi_i \quad (n = 1, 2, \ldots). \tag{80}$$

Since $\xi = \lim F_n(0)$, ξ must be the smallest positive root of (78). Using (79) we can now formulate the following result.

Let μ denote the mean number of offspring per individual in a simple discrete branching process. The probability ξ of ultimate extinction is given by the smallest positive root of the equation $x = G(x)$. When $\mu \leqslant 1$ then $\xi = 1$ and extinction is certain; when $\mu > 1$ then $\xi < 1$ and the probability that the population grows indefinitely is $1 - \xi$.

We have not yet proved the last statement concerning the possibility of indefinite growth of the population when extinction is uncertain. The zero state is an absorbing state and all other states are transient, since they all communicate with the zero state. Thus from (67) we have

$$\lim_{n \to \infty} p_k^{(n)} = 0 \quad (k = 1, 2, \ldots)$$

and

$$\lim_{n \to \infty} p_0^{(n)} = \xi.$$

Now for any K

$$\text{prob}(X_n > K) + \sum_{k=1}^{K} p_k^{(n)} + p_0^{(n)} = 1$$

and

$$\lim_{n \to \infty} \text{prob}(X_n > K) + \xi = 1.$$

Hence for any K, however large, $\text{prob}(X_n > K) \to 1 - \xi$ as $n \to \infty$ and so when $\xi < 1$ there is a positive probability $1 - \xi$ of indefinite growth.

Example 3.16. *Family surnames in the United States.* The problem of the survival of family surnames mentioned earlier in this section was considered by Lotka (1931) in the particular case of males in the United States. From data for white males in 1920 he estimated the probabilities g_k of k male offspring per male and found that the p.g.f. (65) can be approximately represented by

$$G(z) = \frac{0 \cdot 482 - 0 \cdot 041 z}{1 - 0 \cdot 559 z}. \tag{81}$$

The equation $x = G(x)$ thus leads to a quadratic equation whose roots are 1 and 0·86. Hence the probability of extinction of a surname descended from a single male is $\xi = 0 \cdot 86$, perhaps an unexpectedly high value. On the other hand, if initially there are m males with the same surname and if their descendants reproduce independently of one another, then the probability that the surname becomes extinct is ξ^m. This becomes small for m large and consequently the probability of survival of the surname is high.

3.8. Limiting properties of irreducible chains

Throughout this section we shall restrict the discussion to an infinite irreducible system with states 0, 1, 2, ... and for simplicity we shall assume the system to be aperiodic.

When dealing with such systems in practice, one important problem is to determine whether a given system is ergodic, null or transient. We already know that if the system has a finite number of states then it is ergodic with a unique stationary distribution which is also the equilibrium distribution. However, if the system is not finite then it may belong to any one of the three classes, as the simple random walk, Example 3.14, shows. In general, one is given the one-step transition probabilities p_{jk} and it would be a simple matter to classify the system if we could determine directly the limiting behaviour of $p_{jk}^{(n)}$. This is, however, not usually possible and so we must look for criteria based on the p_{jk} themselves. These criteria depend on the existence and nature of the solutions of certain fundamental systems of linear equations associated with the chain.

In what follows we shall have occasion to use *taboo probabilities*. For an arbitrary Markov chain let H be a given sub-set of states. The probability

$$_H p_{jk}^{(n)} = \text{prob}(X_m \notin H,\ m = 1, \ldots, n-1,\ X_n = k | X_0 = j) \quad (82)$$

is called the n-fold transition probability of reaching state k from state j under the taboo H, i.e. it is the conditional probability, given the initial state j, of avoiding states in H at times $1, 2, \ldots, n-1$ and entering state k at time n. First passage and first return probabilities are particular cases, for

$$f_{jk}^{(n)} = {}_k p_{jk}^{(n)} \quad \text{and} \quad f_{jj}^{(n)} = {}_j p_{jj}^{(n)}.$$

If j and k are not states in the sub-set H then the taboo probability (82) can be interpreted as a transition probability in a modified chain in which H has been made a closed set, a particular case of such a modification being one for which all the states in H are made absorbing. This method was used in Section 2.2 in the discussion of first passage problems for the random walk. If the original chain is irreducible then any state $j \notin H$ communicates with H, and so in the modified chain return to j is not a certain event; thus all states outside H in the modified chain are transient. It follows from the limit theorem of Section 3.4 that in this case

$$\lim_{n \to \infty} {}_H p_{jk}^{(n)} = 0 \quad (j \notin H,\ k \notin H). \quad (83)$$

Chung (1960) gives an extensive discussion of taboo probabilities and some of their useful properties and applications.

(i) ERGODIC SYSTEMS

Consider first an infinite ergodic system. We know from the main limit theorem that for each j and k

$$\lim_{n \to \infty} p_{jk}^{(n)} = \pi_k > 0. \quad (84)$$

Thus π_k is the limiting probability of finding the system in the state k. For an initial distribution $\{p_j^{(0)}\}$ let $p_k^{(n)}$ be the probability of finding the system in state k at time n. Then

$$p_k^{(n)} = \sum_{j=0}^{\infty} p_j^{(0)} p_{jk}^{(n)} \quad (k = 0, 1, \ldots) \tag{85}$$

and by the dominated convergence theorem (Titchmarsh, 1939, p. 345; Parzen, 1962, p. 274)

$$\lim_{n \to \infty} p_k^{(n)} = \sum_{j=0}^{\infty} p_j^{(0)} \pi_k = \pi_k \quad (k = 0, 1, \ldots). \tag{86}$$

Thus the limiting probability of finding the system in state k is independent of the initial distribution. In view of the relation

$$p_k^{(n)} = \sum_{j=0}^{\infty} p_j^{(n-1)} p_{jk} \quad (k = 0, 1, \ldots), \tag{87}$$

we expect π_k to satisfy the set of linear equations, given in (63) for a finite chain,

$$\pi_k = \sum_{j=0}^{\infty} \pi_j p_{jk} \quad (k = 0, 1, \ldots). \tag{88}$$

We call (88) the equilibrium equations (strictly, these are defining equations for a stationary distribution). We now prove that an ergodic chain has a unique equilibrium distribution of positive probabilities and that this distribution is also the unique stationary distribution. We may state this as follows. *For an irreducible ergodic chain the limiting probabilities (π_k) form a probability distribution satisfying the equilibrium equations (88) and any solution of the equations*

$$\sum_{j=0}^{\infty} x_j p_{jk} = x_k \quad (k = 0, 1, \ldots), \tag{89}$$

or in matrix notation

$$\mathbf{xP} = \mathbf{x}, \tag{90}$$

satisfying $\sum |x_k| < \infty$, is a scalar multiple of (π_k).

To prove this we note that, since

$$\sum_{k=0}^{K} p_{jk}^{(n)} \leqslant 1,$$

it follows on letting $n \to \infty$ that

$$\sum_{k=0}^{K} \pi_k \leqslant 1$$

and so $\sum \pi_k \leqslant 1$. Next consider the Chapman–Kolmogorov equation

$$p_{jk}^{(n)} = \sum_{i=0}^{\infty} p_{ji}^{(n-1)} p_{ik}. \tag{91}$$

Then for any positive integer M,

$$p_{jk}^{(n)} \geq \sum_{i=0}^{M} p_{ji}^{(n-1)} p_{ik}.$$

Letting $n \to \infty$,

$$\pi_k \geq \sum_{i=0}^{M} \pi_i p_{ik}$$

and letting $M \to \infty$,

$$\pi_k \geq \sum_{i=0}^{\infty} \pi_i p_{ik}. \tag{92}$$

Suppose that for some k strict inequality holds in (92). Then

$$\sum_k \pi_k > \sum_k \sum_i \pi_i p_{ik} = \sum_i \pi_i \sum_k p_{ik} = \sum_i \pi_i$$

where the summations are all taken over $0, 1, \ldots$ This is a contradiction since $\sum \pi_i$ is finite. It follows therefore that

$$\pi_k = \sum_{i=0}^{\infty} \pi_i p_{ik} \quad (k = 0, 1, \ldots). \tag{93}$$

Writing (93) as $\pi P = \pi$, we see that $\pi P^n = \pi$, i.e.

$$\sum_{j=0}^{\infty} \pi_j p_{jk}^{(n)} = \pi_k \quad (k = 0, 1, \ldots). \tag{94}$$

Letting $n \to \infty$, we must have $\pi_k \sum \pi_j = \pi_k$ by the dominated convergence theorem and hence $\sum \pi_k = 1$. Now suppose that $\mathbf{x} = (x_k)$ is an absolutely convergent solution of (89). Then we have, again by the dominated convergence theorem, that as $n \to \infty$

$$x_k = \sum_{j=0}^{\infty} x_j p_{jk}^{(n)} \to \pi_k (\sum x_j)$$

and so the result is proved.

The equilibrium equations (89) are a fundamental set of equations in the theory of Markov chains. We have shown that when the chain is ergodic there is a unique absolutely convergent solution to (89). We show below that the converse is also true, namely that if the equations (89) have an absolutely convergent solution then the chain is ergodic. Consequently if, for a given chain, we can obtain a probability distribution as the solution of the equilibrium equations, then we may be assured that the chain is ergodic. This is a frequently used method of establishing ergodicity and it has the additional merit of providing the equilibrium distribution. This is expressed in the following statement. *An irreducible aperiodic system is ergodic if the equilibrium equations*

$$\sum_{j=0}^{\infty} x_j p_{jk} = x_k \quad (k = 0, 1, \ldots) \tag{95}$$

have a solution (x_k) (x_k not all zero) satisfying

$$\sum |x_k| < \infty, \tag{96}$$

and only if every non-negative solution of (95) satisfies (96).

For suppose (x_j) satisfies (95) and (96). Then

$$|x_k| \leqslant \sum_{j=0}^{\infty} |x_j| \, p_{jk}^{(n)} \leqslant \sum_{j=0}^{J} |x_j| \, p_{jk}^{(n)} + \epsilon$$

for an arbitrary $\epsilon > 0$ and J sufficiently large. Hence, choosing k so that $x_k \neq 0$, we have

$$\sum_{j=0}^{J} |x_j| \lim_{n \to \infty} p_{jk}^{(n)} > \delta > 0$$

and so, for some j and k, $\lim p_{jk}^{(n)} > \delta > 0$. Hence k is an ergodic state and, since the chain is irreducible, all states are ergodic.

Conversely suppose the system is ergodic so that $\lim p_{jk}^{(n)} = \pi_k > 0$. If (x_j) is a non-negative solution of (95) then

$$x_k = \sum_{j=0}^{\infty} x_j p_{jk}^{(n)} \geqslant \sum_{j=0}^{J} x_j p_{jk}^{(n)}.$$

Hence letting $n \to \infty$, we have for every $J > 0$

$$\pi_k \sum_{j=0}^{J} x_j \leqslant x_k$$

and so $\sum x_j$ is convergent.

Example 3.17. *Doubly stochastic matrices.* A double stochastic matrix is a stochastic matrix whose column sums are also unity. For an irreducible aperiodic Markov chain with a doubly stochastic transition matrix, one solution to the equilibrium equations (95) is $x_k = 1$ ($k = 0, 1, \ldots$). In the denumerably infinite case this is not a convergent solution and so such a chain cannot be ergodic, and in the finite case this implies that all states are equally likely in the limit. Clearly for a finite chain a necessary and sufficient condition that all states should be equally likely in the limit is that the transition matrix should be doubly stochastic.

(ii) TRANSIENT SYSTEMS

We now proceed to develop a criterion for determining whether or not a system is transient. For a transient system the first passage probabilities $f_{jk}^{(n)}$ defined at (36) satisfy

$$f_{jk} = \sum_{n=1}^{\infty} f_{jk}^{(n)} < 1. \tag{97}$$

Consider f_{j0}, the probability of ever reaching state 0 when starting from state j. Then

$$f_{j0} = p_{j0} + \sum_{k=1}^{\infty} p_{jk} f_{k0} \quad (j = 1, 2, \ldots). \tag{98}$$

Let $g_j = 1 - f_{j0}$ be the probability of never reaching state 0 from state j. For a transient chain $g_j > 0$ and

$$g_j = \sum_{k=1}^{\infty} p_{jk} g_k \quad (j = 1, 2, \ldots). \tag{99}$$

Hence the system of equations

$$\sum_{k=1}^{\infty} p_{jk} y_k = y_j \quad (j = 1, 2, \ldots) \tag{100}$$

has a non-zero, bounded solution, namely $y_j = g_j$ $(j = 1, 2, \ldots)$.

Suppose conversely that the equations (100) have a non-zero bounded solution (y_j); we may assume $|y_j| \leqslant 1$ $(j = 1, 2, \ldots)$. Let $\mathbf{P}_0$ be the matrix obtained from $\mathbf{P}$ by striking out the row and column corresponding to state 0. Then, if $\mathbf{y} = (y_j)$, we may write (100) as $\mathbf{P}_0 \mathbf{y} = \mathbf{y}$ and therefore

$$\mathbf{P}_0^n \mathbf{y} = \mathbf{y}. \tag{101}$$

The elements of $\mathbf{P}_0^n$ are in fact the taboo probabilities ${}_0 p_{jk}^{(n)}$. Thus (101) may be written

$$y_j = \sum_{k=1}^{\infty} {}_0 p_{jk}^{(n)} y_k \quad (j = 1, 2, \ldots). \tag{102}$$

Hence

$$|y_j| \leqslant \sum_{k=1}^{\infty} {}_0 p_{jk}^{(n)} = g_j^{(n)},$$

where $g_j^{(n)}$ is the probability of not entering zero during n transitions when starting from j. Now

$$1 - f_{j0} = g_j = \lim_{n \to \infty} g_j^{(n)} \geqslant |y_j| > 0 \tag{103}$$

for some j; hence $f_{j0} < 1$. Thus j is a transient state and so, therefore, are all states since the chain is irreducible.

We may express these results as follows. *An infinite irreducible aperiodic chain is transient if and only if the system of equations* (100) *has a non-zero bounded solution.* According to (98) this may be expressed alternatively in the following statement. *An irreducible aperiodic chain is transient if and only if the system of equations*

$$y_j = \sum_{k=0}^{\infty} p_{jk} y_k \quad (j \neq 0) \tag{104}$$

has a bounded non-constant solution.

The proof is left as an exercise.

The implication of this result is that if for any state, which we may take to be state 0 without loss of generality, we can solve the equations (98) for the first passage probabilities into state 0 and obtain a solution with $f_{j0} < 1$ then we may be assured that the system is transient. Similarly, if we can obtain non-zero probabilities g_k from (99) of forever remaining in the states 1, 2, ... when starting from k then again we may be assured that the system is transient.

Example 3.18. *Random walk with variable transition probabilities.* Consider the random walk with variable transition probabilities described by the transition matrix (27). To determine whether the system is transient we examine the equations (100), which take the form

$$\psi_1 y_1 + \theta_1 y_2 = y_1,$$
$$\phi_j y_{j-1} + \psi_j y_j + \theta_j y_{j+1} = y_j \quad (j = 2, 3, \ldots),$$

(105)

and since $\psi_j = 1 - \theta_j - \phi_j$, these may be written

$$y_2 - y_1 = \frac{\phi_1}{\theta_1} y_1,$$
$$y_{j+1} - y_j = \frac{\phi_j}{\theta_j} (y_j - y_{j-1}).$$

(106)

These may be solved recursively for $y_j - y_{j-1}$, obtaining

$$y_j - y_{j-1} = \frac{\phi_1 \ldots \phi_{j-1}}{\theta_1 \ldots \theta_{j-1}} y_1.$$

Hence

$$y_j = \left(\sum_{k=1}^{j-1} \rho_k \right) y_1,$$

(107)

where $\rho_k = (\phi_1 \ldots \phi_{k-1})/(\theta_1 \ldots \theta_{k-1})$. The system is transient if y_j is bounded, i.e. if $\sum \rho_k < \infty$, and recurrent if $\sum \rho_k = \infty$. In particular if $\phi_j = q$, $\theta_j = p$ $(j = 0, 1, \ldots)$ then we have the simple random walk with a reflecting barrier at the origin; in this case $\rho_k = (q/p)^{k-1}$ and the system is transient if $q < p$.

(*iii*) RECURRENT SYSTEMS

We have given criteria by means of which we can show whether a given irreducible aperiodic system is ergodic or transient; if it is neither of these it must be null-recurrent. In addition we now obtain some criteria for recurrent systems. The first of these is a sufficient condition by means of which it may be possible to show that a given system is recurrent; once recurrence has been established the second enables us to distinguish between the ergodic and null-recurrent cases.

An irreducible aperiodic system is recurrent if there is a solution of the inequalities

$$\sum_{k=1}^{\infty} p_{jk} y_k \leqslant y_j \quad (j = 1, 2, \ldots) \tag{108}$$

with the property that $y_j \to \infty$ *as* $j \to \infty$.

To prove this we use the matrix $\mathbf{P}_0$, defined just before equation (101). Then (108) reads $\mathbf{P}_0 \mathbf{y} \leqslant \mathbf{y}$, and since $\{y_j = 1, j = 1, 2, \ldots\}$ satisfies (108) it follows that $(y_j + \alpha)$ is a solution if (y_j) is. Hence we may suppose $y_j \geqslant 0$ without loss of generality. On iteration (108) becomes

$$\mathbf{P}_0^n \mathbf{y} \leqslant \mathbf{y}$$

where the inequality between matrices is defined to hold element-wise. Thus

$$\sum_{k=1}^{\infty} {}_0 p_{jk}^{(n)} y_k \leqslant y_j. \tag{109}$$

Let $z_i = \min(y_{i+1}, y_{i+2}, \ldots)$, which always exists since $y_i \to \infty$ as $i \to \infty$. Then we have

$$z_i \sum_{k=i+1}^{\infty} {}_0 p_{jk}^{(n)} \leqslant y_j \quad (n = 1, 2, \ldots). \tag{110}$$

From (103) we have

$$1 = f_{j0} + \lim_{n \to \infty} \left\{ \sum_{k=1}^{\infty} {}_0 p_{jk}^{(n)} \right\}. \tag{111}$$

Now (110) gives us

$$\sum_{k=1}^{\infty} {}_0 p_{jk}^{(n)} \leqslant \sum_{k=1}^{i} {}_0 p_{jk}^{(n)} + y_j/z_i \quad (n = 1, 2, \ldots). \tag{112}$$

The right-hand side of (112) can be made arbitrarily small in view of (83) by first choosing i sufficiently large and then letting $n \to \infty$. It follows that the limit on the right of (111) is zero. Hence $f_{j0} = 1$, showing that 0 is a recurrent state. By irreducibility, all other states are also recurrent and the result is therefore proved.

If an irreducible aperiodic Markov chain is recurrent then there exists a positive solution of the equilibrium equations

$$\sum_{j=0}^{\infty} x_j p_{jk} = x_k \quad (k = 0, 1, \ldots), \tag{113}$$

which is unique up to a multiplicative constant; the chain is ergodic or null-recurrent according as $\sum x_j < \infty$ *or* $\sum x_j = \infty$.

Although we have already proved this result for the ergodic case, the following proof is quite general.

Suppose the system starts in state 0. Let N denote the time of first return to state 0, and let N_k denote the number of visits to state k during this time. We define $N_0 = 1$. Then

$$N = N_0 + N_1 + N_2 + \ldots \tag{114}$$

We shall show that $x_k = E(N_k)$ $(k = 1, 2, \ldots)$ is finite and satisfies (113). Since $E(N) = \sum E(N_k)$ is finite for an ergodic system and infinite for a null-recurrent system, our result will be proved.

In terms of the taboo probabilities $_0p_{0k}^{(n)}$ we have

$$E(N_k) = \sum_{n=1}^{\infty} {_0}p_{0k}^{(n)} \quad (k = 1, 2, \ldots). \tag{115}$$

Now

$$f_{00}^{(n+m)} \geqslant {_0}p_{0k}^{(n)} f_{k0}^{(m)}$$

and if we choose m so that $f_{k0}^{(m)} > 0$, then $_0p_{0k}^{(n)} = O(f_{00}^{(n+m)})$, as $n \to \infty$, and $E(N_k) < \infty$.

Next we have

$$_0p_{0k}^{(1)} = p_{0k},$$

$$_0p_{0k}^{(n)} = \sum_{j=1}^{\infty} {_0}p_{0j}^{(n-1)} p_{jk} \quad (n > 1),$$

and summing for $n = 1, 2, \ldots$ we have, from (115), that

$$E(N_k) = p_{0k} + \sum_{j=1}^{\infty} p_{jk} E(N_j)$$

$$= \sum_{j=0}^{\infty} p_{jk} E(N_j) \quad (k = 0, 1, \ldots).$$

Thus $x_k = E(N_k)$ satisfies (113). Also no x_k can be zero, for as there is at least one positive x_k, namely $x_0 = 1$, the relation

$$x_k = \sum_{j=0}^{\infty} p_{jk} x_j = \sum_{j=0}^{\infty} p_{jk}^{(n)} x_j = 0$$

would imply that $p_{0k}^{(n)} = 0$ for all n and this contradicts the hypothesis of irreducibility. It now remains to show that (x_k) is unique up to a multiplicative constant.

Let $y_0 = 1, y_1, \ldots$ be a positive solution of (113). Define

$$q_{jk}^{(n)} = \frac{y_k}{y_j} p_{kj}^{(n)} \quad (j, k = 0, 1, \ldots; n = 1, 2, \ldots). \tag{116}$$

Then, writing $q_{jk}^{(1)} = q_{jk}$ for all j, k, we have

$$q_{jk}^{(n)} \geqslant 0, \quad \sum_{k=0}^{\infty} q_{jk}^{(n)} = 1, \quad q_{jk}^{(n+1)} = \sum_{i=0}^{\infty} q_{ji} q_{ik}^{(n)}$$

and so the matrix (q_{jk}) is stochastic and defines a recurrent Markov chain in virtue of (116). If $g_{jk}^{(n)}$ denote the first passage probabilities for this chain, it is not difficult to show that

$$_0p_{0k}^{(n)} = y_k\, g_{k0}^{(n)}$$

which, on summation over n, gives $x_k = y_k$.

Example 3.19. *The single-server queue with Poisson arrivals and general service-times.* Consider the queueing process of Example 3.10. The imbedded Markov chain, which is clearly irreducible, has transition matrix

$$\mathbf{P} = \begin{bmatrix} b_0 & b_1 & b_2 & b_3 & \cdots \\ b_0 & b_1 & b_2 & b_3 & \cdots \\ 0 & b_0 & b_1 & b_2 & \cdots \\ 0 & 0 & b_0 & b_1 & \cdots \\ \cdot & \cdot & \cdot & \cdot & \cdots \end{bmatrix},$$

where the quantities b_i, given by (24), are positive and form a probability distribution. Let $\rho = \sum n b_n$ be the mean of this distribution. We shall show that the system is ergodic if $\rho < 1$, null-recurrent if $\rho = 1$ and transient if $\rho > 1$.

First we attempt to solve the equilibrium equations which in this case take the form

$$\pi_0 b_k + \pi_1 b_k + \pi_2 b_{k-1} + \ldots + \pi_{k+1} b_0 = \pi_k \quad (k = 0, 1, \ldots). \quad (117)$$

The left-hand side is an expression of the convolution type suggesting the use of generating functions. Let

$$\Pi(z) = \sum_{j=0}^{\infty} \pi_j z^j, \qquad G_b(z) = \sum_{j=0}^{\infty} b_j z^j.$$

Note that in virtue of (24)

$$G_b(z) = b^*(\lambda - \lambda z),$$

where

$$b^*(s) = \int_0^{\infty} e^{-st}\, dB(t).$$

Multiplying (117) by z^k and summing over k, we obtain

$$\pi_0 G_b(z) - z^{-1}\pi_0 G_b(z) + z^{-1} G_b(z)\, \Pi(z) = \Pi(z), \quad (118)$$

whence

$$\Pi(z) = \frac{\pi_0 G_b(z)\,(z-1)}{z - G_b(z)} \quad (119)$$

$$= \pi_0 G_b(z) \Big/ \left\{ 1 - \frac{1 - G_b(z)}{1 - z} \right\}. \quad (120)$$

Let

$$C(z) = \frac{1 - G_b(z)}{1 - z} = \sum_{j=0}^{\infty} z^j (b_{j+1} + b_{j+2} + \ldots). \qquad (121)$$

Then

$$C(1) = \sum_{j=0}^{\infty} \sum_{k=j+1}^{\infty} b_k = \sum_{k=0}^{\infty} k b_k = \rho.$$

Thus

$$\Pi(z) = \pi_0 G_b(z) / \{1 - C(z)\} = \pi_0 G_b(z) \sum_{j=0}^{\infty} \{C(z)\}^j \quad (|z| < 1)$$

and it is clear that the π_j are positive if π_0 is. Further, by Abel's Theorem (Theorem 1, Section 3.13)

$$\sum \pi_j = \lim_{z \to 1-} \Pi(z) \quad \begin{array}{l} < \infty \quad (\rho < 1), \\ = \infty \quad (\rho \geqslant 1). \end{array} \qquad (122)$$

Hence by the second result of part (i) of this section the system is ergodic if and only if $\rho < 1$; the equilibrium distribution is given by (119) in the form of a generating function. The quantity π_0 must be chosen so that $\Pi(1) = 1$. Hence

$$\pi_0 = \lim_{z \to 1} \frac{\Pi(z)}{G_b(z)} \left\{ \frac{z - G_b(z)}{z - 1} \right\} = 1 - \rho. \qquad (123)$$

Next consider the question of transience. Here we seek a bounded solution to the equations (100) which, in this case, take the form

$$\sum_{j=1}^{\infty} b_j y_j = y_1,$$

$$\sum_{j=0}^{\infty} b_j y_{j+k-1} = y_k \quad (k = 2, 3, \ldots). \qquad (124)$$

Here we must resort to trial solutions and with one of the form $y_k = \alpha + \beta \xi^k$, (124) reduces to

$$\alpha (1 - b_0) + \beta \{ G_b(\xi) - b_0 \} = \alpha + \beta \xi,$$

$$\alpha + \beta G_b(\xi) \xi^{k-1} = \alpha + \beta \xi^k \quad (k = 2, 3, \ldots). \qquad (125)$$

If we take $\alpha = 1$, $\beta = -1$ we remain with the single equation for ξ,

$$G_b(\xi) = \xi. \qquad (126)$$

From the theory of branching processes we have seen this equation has a root ξ_0 ($0 < \xi_0 < 1$) if and only if $\rho = G_b'(1) > 1$. Hence if $\rho > 1$ there is a bounded solution $y_j = 1 - \xi_0^j$ to the equations (124) and so the system is transient. It remains now to establish the nature of the system when $\rho = 1$.

Consider a trial solution, $y_j = j$, to the inequalities (108). It is not difficult to see that this is a solution if $\rho \leqslant 1$, and the system is therefore recurrent if $\rho \leqslant 1$; it must be null-recurrent when $\rho = 1$ since we have shown that it is ergodic if and only if $\rho < 1$. Alternatively one may argue that the system is null-recurrent for $\rho = 1$ since by (122) the equilibrium equations (113) have a divergent positive solution when $\rho = 1$.

From the definition (24) of the quantities b_r the condition $\rho < 1$ can be seen to be equivalent to

$$\lambda \int_0^\infty t\,dB(t) < 1$$

or
$$\frac{\text{Mean service-time}}{\text{Mean inter-arrival time}} < 1. \tag{127}$$

This is therefore the condition for the queue size to have an equilibrium distribution. The same condition (127) has been shown by Lindley (1952) to hold when both service-times and inter-arrival times have general distributions.

(iv) PERIODIC CHAINS

To simplify the discussion of the present section we have confined our attention to aperiodic irreducible chains. However, if an irreducible chain is periodic all the results of this section still hold with appropriate modifications; in particular wherever the word ergodic occurs we must substitute positive recurrent.

3.9. Absorption problems

We now turn to Markov chains which are not irreducible. According to the decomposition theorem of Section 3.5, a Markov chain will in general be composed of a set T of transient states and a set Q of recurrent states, the latter being uniquely decomposable into closed sets C_1, C_2, ... Once one of the closed sets C_i is entered it is never vacated and we may call such an event absorption in C_i. We consider the problem of determining the probability, given that we start in a transient state, of being absorbed in Q or, more specifically, in a particular C_i. We assume that T does not contain an irreducible sub-set for such a sub-set would itself form an irreducible transient chain; in other words we assume that every state in T communicates with some recurrent state.

For any state $j \in T$ and any closed set C of recurrent states we define f_{jC} to be the probability of eventually entering C. Let $f_{jC}^{(n)}$ be the probability that passage from j to C occurs for the first time at time n. Then

$$f_{jC} = \sum_{n=1}^\infty f_{jC}^{(n)}.$$

Further,

$$f_{jC}^{(n)} = \sum_{k \in T} p_{jk} f_{kC}^{(n-1)} \quad (j \in T, \; n = 2, 3, \ldots),$$

$$f_{jC}^{(1)} = \sum_{k \in C} p_{jk} \quad (j \in T). \tag{128}$$

On summing (128) over $n = 1, 2, \ldots$ we obtain

$$f_{jC} = \sum_{k \in C} p_{jk} + \sum_{k \in T} p_{jk} f_{kC} \quad (j \in T). \tag{129}$$

Thus if j is a transient state and C a closed set of recurrent states, then the probability f_{jC} of absorption in C when starting in j satisfies the system of equations (129).

Example 3.20. Simple random walk with absorbing barriers. In Section 2.2(ii) we considered a simple random walk with absorbing barriers at $-b$ and a and with probabilities q, $1-p-q$, p of steps -1, 0, 1 respectively. In this case T consists of the states $-b+1, \ldots, a-1$ and the two absorbing states each form a closed set consisting of a single state. Hence for absorption at the upper barrier a, say, the equations (129) become

$$f_{ja} = q f_{j-1,a} + (1-p-q) f_{ja} + p f_{j+1,a} \quad (j = -b+2, \ldots, a-2), \tag{130}$$

$$f_{-b+1,a} = (1-p-q) f_{-b+1,a} + p f_{-b+2,a}, \tag{131}$$
$$f_{a-1,a} = p + q f_{a-2,a} + (1-p-q) f_{a-1,a}.$$

These equations are of course equivalent to (2.14) and its boundary conditions with $s = 1$.

3.10. Non-negative square matrices

The present and following two sections are rather specialized and may be omitted on a first reading. We give for the finite chain a demonstration of most of the preceding results from a different point of view. The treatment is based on the properties of non-negative square matrices which we describe in this section. At the beginning of Section 3.11, a brief summary is given of the connexion between the properties of the Markov chain and those of its transition matrix.

For non-negative square matrices, the main result is the theorem of Perron and Frobenius which states, broadly speaking, that a non-negative square matrix has a maximal non-negative eigenvalue which is not exceeded in absolute value by any other eigenvalue and corresponding to which there is a non-negative eigenvector. Proofs of the results are given, for example, by Gantmacher (1959).

(*i*) SOME DEFINITIONS

Let $\mathbf{A} = (a_{ij})$ be a matrix with non-negative elements a_{ij}. We call $\mathbf{A}$ *positive* and write $\mathbf{A} > 0$ if each of the elements is strictly positive. Otherwise we write $\mathbf{A} \geqslant 0$ and call $\mathbf{A}$ *non-negative*. Similarly, inequalities $\mathbf{A} \leqslant \mathbf{B}$ or $\mathbf{A} < \mathbf{B}$ between non-negative matrices of the same order mean that the inequalities hold element-wise. If $\mathbf{C} = (c_{ij})$ is an arbitrary matrix of complex-valued elements, then we define $\mathbf{C}^*$ to be the matrix obtained from $\mathbf{C}$ by replacing each element by its modulus. Thus

$$\mathbf{C}^* = (|c_{ij}|). \tag{132}$$

An immediate consequence of non-negativity is the following.

If $\mathbf{A} > 0$ *and* $\mathbf{B} \geqslant 0$ *then* $\mathbf{AB} = 0$ *if and only if* $\mathbf{B} = 0$.

For square matrices we can compute the powers $\mathbf{A}^2$, $\mathbf{A}^3$, ... and as in the case of transition probabilities we use the notation $\mathbf{A}^n = (a_{ij}^{(n)})$. We define $\mathbf{A}^0 = \mathbf{I}$. Clearly if $\mathbf{A} \geqslant 0$, then $\mathbf{A}^n \geqslant 0$ for every positive integer n.

Suppose $\mathbf{A}$ is a square matrix of order $h \times h$. If, by permuting the rows and columns of $\mathbf{A}$ in the same way (for a transition matrix this means relabelling the states), we can obtain a matrix of the form

$$\begin{bmatrix} \mathbf{B} & 0 \\ \mathbf{C} & \mathbf{D} \end{bmatrix}, \tag{133}$$

where $\mathbf{B}$ is a square matrix, then $\mathbf{A}$ is said to be *reducible*. Here and throughout it is convenient to use 0 to denote a matrix of zero elements. Otherwise it is called *irreducible*. Thus a square matrix is irreducible if there is no permutation which, when applied to rows and columns, changes $\mathbf{A}$ to the form (133). Note that this corresponds to the concept of irreducibility for Markov chains; for if a finite stochastic matrix has the form (133) when the states are suitably labelled, then the states corresponding to $\mathbf{B}$ form a closed set from which the remaining states cannot be reached and the chain is therefore not irreducible. Observe that if a matrix is of the form (133) then the zero sub-matrix is preserved through all powers of the matrix.

The irreducible square matrices are the important elements in the theory of non-negative square matrices. For if $\mathbf{A}$ is any square matrix then by a suitable permutation applied to both rows and columns we can reduce $\mathbf{A}$ to the form,

$$\mathbf{A} = \begin{bmatrix} \mathbf{A}_{11} & 0 & 0 & \cdots & 0 \\ \mathbf{A}_{21} & \mathbf{A}_{22} & 0 & \cdots & 0 \\ \vdots & \vdots & \vdots & & \vdots \\ \mathbf{A}_{m1} & \mathbf{A}_{m2} & \mathbf{A}_{m3} & \cdots & \mathbf{A}_{mm} \end{bmatrix}, \tag{134}$$

where the matrices A_{ii} on the diagonal are square and irreducible. If A is itself irreducible then, of course, $A = A_{11}$. On account of the determinantal relation

$$|\lambda I - A| = \prod_{i=1}^{m} |\lambda I - A_{ii}| \tag{135}$$

any eigenvalue of A is an eigenvalue of one of the irreducible matrices A_{ii}. We proceed therefore to study irreducible matrices.

(*ii*) IRREDUCIBLE NON-NEGATIVE SQUARE MATRICES

The basic result is the following.

The Perron–Frobenius theorem. Suppose $A \geqslant 0$ *and irreducible.*

(a) *Then*

  (i) A *has a real positive eigenvalue* λ_1 *with the following properties;*
  (ii) *corresponding to* λ_1 *there is an eigenvector* x *all of whose elements may be taken as positive, i.e. there exists a vector* $x > 0$ *such that*

$$Ax = \lambda_1 x;$$

  (iii) *if* α *is any other eigenvalue of* A *then*

$$|\alpha| \leqslant \lambda_1; \tag{136}$$

  (iv) λ_1 *increases when any element of* A *increases;*
  (v) λ_1 *is a simple root of the determinantal equation*

$$|\lambda I - A| = 0. \tag{137}$$

  (vi) $$\lambda_1 \leqslant \max_j \left(\sum_k a_{jk} \right), \qquad \lambda_1 \leqslant \max_k \left(\sum_j a_{jk} \right). \tag{138}$$

(b) *If* B *is a matrix of complex-valued elements such that* $B^* \leqslant A$, *then any eigenvalue* μ *of* B *satisfies*

$$|\mu| \leqslant \lambda_1.$$

If for some μ, $|\mu| = \lambda_1$, *then* $B^* = A$. *More precisely if* $\mu = \lambda_1 e^{i\phi}$, *then*

$$B = e^{i\phi} DAD^{-1}, \tag{139}$$

where $D^* = I$.

(c) *If* A *is of order* $h \times h$ *and if* A *has exactly* t *eigenvalues equal in modulus to* λ_1, *then these numbers are all different and are the roots of the equation*

$$\lambda^t - \lambda_1^t = 0, \tag{140}$$

and in general the set of h *eigenvalues, when plotted as points in the complex* λ*-plane, is invariant under a rotation of the plane through the angle* $2\pi/t$ *but* *not through smaller angles. When $t > 1$, then A can be reduced to the following cyclic form by a permutation applied to both rows and columns.*

$$
A = \begin{bmatrix}
0 & A_{12} & 0 & \ldots & 0 \\
0 & 0 & A_{23} & \ldots & 0 \\
\vdots & \vdots & \vdots & & \vdots \\
0 & 0 & 0 & \ldots & A_{t-1,t} \\
A_{t1} & 0 & 0 & \ldots & 0
\end{bmatrix}, \tag{141}
$$

where the sub-matrices on the main diagonal are square.

A proof of this result is given also by Debreu and Herstein (1953).

Clearly if A is irreducible so is its transpose A' and conversely. It follows that A also has a positive row or left eigenvector corresponding to λ_1. If λ_1 itself is the only eigenvalue of modulus λ_1 then A is said to be *primitive*. Note that (a) (vi) gives the largest row sum or largest column sum as an upper bound to λ_1.

(*iii*) JORDAN CANONICAL FORM

Let A be an $h \times h$ matrix. It is well known that if the eigenvalues $\lambda_1, \ldots, \lambda_h$ of A are distinct then there is a non-singular matrix H which transforms A to diagonal form, i.e.

$$
HAH^{-1} = \Lambda, \tag{142}
$$

where $\Lambda = \operatorname{diag}(\lambda_i)$. The usefulness of this representation is that powers of A can simply be expressed in terms of powers of the eigenvalues:

$$
A^n = H^{-1}\Lambda^n H. \tag{143}
$$

In general, however, when the eigenvalues are not distinct, A need not have a diagonal form, as for example, if

$$
A = \begin{bmatrix} 1 & 1 \\ 0 & 1 \end{bmatrix}.
$$

There is however a nearly diagonal form called the Jordan canonical form.

Let $J_1(\lambda) = \lambda$ and for $k \geqslant 2$ let $J_k(\lambda)$ be a $k \times k$ matrix of the form

$$
J_k(\lambda) = \begin{bmatrix}
\lambda & 1 & 0 & 0 & \ldots & 0 \\
0 & \lambda & 1 & 0 & \ldots & 0 \\
0 & 0 & \lambda & 1 & \ldots & 0 \\
\vdots & \vdots & \vdots & \vdots & & \vdots \\
0 & 0 & 0 & 0 & & \lambda
\end{bmatrix} \tag{144}
$$

$$
= \lambda I + M,
$$

where $\mathbf{M}$ is matrix with 1's on the first super-diagonal and zero elsewhere. It is not difficult to see that $\mathbf{M}^k = 0$ and it follows that if we take the nth power $(n \geqslant k)$ of $\mathbf{J}_k(\lambda)$ we obtain

$$\mathbf{J}_k^n(\lambda) = \lambda^n \mathbf{I} + \binom{n}{1} \lambda^{n-1} \mathbf{M} + \ldots + \binom{n}{k-1} \lambda^{n-k+1} \mathbf{M}^{k-1}. \tag{145}$$

The Jordan canonical form is built up of matrices of the type $\mathbf{J}_k(\lambda)$. Let $\mathbf{A}$ be an $h \times h$ square matrix. Then there is a non-singular matrix $\mathbf{H}$ such that

$$\mathbf{HAH}^{-1} = \begin{bmatrix} \mathbf{J}_{k_1}(\lambda_1) & 0 & \ldots & 0 \\ 0 & \mathbf{J}_{k_2}(\lambda_2) & \ldots & 0 \\ \vdots & \vdots & & \vdots \\ 0 & 0 & \ldots & \mathbf{J}_{k_m}(\lambda_m) \end{bmatrix}, \tag{146}$$

where $k_1 + \ldots + k_m = h$. The λ_ν are eigenvalues of $\mathbf{A}$, not necessarily distinct, and to every eigenvalue of $\mathbf{A}$ there corresponds at least one $\mathbf{J}$. If an eigenvalue is simple there is exactly one corresponding $\mathbf{J}$ of order 1×1. The representation (146) is called the Jordan canonical form of $\mathbf{A}$; it reduces to the diagonal form (142) when the eigenvalues are distinct. Let us call a matrix of the form displayed on the right-hand side of (146) a *Jordan matrix*.

Returning now to non-negative matrices, we state a result concerning the Jordan canonical form of an irreducible non-negative matrix. *If the irreducible non-negative matrix $\mathbf{A}$ has exactly t eigenvalues of maximum modulus λ_1 then there is a non-singular matrix $\mathbf{H}$ such that*

$$\mathbf{HAH}^{-1} = \begin{bmatrix} \mathbf{\Lambda} & 0 \\ 0 & \mathbf{J} \end{bmatrix} \tag{147}$$

where $\mathbf{\Lambda}$ is a $t \times t$ diagonal matrix, $\mathbf{\Lambda} = \mathrm{diag}\,\{\lambda_1 e^{2i\pi k/t}\}$ $(k = 0, \ldots, t-1)$, and $\mathbf{J}$ is a Jordan matrix whose diagonal elements are all less than λ_1 in modulus. Note that the diagonal elements of $\mathbf{\Lambda}$ are the t eigenvalues of modulus λ_1. When $t = 1$ then $\mathbf{A}$ is primitive and the result becomes

$$\mathbf{HAH}^{-1} = \begin{bmatrix} \lambda_1 & 0 \\ 0 & \mathbf{J} \end{bmatrix}. \tag{148}$$

(iv) LIMITING BEHAVIOUR OF $\mathbf{A}^n$

The previous results enable us to give the following result concerning the limiting behaviour of the powers $\mathbf{A}^n$ as $n \to \infty$. *If $\mathbf{A}$ is non-negative and irreducible with exactly t eigenvalues of modulus λ_1, and if $\mathbf{x} = (x_j)$ and*

$\mathbf{y} = (y)_j$ *are positive column and row eigenvectors corresponding to* λ_1, *normalized so that* $\mathbf{yx} = \sum x_j y_j = 1$, *then*

$$\lim_{n \to \infty} \frac{1}{t} \{(1/\lambda_1^n) \mathbf{A}^n + \ldots + (1/\lambda_1^{n+t-1}) \mathbf{A}^{n+t-1}\} = \mathbf{xy}, \qquad (149)$$

the limit being approached geometrically fast and uniformly for all elements. In particular if $\mathbf{A}$ *is primitive* ($t = 1$) *then*

$$\lim_{n \to \infty} (1/\lambda_1)^n \mathbf{A}^n = \mathbf{xy}. \qquad (150)$$

Note in particular the result (150) for the primitive case, which in terms of the elements of $\mathbf{A}^n$ becomes

$$\lim_{n \to \infty} a_{jk}^{(n)}/\lambda_1^n = x_j y_k \qquad (151)$$

and these limits are positive for all j and k.

3.11. Finite Markov chains

The results of the previous section enable us to deduce many of the properties of finite chains from those of their stochastic matrices. Firstly we note that any stochastic matrix $\mathbf{P}$ has an eigenvalue 1 since $\mathbf{P1} = \mathbf{1}$, where $\mathbf{1}$ is a column vector of 1's. By a suitable labelling of the states we can always express $\mathbf{P}$ in the form (134) and it follows from (135), (136) and (138) that no eigenvalue can exceed 1 in modulus. Secondly we shall assume the truth of part A of the limit theorem (Section 3.4) since these results are simple consequences of the properties of power series with non-negative coefficients.

One of the main points emerging from the ensuing analysis is that the nature of a finite chain is determined by the properties of the eigenvalues of $\mathbf{P}$ which have unit modulus. Another point is that limiting values of transition probabilities are approached geometrically fast, the rate of approach being determined in general by the eigenvalue of largest modulus less than unity. For the reader who does not wish to pursue the detailed discussion we summarize here the results for the three cases distinguished in this section.

(i) If the finite Markov chain is ergodic then its transition matrix is primitive and irreducible, i.e. $\mathbf{P}$ has a simple eigenvalue 1 which exceeds all other eigenvalues in modulus, and $\mathbf{P}$ is not of the form (133). Conversely, if $\mathbf{P}$ is primitive and irreducible then the system is ergodic.

(ii) If the chain is irreducible, i.e. consists of a single closed set, and periodic with period t, then the transition matrix $\mathbf{P}$ is irreducible and has exactly t eigenvalues of unit modulus, the remaining eigenvalues being of less than unit modulus. The converse is also true. Further, the states of the chain divide into mutually exclusive cyclic sub-sets, as described at the end of Section 3.6.

(iii) If the chain has transient states, then **P** is reducible and conversely. The number of closed sets of recurrent states is equal to the multiplicity of the eigenvalue 1. Formula (174) below is derived for determining the probability of eventually entering a given closed set when starting from a given transient state.

In what follows we suppose that the chain has h states and we distinguish three cases.

(i) P IRREDUCIBLE AND PRIMITIVE

We show that in this case the system is ergodic.

The eigenvalue $\lambda_1 = 1$ is simple and all other eigenvalues are strictly less than 1 in modulus. According to the theorem of Perron and Frobenius there is a positive row eigenvector $\pi = (\pi_j)$ satisfying $\pi P = \pi$ and we can normalize this vector so that $\sum \pi_j = 1$. We now have the following result.

If **P** *is primitive and irreducible then the system is ergodic and*

$$\lim_{n \to \infty} p_{jk}^{(n)} = \pi_k > 0, \tag{152}$$

the limit being approached geometrically fast and uniformly for all j and k. Conversely if the system is ergodic then **P** *is primitive and irreducible.*

The result (152) is a direct consequence of (150); if (152) holds then all the states are aperiodic and, by the limit theorem, part A(ii) (Section 3.4), ergodic. Conversely suppose that the system is ergodic. Then there is a positive probability distribution (π_j) such that $p_{jk}^{(n)} \to \pi_j$, or in matrix notation

$$\lim_{n \to \infty} P^n = 1\pi,$$

where π is the row vector (π_j). It follows from (63) that π is an eigenvector of **P** corresponding to the eigenvalue 1. Since $\lim P^n > 0$ we must have $P^n > 0$ for n sufficiently large and so **P** must be irreducible, for if not then **P** would be of the form (133) and the zero sub-matrix would persist through all powers of **P**. If **P** has any other eigenvalues of modulus 1 then it would be expressible in the cyclic form (141) and we would not have $P^n > 0$ for any values of n. It follows that **P** is primitive.

(ii) P IRREDUCIBLE BUT NOT PRIMITIVE

In this case we show that the chain is irreducible, positive-recurrent and periodic with period equal to the number of eigenvalues of unit modulus.

If **P** *is irreducible and has exactly t eigenvalues of unit modulus then each state is periodic with period t and the states can be divided into t mutually exclusive and exhaustive cyclic sub-sets $S_0, \ldots, S_{t-1}$ such that one-step* *transitions from a state in S_α can lead only to a state in $S_{\alpha+1}$ (if $\alpha = t-1$ then $\alpha+1$ is defined to be 0); moreover if $j \in S_\alpha$ and $k \in S_\beta$ then*

$$\lim_{n \to \infty} p_{jk}^{(nt+\beta-\alpha)} = t\pi_k > 0, \qquad (153)$$

the limit being approached geometrically fast and uniformly in j and k.

Conversely if each state has period $t > 1$ and if for each j and k there is an integer r_{jk} such that

$$\lim_{n \to \infty} p_{jk}^{(nt+r_{jk})} > 0 \qquad (154)$$

then $\mathbf{P}$ is irreducible and has exactly t eigenvalues of unit modulus.

If $\mathbf{P}$ is irreducible and has exactly t eigenvalues of modulus 1 then it can be represented in the form (141) and so in a linear combination of t successive powers of $\mathbf{P}$

$$\mathbf{Q} = a_0 \mathbf{P}^n + a_1 \mathbf{P}^{n+1} + \ldots + a_{t-1} \mathbf{P}^{n+t-1} \quad (a_j > 0,\ j = 0, \ldots, t-1),$$

the non-zero elements of $\mathbf{Q}$ are contributed by only one of the terms $a_j \mathbf{P}^j$. This can be seen by examining the successive powers of a matrix of the form (141). Further details of the proof are left to the reader.

(iii) $\mathbf{P}$ REDUCIBLE: ABSORPTION PROBLEMS

Suppose firstly that $\mathbf{P}$ has the form

$$\mathbf{P} = \begin{array}{c} \\ C \\ T \end{array} \begin{array}{cc} C & T \\ \begin{bmatrix} \mathbf{P}_1 & 0 \\ \mathbf{R}_{21} & \mathbf{Q}_2 \end{bmatrix} \end{array} \quad (\mathbf{R}_{21} \neq 0) \qquad (155)$$

where $\mathbf{P}_1$ (of order $m \times m$, say) and $\mathbf{Q}_2$ (of order $(h-m) \times (h-m)$) are irreducible square matrices, $\mathbf{P}_1$ being primitive, and C and T represent two sets of states. Then $\mathbf{P}_1$ represents transition probabilities within C, $\mathbf{Q}_2$ within T and $\mathbf{R}_{21}$ from T to C. Transitions out of C are impossible and so C is a closed set. We shall show that the states in C are ergodic and those in T transient, and that whatever the initial conditions, ultimately the set C is entered and there is an equilibrium distribution over the states of C.

The row sums of $\mathbf{Q}_2$ are bounded by 1 and at least one row sum is strictly less than 1. It follows from the Perron–Frobenius theorem, part (b), that all the eigenvalues of $\mathbf{Q}_2$ are strictly less than 1 and the result (149) tells us that $\mathbf{Q}_2^n \to 0$. From the theory of the ergodic finite chain

$$\lim_{n \to \infty} \mathbf{P}_1^n = \mathbf{\Pi} = \mathbf{1}\boldsymbol{\pi} \qquad (156)$$

exists, since $\mathbf{P}_1$ is itself a primitive, irreducible, stochastic matrix. We have

$$\mathbf{P}^n = \begin{bmatrix} \mathbf{P}_1^n & 0 \\ \mathbf{V}_n & \mathbf{Q}_2^n \end{bmatrix}, \tag{157}$$

and since all but one of the eigenvalues of $\mathbf{P}$ are less than unity in absolute value the Jordan canonical form of $\mathbf{P}$ is

$$\mathbf{P} = \mathbf{H}^{-1} \begin{bmatrix} 1 & 0 \\ 0 & \mathbf{J} \end{bmatrix} \mathbf{H}, \tag{158}$$

where $\mathbf{J}$ is a Jordan matrix whose diagonal elements are less than unity. Thus $\mathbf{J}^n \to 0$ by (145) and so $\lim \mathbf{P}^n$ exists, and therefore $\lim \mathbf{V}_n = \mathbf{V}$ exists. Moreover (145) also tells us that the elements of $\mathbf{J}^n$ are all $O(\rho^n)$ for some ρ $(0 < \rho < 1)$. Hence the elements of $\mathbf{P}^n$ approach their limits geometrically fast since, according to (158),

$$\mathbf{P}^n = \mathbf{H}^{-1} \begin{bmatrix} 1 & 0 \\ 0 & 0 \end{bmatrix} \mathbf{H} + \mathbf{H}^{-1} \begin{bmatrix} 0 & 0 \\ 0 & \mathbf{J}^n \end{bmatrix} \mathbf{H}. \tag{159}$$

From (157) we have

$$\lim_{n \to \infty} \mathbf{P}^n = \begin{bmatrix} \mathbf{\Pi} & 0 \\ \mathbf{V} & 0 \end{bmatrix}. \tag{160}$$

We know that $\mathbf{\Pi} = \mathbf{1}\pi$ where $\pi = (\pi_1, \ldots, \pi_m)$ is the limiting probability vector for $\mathbf{P}_1$. To evaluate $\mathbf{V}$ we note that $\mathbf{V}_n$ satisfies the recurrence relation

$$\mathbf{V}_{n+1} = \mathbf{R}_{21} \mathbf{P}_1^n + \mathbf{Q}_2 \mathbf{V}_n. \tag{161}$$

Hence $\mathbf{V} = \mathbf{R}_{21}\mathbf{\Pi} + \mathbf{Q}_2 \mathbf{V}$ and therefore

$$\mathbf{V} = (\mathbf{I} - \mathbf{Q}_2)^{-1} \mathbf{R}_{21} \mathbf{\Pi}. \tag{162}$$

The inverse of $\mathbf{I} - \mathbf{Q}_2$ exists since 1 is not an eigenvalue of $\mathbf{Q}_2$. In fact, since $\mathbf{R}_{21}\mathbf{1} = (\mathbf{I} - \mathbf{Q}_2)\mathbf{1}$, it follows from (158) and (162) that

$$\mathbf{V} = \mathbf{1}\pi. \tag{163}$$

Thus

$$\lim_{n \to \infty} \mathbf{P}^n = \begin{pmatrix} 1 \\ \vdots \\ 1 \end{pmatrix} (\pi_1 \ldots \pi_m \; 0 \; 0 \ldots 0), \tag{164}$$

where m is the number of states in C. For j in C, $\lim p_{jj}^{(n)} = \pi_j > 0$ and so j is ergodic. For j in T, $p_{jj}^{(n)} = O(\rho^n)$ $(0 < \rho < 1)$ which implies that $\sum_n p_{jj}^{(n)}$ is convergent so that j is transient.

Thus when $\mathbf{P}$ is of the form (155) we have completely evaluated $\lim \mathbf{P}^n$ and determined the nature of the states of the system. Moreover, it is clear that we can apply the same sort of reasoning to the case where **P** is of the form

$$
\mathbf{P} = \begin{bmatrix}
\mathbf{P}_1 & \cdots & 0 & 0 & \cdots & 0 \\
\vdots & & \vdots & \vdots & & \vdots \\
0 & \cdots & \mathbf{P}_r & 0 & \cdots & 0 \\
\mathbf{R}_{r+1,1} & \cdots & \mathbf{R}_{r+1,r} & \mathbf{Q}_{r+1} & \cdots & 0 \\
\vdots & & \vdots & \vdots & & \vdots \\
\mathbf{R}_{s1} & \cdots & \mathbf{R}_{sr} & 0 & \cdots & \mathbf{Q}_s
\end{bmatrix},
\tag{165}
$$

where the $\mathbf{P}_\alpha$ are irreducible and primitive, the $\mathbf{Q}_\beta$ irreducible and

$$
\mathbf{R}_{\alpha 1} + \ldots + \mathbf{R}_{\alpha,r} \neq 0 \quad (\alpha = r+1,\ldots,s).
\tag{166}
$$

We may write

$$
\mathbf{P} = \begin{bmatrix} \mathbf{A} & 0 \\ \mathbf{B} & \mathbf{C} \end{bmatrix}, \qquad
\mathbf{P}^n = \begin{bmatrix} \mathbf{A}^n & 0 \\ \mathbf{B}_n & \mathbf{C}^n \end{bmatrix},
\tag{167}
$$

where

$$
\mathbf{A} = \begin{bmatrix} \mathbf{P}_1 & \cdots & 0 \\ \vdots & & \vdots \\ 0 & \cdots & \mathbf{P}_r \end{bmatrix}.
$$

Then

$$
\lim_{n \to \infty} \mathbf{A}^n = \begin{bmatrix} \mathbf{\Pi}_1 & & 0 \\ & \ddots & \\ 0 & & \mathbf{\Pi}_r \end{bmatrix} = \mathbf{\Pi},
\tag{168}
$$

say, where

$$
\mathbf{\Pi}_\alpha = \lim_{n \to \infty} \mathbf{P}_\alpha^n > 0, \qquad \lim_{n \to \infty} \mathbf{C}^n = 0
\tag{169}
$$

and

$$
\lim_{n \to \infty} \mathbf{B}_n = \mathbf{V} = \begin{bmatrix} \mathbf{V}_{r+1,1} & \cdots & \mathbf{V}_{r+1,r} \\ \vdots & & \vdots \\ \mathbf{V}_{s1} & \cdots & \mathbf{V}_{sr} \end{bmatrix},
$$

where

$$
\mathbf{V} = (\mathbf{I} - \mathbf{C})^{-1} \mathbf{B} \mathbf{\Pi}.
\tag{170}
$$

In this case the sets of states $C_1, \ldots, C_r$ corresponding to $\mathbf{P}_1, \ldots, \mathbf{P}_r$ are closed sets of recurrent states while the sets $T_{r+1}, \ldots, T_s$ corresponding to $\mathbf{Q}_{r+1}, \ldots, \mathbf{Q}_s$ are sets of transient states. We can in fact express $\mathbf{V}$ more simply, for if we consider the sub-matrix of $\mathbf{P}$

$$
\mathbf{U}_{\alpha\beta} = \begin{array}{c} \\ C_\beta \\ T_\alpha \end{array} \!\! \begin{array}{cc} C_\beta & T_\alpha \\ \begin{bmatrix} \mathbf{P}_\beta & 0 \\ \mathbf{R}_{\alpha\beta} & \mathbf{Q}_\alpha \end{bmatrix} \end{array},
\tag{171}
$$

then $\mathbf{U}_{\alpha\beta}$ is of the form (155) with, possibly, $\mathbf{R}_{\alpha\beta} = 0$, and the elements of $\mathbf{U}_{\alpha\beta}^n$ are given by the corresponding sub-matrix of $\mathbf{P}^n$. Hence if we write

$$\mathbf{B}_n = \begin{bmatrix} \mathbf{B}_{r+1,1}^{(n)} & \cdots & \mathbf{B}_{r+1,r}^{(n)} \\ \vdots & & \vdots \\ \mathbf{B}_{s,1}^{(n)} & \cdots & \mathbf{B}_{s,r}^{(n)} \end{bmatrix}, \tag{172}$$

then

$$\lim_{n \to \infty} \mathbf{B}_{\alpha\beta}^{(n)} = \mathbf{V}_{\alpha\beta}$$

exists and by (162) is given by

$$\mathbf{V}_{\alpha\beta} = (\mathbf{I} - \mathbf{Q}_\alpha)^{-1} \mathbf{R}_{\alpha\beta} \mathbf{\Pi}_\beta \quad (\beta = 1, \dots, r; \ \alpha = r+1, \dots, s). \tag{173}$$

If a finite Markov chain has a transition matrix of the form (165) *where the* $\mathbf{P}_\beta$ $(\beta = 1, \dots, r)$ *are irreducible and primitive, the* $\mathbf{Q}_\alpha$ $(\alpha = r+1, \dots, s)$ *irreducible and the* $\mathbf{R}_{\alpha\beta}$ *satisfy* (166), *then* $\lim_{n \to \infty} p_{jk}^{(n)}$ *exists for each* j *and* k *and is given in matrix form by* (167)–(170). *The sets* C_β $(\beta = 1, \dots, r)$ *are closed sets of ergodic states and* T_α $(\alpha = r+1, \dots, s)$ *sets of transient states. Given that the state* $j \in T_\alpha$ *is occupied initially, the probability of ultimately being in state* $k \in C_\beta$ *is given by the appropriate element of* $\mathbf{V}_{\alpha\beta}$ *at* (173).

Note that to obtain the total probability of absorption in the closed set C_β when starting from $j \in T_\alpha$ we sum over the states of C_β in the matrix $\mathbf{V}_{\alpha\beta}$ of limiting probabilities, i.e. we take the element corresponding to j in

$$\mathbf{V}_{\alpha\beta} \mathbf{1} = \{ \text{prob}(X_\infty \in C_\beta | X_0 = j \in T_\alpha) \} = (\mathbf{I} - \mathbf{Q}_\alpha)^{-1} \mathbf{R}_{\alpha\beta} \mathbf{1}. \tag{174}$$

We have assumed so far that the stochastic matrices $\mathbf{P}_1, \dots, \mathbf{P}_r$ within closed sets are primitive. Now whereas it is clear that any finite Markov chain, after a suitable relabelling of the states, can have its transition matrix written in the form (165), it is only when each of the $\mathbf{P}_\beta$ are primitive that $\mathbf{P}^n$ will tend to a limit as $n \to \infty$. When any one of the matrices $\mathbf{P}_1, \dots, \mathbf{P}_r$ has more than one eigenvalue of unit modulus then we get periodic phenomena which, however, may be smoothed out by, for example, taking simple arithmetic averages

$$\mathbf{Q}_n = \frac{1}{n} (\mathbf{P} + \mathbf{P}^2 + \dots + \mathbf{P}^n), \tag{175}$$

and it is not difficult to show now that whatever the form of the matrix $\mathbf{P}$, the average $\mathbf{Q}_n$ tends to a limit as $n \to \infty$. We leave it as an exercise to the reader to prove the following.

For any finite Markov chain with transition matrix $\mathbf{P}$ written in the form (165) where the $\mathbf{P}_\beta$ and $\mathbf{Q}_\alpha$ are irreducible and the $\mathbf{R}_{\alpha\beta}$ satisfy (166), the limit

$$\lim_{n \to \infty} \frac{1}{n} (\mathbf{P} + \mathbf{P}^2 + \ldots + \mathbf{P}^n)$$

exists and is equal to

$$\begin{bmatrix} \mathbf{\Pi}_1 & \ldots & 0 & \ldots & 0 \\ \vdots & & \vdots & & \vdots \\ 0 & \ldots & \mathbf{\Pi}_r & \ldots & 0 \\ \mathbf{V}_{r+1,1} & \ldots & \mathbf{V}_{r+1,r} & \ldots & 0 \\ \vdots & & \vdots & & \vdots \\ \mathbf{V}_{s1} & \ldots & \mathbf{V}_{sr} & \ldots & 0 \end{bmatrix}, \tag{176}$$

where

$$\mathbf{\Pi}_\beta = \lim \frac{1}{n} (\mathbf{P}_\beta + \ldots + \mathbf{P}_\beta^n) \tag{177}$$

and the $\mathbf{V}_{\alpha\beta}$ are given by (173).

Example 3.21. *The Ehrenfest model.* Following Kac (1947), we now apply the matrix methods for finite Markov chains developed in this section to the following model. A particle performs a discrete random walk between two barriers of the reflecting type placed at the points $-a$ and a; at each stage the particle can move one unit up or down, the probability of moving one unit towards the origin being greater than the probability of moving one unit away from the origin by an amount proportional to the particle's distance from the origin. If the particle reaches one of the barrier points then it is certain to move to the adjacent interior position at the next step. The process so described is clearly an irreducible Markov chain of period 2 with $2a+1$ states and transition probabilities

$$p_{j,j+1} = p_{-j,-j-1} = \frac{a-j}{2a}$$
$$\qquad\qquad\qquad\qquad\qquad (j = 0.1, \ldots, a-1),$$
$$p_{j,j-1} = p_{-j,-j+1} = \frac{a+j}{2a} \tag{178}$$

$$p_{a,a-1} = p_{-a,-a+1} = 1, \qquad p_{jk} = 0, \text{ otherwise.}$$

The transition matrix is therefore

$$
\begin{array}{c}
 \\
\end{array}
\begin{array}{c}
\quad a \quad a-1 \quad a-2 \quad a-3 \; \ldots \; -a+2 \; -a+1 \; -a
\end{array}
$$

$$
\mathbf{P} =
\begin{array}{c}
a \\ a-1 \\ a-2 \\ \vdots \\ -a+1 \\ -a
\end{array}
\left[
\begin{array}{ccccccc}
0 & 1 & 0 & 0 & \ldots & 0 & 0 & 0 \\
\dfrac{1}{2a} & 0 & \dfrac{2a-1}{2a} & 0 & \ldots & 0 & 0 & 0 \\
0 & \dfrac{2}{2a} & 0 & \dfrac{2a-2}{2a} & \ldots & 0 & 0 & 0 \\
\vdots & \vdots & \vdots & \vdots & & \vdots & \vdots & \vdots \\
0 & 0 & 0 & 0 & \ldots & \dfrac{2a-1}{2a} & 0 & \dfrac{1}{2a} \\
0 & 0 & 0 & 0 & \ldots & 0 & 1 & 0
\end{array}
\right] .
\tag{179}
$$

The process may be described as a random walk with a central restoring tendency.

The transition matrix (179) has a second interpretation in terms of the Ehrenfest model of diffusion, describing the stochastic motion of molecules between two connected containers C and D. Suppose there are $2a$ molecules distributed between the two containers. At time n a molecule is chosen at random from among the $2a$ molecules and removed from its container to the other. We identify the state j of the system with the number of molecules $a+j$ in C $(-a \leqslant j \leqslant a)$. Thus the transition probabilities are given by (178) and we arrive at the transition matrix (179).

We now look for eigenvalues λ and corresponding left eigenvectors $\mathbf{x} = (x_0, \ldots, x_{2a})$ which satisfy $\mathbf{xP} = \lambda \mathbf{x}$ or

$$
\begin{aligned}
x_1/(2a) &= \lambda x_0, \\
2x_2/(2a) + 2ax_0/(2a) &= \lambda x_1, \\
3x_3/(2a) + (2a-1)x_1/(2a) &= \lambda x_2, \\
&\;\;\vdots \qquad\qquad \vdots \\
2ax_{2a}/(2a) + 2x_{2a-2}/(2a) &= \lambda x_{2a-1}, \\
x_{2a-1}/(2a) &= \lambda x_{2a}.
\end{aligned}
\tag{180}
$$

If we take a generating function

$$
G(z) = \sum_{j=0}^{2a} x_j z^j
$$

then we can write the system (180) as

$$
\sum_{j=0}^{2a} jx_j z^{j-1} + \sum_{j=0}^{2a} (2a-j)x_j z^{j+1} = 2a\lambda G(z)
\tag{181}
$$

or

$$
(1 - z^2) G'(z) + 2a(z - \lambda) G(z) = 0.
\tag{182}
$$

The general solution to this differential equation is

$$G(z) = C(1+z)^{(1+\lambda)a}(1-z)^{(1-\lambda)a}, \tag{183}$$

where C is an arbitrary constant. We are interested only in solutions $G(z)$ which are polynomials of degree $2a$ in z and clearly the values

$$\lambda = j/a \quad (j = -a, \ldots, a) \tag{184}$$

furnish $2a+1$ such polynomials. Hence the $2a+1$ numbers (184) are the (distinct) eigenvalues of **P** and elements of the eigenvector corresponding to $\lambda = j/a$ are proportional to the coefficients of $1, z, \ldots, z^{2a}$ in

$$(1+z)^{a+j}(1-z)^{a-j}. \tag{185}$$

Since the chain is periodic there is no equilibrium distribution but there is a unique limiting probability vector obtained by setting $j = a$ (i.e. $\lambda = 1$) in (185) and normalizing so that the coefficients sum to 1. We obtain $2^{-2a}(1+z)^{2a} = (1/2+z/2)^{2a}$, a symmetric binomial distribution. Let

$$H_j(z) = 2^{-2a}z^{-a}(1+z)^{a+j}(1-z)^{a-j} = \sum_{k=-a}^{a} h_{jk}z^k. \tag{186}$$

The eigenvector corresponding to $\lambda = j/a$ is now given by the coefficients of $z^{-a}, \ldots, z^a$ in $H_j(z)$ and $H_a(z)$ gives the limiting probability vector. The matrix $\mathbf{H} = (h_{jk})$ clearly satisfies

$$\mathbf{HP} = \mathbf{\Lambda H}, \tag{187}$$

where $\qquad \mathbf{\Lambda} = \operatorname{diag}(j/a) \quad (j = -a, \ldots, a).$

We then have

$$\mathbf{P} = \mathbf{H}^{-1}\mathbf{\Lambda H} \tag{188}$$

and to complete our diagonal representation (188) we require to find $\mathbf{H}^{-1}$, which in this case turns out to be simply related to $\mathbf{H}$. Let $(h^{jk}) = \mathbf{H}^{-1}$. Then

$$\delta_{jk} = \sum_{i=-a}^{a} h^{ji} h_{ik}$$

and

$$z^j = \sum_{k=-a}^{a} \delta_{jk} z^k = \sum_{i=-a}^{a} h^{ji} \sum_{k=-a}^{a} h_{ik} z^k$$

$$= \left(\frac{1-z^2}{4z}\right)^a \sum_{i=-a}^{a} h^{ji} \left(\frac{1+z}{1-z}\right)^i \tag{189}$$

on substituting (186). Let $w = (1+z)/(1-z)$. Then (189) may be written

$$\sum_{i=-a}^{a} h^{ji} w^i = (-1)^{a+j}(1-w)^{a+j}(1+w)^{a-j} w^{-a}$$

$$= (-1)^{a+j} 2^{2a} H_{-j}(w),$$

from which it follows that

$$h^{jk} = (-1)^{a+j} 2^{2a} h_{-j,k}. \tag{190}$$

We have therefore determined the rows of $\mathbf{H}$ and $\mathbf{H}^{-1}$ in terms of generating functions. Another method of finding $\mathbf{H}^{-1}$ is to determine the right eigenvectors of $\mathbf{P}$ in view of the relation $\mathbf{PH}^{-1} = \mathbf{H}^{-1}\mathbf{\Lambda}$ but the method used above for the left eigenvectors does not work in this case.

We can also determine the mean and variance of the particle's displacement after n steps. Suppose we represent the probability distribution of the particle's displacement at time n by the row vector $\mathbf{p}^{(n)} = (p_{-a}^{(n)}, \ldots, p_a^{(n)})$, and let μ_n and σ_n^2 denote the mean and variance of this displacement. Then we have $\mathbf{p}^{(n+1)} = \mathbf{p}^{(n)}\mathbf{P}$ and a simple calculation analogous to that used to obtain (182) will show that the moment generating function

$$M_n(\theta) = \sum_{k=-a}^{a} p_k^{(n)} e^{-k\theta}$$

satisfies

$$a(e^{-\theta} + e^{\theta}) M_n(\theta) + (e^{-\theta} - e^{\theta}) M_n'(\theta) = 2aM_{n+1}(\theta) \tag{191}$$

or

$$aM_n(\theta) \cosh \theta - M_n'(\theta) \sinh \theta = aM_{n+1}(\theta). \tag{192}$$

Remembering that $M_n'(0) = -\mu_n$ and $M_n''(0) = \sigma_n^2 + \mu_n^2$ we may differentiate (192) with respect to θ and obtain the recurrence relations

$$\mu_{n+1} = (1 - 1/a)\mu_n, \tag{193}$$

$$\sigma_{n+1}^2 = \left(1 - \frac{2}{a}\right)\sigma_n^2 - \mu_n^2/a^2 + 1. \tag{194}$$

If the particle is initially in the position j_0 the solutions to (193) and (194) are

$$\mu_n = (1 - 1/a)^n j_0, \tag{195}$$

$$\sigma_n^2 = \tfrac{1}{2}a\{1 - (1 - 2/a)^n\} + j_0^2\{(1 - 2/a)^n - (1 - 1/a)^{2n}\}. \tag{196}$$

When $n \to \infty$, μ_n and σ_n^2 tend respectively to 0 and $\tfrac{1}{2}a$, corresponding to the binomial distribution with p.g.f. $H_a(z)$ as given by (186).

3.12. Further topics

(i) m-DEPENDENT CHAINS

For a Markov chain we have essentially one-step dependence, i.e. given the value of X_{n-1} we can find the distribution of X_n. However, processes may be encountered in which the dependence goes back more than one time unit and in such cases we can reduce the process to a Markov chain by appropriately redefining the state space. This has already been illustrated in Example 1.7. As another example, consider a sequence of dependent Bernoulli trials in which the probability of success, denoted by 1, or failure, denoted by 0, at any given trial depends on the outcome of the two preceding trials. We redefine the state space according to the outcome of two successive trials

00	state 0
01	state 1
10	state 2
11	state 3.

Thus if trials $n-1$ and n give rise, for example, to 1 and 0, then we say that the process is in state 2 at time n. It is easy to see that the process is now a Markov chain with four states.

Multivariate discrete Markov processes can be handled in a similar way. We reduce the process to a Markov chain by appropriately redefining the state space. For example consider a trivariate discrete Markov process $\mathbf{X}_n = (X_n^{(1)}, X_n^{(2)}, X_n^{(3)})$ such that the distribution of $\mathbf{X}_n$ depends on the value of the vector $\mathbf{X}_{n-1}$. If the component $X_n^{(i)}$ takes the possible values $1, 2, \ldots, h_i$ ($i = 1, 2, 3$), then we construct a single state space with $h_1 h_2 h_3$ states where each state is a triplet $(x^{(1)}, x^{(2)}, x^{(3)})$ and the process so defined is a Markov chain. For an application of this method to a reservoir model see Lloyd (1963).

(ii) MARKOV PROCESSES WITH CONTINUOUS STATES IN DISCRETE TIME

Throughout this chapter so far we have been concerned with Markov processes with a discrete state space in discrete time. We now discuss briefly the discrete time Markov process for which the state space is continuous. One special process of this type with which we have already dealt in some detail in Chapter 2 is the general random walk. Suppose that the steps of the random walk are mutually independent, continuous random variables $Z_1, Z_2, \ldots$ with common p.d.f. $f(z)$. If the walk is unrestricted then the p.d.f. $f_n(x)$ of

$$X_n = Z_1 + \ldots + Z_n$$

is the n-fold convolution of $f(z)$ and clearly satisfies the recurrence relation

$$f_n(x) = \int_{-\infty}^{\infty} f_{n-1}(y) f(x-y) \, dy. \tag{197}$$

This is the one-step Chapman–Kolmogorov relation in this case. Clearly the kernel function $f(x-y)$ is the p.d.f. of X_n conditional on $X_{n-1} = y$ and therefore has the nature of a transition probability. In the particular case of the random walk the transition p.d.f. is a difference or translation kernel, expressing the fact that the increment in the process is independent of the current state of the process.

More generally, suppose we have a Markov process $X_1, X_2, \ldots$ in which $X_1, X_2, \ldots$ are continuous random variables such that the p.d.f. of X_n conditional on $X_{n-1} = y$ is a given kernel function $f(y, x)$, a function of two continuous variables satisfying

$$\int_{-\infty}^{\infty} f(y, x) \, dx = 1 \quad (-\infty < y < \infty). \tag{198}$$

The function $f(y, x)$ corresponds to a transition probability matrix (p_{jk}) which is essentially a function of two discrete variables j and k. If $f_n(x)$ denotes the p.d.f. of X_n then the one-step Chapman–Kolmogorov relation takes the form

$$f_n(x) = \int_{-\infty}^{\infty} f_{n-1}(y) f(y, x) \, dy. \tag{199}$$

Such a process is clearly homogeneous in time since the transition p.d.f. $f(y, x)$ is independent of n. More generally, for a homogeneous process the transition p.d.f. $f_n(y, x)$ of X_n given $X_0 = y$ satisfies the Chapman–Kolmogorov relation

$$f_{n+m}(y, x) = \int_{-\infty}^{\infty} f_n(y, z) f_m(z, x) \, dz, \tag{200}$$

which corresponds to the relation

$$p_{jk}^{(m+n)} = \sum_l p_{jl}^{(m)} p_{lk}^{(n)}$$

for Markov chains. Corresponding relations can be written down for non-homogeneous processes.

The random variables $X_1, X_2, \ldots$ need not be purely continuous. For example in the case of the random walk with reflecting barriers described in Section 2.3(vii) the distribution of X_n has discrete probabilities on the barriers and is continuous between the barriers. In such cases the process is more aptly described in terms of distribution functions rather than p.d.f.'s. The general homogeneous Markov process in discrete time will be governed by a transition distribution function

$$F(y, x) = \text{prob}(X_n \leqslant x | X_{n-1} = y).$$

The theory and application of Markov processes in discrete time with continuous state space is not as fully developed as, for example, that of Markov chains, except in the case of special processes such as the random walk. Such processes encountered in practice would be dealt with by more or less *ad hoc* methods. For example, if a process governed by equation (199) has an equilibrium distribution $f(x)$ then $f(x)$ will satisfy the integral equation, obtained by letting $n \to \infty$ in (199),

$$f(x) = \int_{-\infty}^{\infty} f(y) f(y, x) \, dy. \qquad (201)$$

In particular instances it may be possible to solve such an equation analytically; otherwise one may need to use numerical methods. For the random walk the method of moment generating functions often works because Laplace or Fourier transforms are particularly suited to linear relations involving difference kernels.

(iii) ADDITIVE PROCESSES DEFINED ON A FINITE MARKOV CHAIN

Let X_0, X_1, X_2, ... denote a realization of a finite Markov chain with transition probability matrix $\mathbf{P} = (p_{jk})$. For simplicity we suppose that the chain is ergodic. With each one-step transition we associate a score in the following manner. If, say, the nth transition is from state j to state k then the score associated with that transition is a random variable W_n whose distribution $F_{jk}(w)$ depends only on j and k. Further, conditionally on a given sequence of transitions, W_1, W_2, ... are mutually independent. We are interested in the distribution of the cumulative score

$$Y_n = W_1 + W_2 + \ldots + W_n, \qquad (202)$$

conditional on the initial state X_0, and possibly also on the final state X_n. The vector process (X_n, Y_n) is a two-dimensional Markov process in discrete time. The first component X_n takes values from among a finite discrete set and the second component Y_n may be continuous or discrete.

For example, let the Markov chain $\{X_n\}$ be imbedded in continuous time and let W_n be the time taken to effect the nth transition, this time being a random variable whose distribution depends on the previously occupied state or on the state next occupied or on both. Such a process is called a semi-Markov process or Markov renewal process. A more detailed discussion from a different point of view is given in Section 9.3. Thus for the semi-Markov process the random variables W_i are positive but in general, the scores W_i in (202) need not necessarily be positive.

If we focus attention say on the kth state of the Markov chain and score 1 each time the state is occupied and zero otherwise, then the cumulative score Y_n is the number of times state k is occupied among times $1, \ldots, n$. Similarly, we may score 1 each time a transition between two particular states occurs and zero otherwise; then the cumulative score Y_n is the number of times that particular transition occurs during n transitions of the chain. More generally, the process (202) may be regarded as a random walk in which the steps possess a particular kind of dependence.

We shall derive a method for determining the distribution of Y_n. Let $F_{jk}(x)$ be the distribution function of the score associated with a transition from state j to state k, and let

$$\phi_{jk}(\theta) = \int_{-\infty}^{\infty} e^{-\theta x} dF_{jk}(x)$$

be its moment generating function. Let $F_{jk}^{(r)}(x)$ be the distribution function of the cumulative score associated with an r-fold transition from state j to state k with corresponding m.g.f. $\phi_{jk}^{(r)}(\theta)$. Thus

$$\phi_{jk}^{(r)}(\theta) = E(e^{-\theta Y_r} | X_0 = j, X_r = k). \tag{203}$$

We define the matrices

$$\mathbf{P}(\theta) = \{p_{jk}\,\phi_{jk}(\theta)\} \tag{204}$$

and

$$\mathbf{P}^{(r)}(\theta) = \{p_{jk}^{(r)}\,\phi_{jk}^{(r)}(\theta)\}. \tag{205}$$

We shall show that $\mathbf{P}^{(r)}(\theta) = \{\mathbf{P}(\theta)\}^r$.

Conditional on the event $\{X_0 = j, X_{n-1} = l, X_n = k\}$, the random variables Y_{n-1} and W_n are independent with m.g.f.'s $\phi_{jl}^{(n-1)}(\theta)$ and $\phi_{lk}(\theta)$ respectively. Hence

$$E(e^{-\theta Y_n} | X_0 = j, X_{n-1} = l, X_n = k) = \phi_{jl}^{(n-1)}(\theta)\,\phi_{lk}(\theta). \tag{206}$$

If we remove the condition $X_{n-1} = l$, we obtain

$$E(e^{-\theta Y_n} | X_0 = j, X_n = k)$$

$$= \frac{1}{p_{jk}^{(n)}} \sum_l p_{jl}^{(n-1)}\, p_{lk}\, E(e^{-\theta Y_n} | X_0 = j, X_{n-1} = l, X_n = k)$$

$$= \frac{1}{p_{jk}^{(n)}} \sum_l p_{jl}^{(n-1)}\, p_{lk}\, \phi_{jl}^{(n-1)}(\theta)\,\phi_{lk}(\theta), \tag{207}$$

or

$$p_{jk}^{(n)}\,\phi_{jk}^{(n)}(\theta) = \sum_l p_{jl}^{(n-1)}\,\phi_{jl}^{(n-1)}(\theta)\, p_{lk}\,\phi_{lk}(\theta). \tag{208}$$

In matrix notation this reads

$$\mathbf{P}^{(n)}(\theta) = \mathbf{P}^{(n-1)}(\theta)\,\mathbf{P}(\theta)$$

and since $\mathbf{P}^{(1)}(\theta) = \mathbf{P}(\theta)$ we have

$$\mathbf{P}^{(n)}(\theta) = \{\mathbf{P}(\theta)\}^n. \tag{209}$$

By summing the elements in any given row of $\{\mathbf{P}(\theta)\}^n$ we obtain the m.g.f. of Y_n conditional on the initial state corresponding to that row, i.e.

$$E(e^{-\theta Y_n} | X_0 = j) = \sum_k p_{jk}^{(n)}\,\phi_{jk}^{(n)}(\theta). \tag{210}$$

Thus for the process (202), which we may describe as an additive process defined on a Markov chain, the matrix $\mathbf{P}(\theta)$ plays a similar role to that of the m.g.f. in the summation of independent random variables, in that the nth power of $\mathbf{P}(\theta)$ determines the distribution of the cumulative sum $Y_n = W_1 + \ldots + W_n$.

Example 3.22. The distribution of the number of successes in a sequence of dependent Bernoulli trials. Consider the two-state Markov chain described in Section 3.2 with states 0 and 1 which we may take to represent failure and success respectively. This chain describes a sequence of Bernoulli trials in which the probability of success or failure at any given trial depends on the outcome of the preceding trial. We have

$$\mathbf{P} = \begin{bmatrix} 1-\alpha & \alpha \\ \beta & 1-\beta \end{bmatrix}.$$

Suppose we wish to determine the m.g.f. of the number Y_n of successes in n trials. We score 1 for a transition into state 1, and 0 for a transition into state 0. Accordingly we take

$$\phi_{00}(\theta) = 1, \qquad \phi_{01}(\theta) = e^{-\theta}, \qquad \phi_{10}(\theta) = 1, \qquad \phi_{11}(\theta) = e^{-\theta}, \quad (211)$$

and so we have that

$$\mathbf{P}(\theta) = \begin{bmatrix} 1-\alpha & \alpha e^{-\theta} \\ \beta & (1-\beta)e^{-\theta} \end{bmatrix}. \quad (212)$$

To evaluate $\{\mathbf{P}(\theta)\}^n$ we use the method of spectral or diagonal representation. The eigenvalues $\lambda_1(\theta)$, $\lambda_2(\theta)$ of $\mathbf{P}(\theta)$ are the solutions of the determinantal equation $|\mathbf{P}(\theta) - \lambda \mathbf{I}| = 0$, i.e.

$$\lambda^2 - \lambda\{(1-\alpha) + (1-\beta)e^{-\theta}\} + (1-\alpha-\beta)e^{-\theta} = 0. \quad (213)$$

This is a quadratic equation in λ having the roots

$$\lambda_1(\theta), \lambda_2(\theta) = \tfrac{1}{2}\{(1-\alpha) + (1-\beta)e^{-\theta}\} \pm \tfrac{1}{2}[\{(1-\alpha) + (1-\beta)e^{-\theta}\}^2 \\ - 4(1-\alpha-\beta)e^{-\theta}]^{\frac{1}{2}}, \quad (214)$$

where $\lambda_1(0) = 1$ and $\lambda_2(0) = 1-\alpha-\beta$. Hence we may write $\mathbf{P}(\theta)$ in the form $\mathbf{Q}(\theta)\mathbf{\Lambda}(\theta)\mathbf{Q}^{-1}(\theta)$, where $\mathbf{\Lambda}(\theta) = \mathrm{diag}\{\lambda_i(\theta)\}$ and the columns $\mathbf{q}_i(\theta)$ of $\mathbf{Q}(\theta)$ are solutions of the equations $\mathbf{P}(\theta)\mathbf{q}_i(\theta) = \lambda_i(\theta)\mathbf{q}_i(\theta)$ $(i = 1, 2)$. Thus

$$\{\mathbf{P}(\theta)\}^n = \frac{1}{\alpha e^{-\theta}\{\lambda_2(\theta) - \lambda_1(\theta)\}} \begin{bmatrix} \alpha e^{-\theta} & \alpha e^{-\theta} \\ \lambda_1(\theta) + \alpha - 1 & \lambda_2(\theta) + \alpha - 1 \end{bmatrix}$$
$$\times \begin{bmatrix} \lambda_1^n(\theta) & 0 \\ 0 & \lambda_2^n(\theta) \end{bmatrix} \begin{bmatrix} \lambda_2(\theta) + \alpha - 1 & -\alpha e^{-\theta} \\ -\{\lambda_1(\theta) + \alpha - 1\} & \alpha e^{-\theta} \end{bmatrix}. \quad (215)$$

By (210), the sum of the elements in the first row of $\{\mathbf{P}(\theta)\}^n$ is the m.g.f.

$\phi_0^{(n)}(\theta)$ of the number of successes in n trials which succeed an initial failure while the sum of elements in the second row gives the corresponding m.g.f. $\phi_1^{(n)}(\theta)$ conditional on an initial success. Hence, for example, the m.g.f. of the number of successes in n trials following an initial success is given by

$$\phi_1^{(n)}(\theta) = E(e^{-\theta Y_n}|X_0 = 1) = \frac{\lambda_1^{n+1} - \lambda_2^{n+1} - (1-\alpha-\beta)(\lambda_1^n - \lambda_2^n)}{\lambda_1 - \lambda_2}, \quad (216)$$

where, for the sake of brevity, we omit the argument θ in $\lambda_1(\theta)$, $\lambda_2(\theta)$. In obtaining the simplification from (215) to (216) we use the fact that

$$\lambda_1(\theta) + \lambda_2(\theta) = 1 - \alpha + (1-\beta)e^{-\theta},$$
$$\lambda_1(\theta)\lambda_2(\theta) = (1-\alpha-\beta)e^{-\theta}. \quad (217)$$

Note that if $\beta = 1 - \alpha$ then we have independent trials; in this case $\lambda_1(\theta) = (1-\alpha) + \alpha e^{-\theta}$, $\lambda_2(\theta) \equiv 0$ and we find that

$$E(e^{-\theta Y_n}) = \phi_1^{(n)}(\theta) = \phi_0^{(n)}(\theta) = (1-\alpha+\alpha e^{-\theta})^n, \quad (218)$$

the m.g.f. of the familiar binomial distribution.

We may obtain the moments of Y_n conditional on, say, $X_0 = 1$ in a straightforward, if laborious, manner by differentiating (216) with respect to θ. The derivatives of $\lambda_1(\theta)$, $\lambda_2(\theta)$ at $\theta = 0$ may be found by differentiating the relations (217) and using the fact that $\lambda_1(0) = 1$, $\lambda_2(0) = 1 - \alpha - \beta$.

Asymptotic results may be obtained by noting that in a neighbourhood of $\theta = 0$ we have $|\lambda_1(\theta)| > |\lambda_2(\theta)|$ since $\lambda_1(0) > |\lambda_2(0)|$. Hence writing (216) in the form

$$\phi_1^{(n)}(\theta) = \frac{\lambda_1^n(\lambda_1 - 1 + \alpha + \beta) - \lambda_2^n(\lambda_2 - 1 + \alpha + \beta)}{\lambda_1 - \lambda_2}, \quad (219)$$

we see that as $n \to \infty$

$$\log \phi_1^{(n)}(\theta) \sim n \log \lambda_1(\theta).$$

Thus asymptotically Y_n behaves like a sum of independent random variables. A familiar central limit argument shows that Y_n is asymptotically normal with mean $-n\lambda_1'(0)$ and variance $n[\lambda_1''(0) - \{\lambda_1'(0)\}^2]$, and is independent of initial conditions. We find that

$$E(Y_n) \sim n\frac{\alpha}{\alpha+\beta}, \qquad V(Y_n) \sim n\frac{\alpha\beta(2-\alpha-\beta)}{(\alpha+\beta)^3}. \quad (220)$$

These results may also be obtained by solving recurrence relations for $E(Y_n|X_0 = 1)$ and $E(Y_n^2|X_0 = 1)$ followed by an asymptotic argument.

Example 3.23. *Rainfall in Tel-Aviv.* Using the asymptotic results of the previous example we can find approximately the distribution of the number of wet days in the rainfall model of Example 3.2. We have $\alpha = 0.250$, $\beta = 0.338$. If we take a period of say $n = 90$ days in the rainy season, then using (220) we can say that the number of wet days in this period is approximately normally distributed with mean 38·3 and standard deviation 7·3.

Returning to the general additive process defined on a finite Markov chain we can obtain similar asymptotic results to those of Example 3.22. If the Markov chain is ergodic then the eigenvalue $\lambda_1 = 1$ of $\mathbf{P}$ will exceed all other eigenvalues in absolute value. Hence, by continuity, the eigenvalue $\lambda_1(\theta)$ $(\lambda_1(0) = 1)$ of $\mathbf{P}(\theta)$ will exceed in absolute value all other eigenvalues in a neighbourhood of $\theta = 0$. Hence, denoting the number of states in the chain by h and writing $\{\mathbf{P}(\theta)\}^n$ in the form

$$\{\mathbf{P}(\theta)\}^n = \mathbf{Q}(\theta) \begin{bmatrix} \lambda_1^n(\theta) & \cdots & 0 \\ \vdots & & \vdots \\ 0 & \cdots & \lambda_h^{(n)}(\theta) \end{bmatrix} \mathbf{Q}^{-1}(\theta),$$

we find that the elements of $\{\mathbf{P}(\theta)\}^n$ are of the form

$$\sum_{j=1}^{h} \psi_j(\theta) \lambda_j^n(\theta).$$

Hence asymptotically

$$\log E(e^{-Y_n \theta}) \sim n \log \lambda_1(\theta)$$

and we see that Y_n is asymptotically normal with mean $-n\lambda_1'(0)$ and variance $n[\lambda_1''(0) - \{\lambda_1'(0)\}^2]$, and is independent of the initial state.

For further discussion of these processes the reader is referred to Miller (1961b, 1962a, 1962b), Keilson and Wishart (1964), and Gabriel (1959).

3.13. Appendix on power series with non-negative coefficients

We quote here three theorems on power series with non-negative coefficients used in the proof of the limit theorem for Markov chains.

Theorem 1. Let $\{c_k\}$ be a sequence of non-negative terms and suppose that the power series

$$C(s) = \sum_{k=0}^{\infty} c_k s^k$$

is convergent for $|s| < 1$. Then we have

$$\lim_{s \to 1-} C(s) = \sum_{k=0}^{\infty} c_k$$

in the sense that if one side is finite, then so is the other and they are equal, or if one side is infinite, then so is the other.

This is Abel's theorem and its converse for power series with non-negative coefficients; see, for example, Parzen (1962, p. 218).

The following theorem is the analogue in discrete time of the result (9.17) for renewal processes in continuous time.

Theorem 2. Let

$$F(s) = \sum_{n=0}^{\infty} f_n s^n$$

be a probability generating function and suppose that $F(s)$ is not a power series in s^t for some integer $t > 1$. Let

$$\mu = \sum_{n=0}^{\infty} n f_n,$$

$$P(s) = \frac{1}{1 - F(s)} = \sum_{n=0}^{\infty} p_n s^n.$$

Then $\qquad \lim_{n \to \infty} p_n = 1/\mu$ (where $1/\mu = 0$ if $\mu = \infty$).

This theorem is due to Erdös, Feller and Pollard (1949) and for a proof we refer to this article or to Feller (1957, p. 306). See also Exercise 12 where a proof is indicated for the case when $P(s)$ is regular at $s = 1$.

Corollary. If f_n vanishes except when n is a multiple of an integer $t > 1$, then $F(s)$ is a power series in s^t and, with p_n and μ defined as in Theorem 2,

$$\lim_{n \to \infty} p_{nt} = \frac{t}{\mu} \quad \text{(where } 1/\mu = 0 \text{ if } \mu = \infty\text{)}.$$

The corollary is an immediate consequence of Theorem 2.

Theorem 3. If $F(s)$ satisfies the hypothesis of Theorem 2, if

$$G(s) = \sum_{n=0}^{\infty} g_n s^n$$

is a power series of non-negative terms such that $G(1) = \Sigma g_n < \infty$, and if

$$Q(s) = \frac{G(s)}{1 - F(s)} = \sum_{n=0}^{\infty} q_n s^n$$

then

$$\lim_{n \to \infty} q_n = \frac{G(1)}{\mu} \quad \text{(where } 1/\mu = 0 \text{ if } \mu = \infty\text{)}.$$

Proof. We have $Q(s) = G(s)P(s)$ where $P(s) = 1/\{1-F(s)\}$. Hence

$$q_n = \sum_{r=0}^{n} g_r\, p_{n-r}$$

and for $n > m$

$$q_n = \sum_{r=0}^{m-1} g_r\, p_{n-r} + \sum_{r=m}^{n} g_r\, p_{n-r}.$$

Since $\lim p_n$ is finite by Theorem 2, p_n is bounded, say $p_n \leqslant p$. Hence

$$\sum_{r=0}^{m-1} g_r\, p_{n-r} \leqslant q_n \leqslant \sum_{r=0}^{m-1} g_r\, p_{n-r} + p \sum_{r=m}^{\infty} g_r.$$

Letting $n \to \infty$, we have that

$$\frac{1}{\mu} \sum_{r=0}^{m-1} g_r \leqslant \liminf_{n \to \infty} q_n \leqslant \limsup_{n \to \infty} q_n \leqslant \frac{1}{\mu} \sum_{r=0}^{m-1} g_r + p \sum_{r=m}^{\infty} g_r$$

and now letting $m \to \infty$, we have

$$\lim_{n \to \infty} q_n = \frac{1}{\mu} \sum_{r=0}^{\infty} g_r.$$

Bibliographic Notes

Markov processes are named after A. A. Markov who introduced the finite Markov chain in 1907. Modern works on the theory of Markov chains are by Feller (1957) and, more extensively, Chung (1960). These authors may be consulted for further references. Chung, amongst other authors, calls the processes of the present chapter discrete parameter Markov chains and those of Chapter 4 continuous parameter Markov chains. Properties of non-homogeneous Markov chains are examined by Hajnal (1956, 1958). The concept of the imbedded Markov chain and its application in Example 3.10 are due to Kendall (1951, 1953). The general results of Sections 3.4 and 3.5 are due mainly to Kolmogorov (1936). Feller (1957) obtained these results using his theory of recurrent events. Branching processes are treated extensively by Harris (1963). The results of Section 3.8(i) and 3.8(ii) are due to Feller (1957). The first result of Section 3.8(iii) is due to Foster (1953) and the second to Derman (1954). Further results on Example 3.18 are given by Chung (1960). Example 3.19 follows Foster (1953). Algebraic approaches to the theory of finite Markov chains are discussed by Fréchet (1938) and Feller (1957). The approach of Section 3.11 is similar to that of Gantmacher (1959). For yet another point of view on finite chains, see Kemeny and Snell (1960). The geometrically fast approach to limiting values in the finite chain has been investigated for more general chains by Kendall (1960) and Vere-Jones (1962). Kendall calls the property *geometric ergodicity*.

Exercises

1. From a realization of a two-state Markov chain with $p_{00} = p_0, p_{11} = p_1$, a new chain is constructed by calling the pair 01 state 0 and the pair 10 state 1, ignoring 00 and 11. Only non-overlapping pairs are considered. Show that for the new chain

$$p_{00} = p_{11} = (1 + p_0 + p_1)^{-1}.$$

(Cambridge Diploma, 1954)

2. A Markov chain with three states has transition matrix

$$\mathbf{P} = \begin{bmatrix} 1-p & p & 0 \\ 0 & 1-p & p \\ p & 0 & 1-p \end{bmatrix} \quad (0 < p < 1).$$

Prove that

$$\mathbf{P}^n = \begin{bmatrix} a_{1n} & a_{2n} & a_{3n} \\ a_{3n} & a_{1n} & a_{2n} \\ a_{2n} & a_{3n} & a_{1n} \end{bmatrix},$$

where $a_{1n} + \omega a_{2n} + \omega^2 a_{3n} = (1 - p + p\omega)^n$, ω being a primitive cube root of unity.

(Cambridge Diploma, 1956)

3. A ball is drawn from an urn containing 4 balls, numbered 1 to 4, and then replaced. In repeated independent trials let Y_n be the number of the ball drawn at the nth trial. For $j = 1, 2, 3$, let $A_n^{(j)}$ be the event $\{Y_n = j$ or $Y_n = 4\}$, and for $m = 1, 2, \ldots$, let $X_{3(m-1)+j} = 1$ or 0 according as $A_m^{(j)}$ occurs or not. Show that, for k_1, k_2, k_3 equal to 0 or 1,

$$\text{prob}(X_n = k_1) = \text{prob}(X_n = k_2 | X_m = k_3) = \tfrac{1}{2} \quad (n > m)$$

whereas

$$\text{prob}(X_{3m+3} = 1 | X_{3m+2} = 1, X_{3m+1} = 1) = 1 \quad (m = 0, 1, \ldots).$$

Hence show that the process X_n satisfies the Chapman–Kolmogorov relation (33) but is not a Markov chain.

(Parzen, 1962, p. 203)

4. Consider a dam of finite capacity h units, where h is a positive integer. Let X_n be an integral random variable representing the content of the dam at the end of the nth time period. For the sake of convenience we

suppose each time period to be 1 day. The daily inputs are independent, identically distributed, integral random variables with p.g.f.

$$G(z) = \sum_{j=0}^{\infty} g_j z^j.$$

Provided the dam is not empty, one unit is released at the end of each day. Otherwise there is no release and overflow is regarded as lost. The content X_n is reckoned after the release. Show that X_n is a Markov chain with h states and write down the transition matrix.

If $\pi_i^{(h)}$ $(i = 0, \ldots, h-1)$ denotes the equilibrium distribution, prove that the ratios $v_i = \pi_i^{(h)}/\pi_0^{(h)}$ are independent of h and satisfy

$$\sum_{i=0}^{\infty} v_i z^i = \frac{g_0(1-z)}{G(z)-z}.$$

(Moran, 1956; Prabhu, 1958)

5. In the previous exercise, show that if the input has a geometric distribution then the equilibrium distribution is a truncated geometric distribution with a modified initial term.

6. An individual at time n in a population becomes independently a family of size R at time $n+1$, where R has the p.g.f. $G(z)$. Immigration in the time interval $(n, n+1)$ occurs independently with p.g.f. $F(z)$. Show that $\Pi_n(z)$, the p.g.f. of population size at time n, satisfies

$$\Pi_{n+1}(z) = F(z)\, \Pi_n\{G(z)\}.$$

7. Use the result of the previous exercise in the following problem. A man founds a society for the upholding of moral standards. Each year he admits as a member one person of sufficiently high moral standard and the probability of this is p. A member who is guilty of a moral lapse must resign and the probability of this happening to a member in any year is λ. The founder himself is not considered a member. If $\Pi_n(z)$ is the p.g.f. of the number of members at the end of the nth year, show that, neglecting the possibility of death,

$$\Pi_n(z) = \{pz + (1-p)\}\, \Pi_{n-1}\{(1-\lambda)z + \lambda\}.$$

Hence show that

$$\Pi_n(0) = \prod_{k=0}^{n-1} \{1 - p(1-\lambda)^k\}$$

and that if p is small, the probability that the society contains no members throughout a particular year will approximate to $\exp(-p/\lambda)$ after a long time.

(*Cambridge Diploma*, 1955)

8. In the random walk with variable transition probabilities, Examples 3.11 and 3.18, show that, provided $\theta_i > 0$, a first passage from any state j to state $j + 1$ is certain. Show that the mean first passage time is

$$\{1/\theta_0 + 1/(\rho_2\theta_1) + \cdots + 1/(\rho_{j+1}\theta_j)\} \qquad \rho_{j+1}^{-1}$$

9. Consider a single-server queueing system for which service-times are independent with a common exponential distribution and inter-arrival times are independent and identically distributed with a given p.d.f. Show that the number of individuals in the queue, considered at the epochs of arrival, forms an imbedded Markov chain with transition matrix of the form

$$\begin{bmatrix} A_0 & a_0 & 0 & 0 & 0 & \cdots \\ A_1 & a_1 & a_0 & 0 & 0 & \cdots \\ A_2 & a_2 & a_1 & a_0 & 0 & \cdots \\ \vdots & \vdots & \vdots & \vdots & \vdots & \end{bmatrix}$$

where $A_j = a_{j+1} + \cdots$ and $\sum a_i = 1$. Define ρ by $\rho^{-1} = \sum n a_n$. Use the results of Section 3.8 to show that the system is ergodic if $\rho < 1$, null-recurrent if $\rho = 1$ and transient if $\rho > 1$. Interpret these conditions in terms of the mean inter-arrival and service-times.

(Foster, 1953; Kendall, 1953)

10. A number, h, of points is arranged on the circumference of a circle and a particle pursues a random walk on the points. The probability that a particle will move in a clockwise direction to the next point is p, otherwise it will move to its anticlockwise neighbouring point. All such steps are independent and p is the same for all points. The particle starts at one of the points, A say. It is said to pursue a 'tour' starting from A if its first return to A is from the other neighbouring point to A than that which it first visited on leaving A. In doing so it must visit all other points on the circle. Show that the probability of completing a tour is, for $p \neq q$, and $p + q = 1$,

$$\frac{(p-q)(p^h + q^h)}{p^h - q^h}.$$

Determine this probability when $p = q = \frac{1}{2}$.

(*Cambridge Diploma*, 1963)

11. For $n = 1, 2, \ldots$ let Y_n be a Markov chain with states $-1, 0, 1$ and transition matrix

$$\begin{array}{c} \\ -1 \\ 0 \\ 1 \end{array} \begin{array}{ccc} -1 & 0 & 1 \\ \begin{bmatrix} \alpha_1 & \alpha_2 & \alpha_3 \\ \beta_3 & \beta_1 & \beta_2 \\ \gamma_2 & \gamma_3 & \gamma_1 \end{bmatrix}, \end{array}$$

and let X_n be a random walk with steps Y_n. Thus

$$X_n = Y_1 + \ldots + Y_n \quad (n = 1, 2, \ldots)$$

is an additive process defined on the Markov chain. Suppose that there is an absorbing barrier at $a > 0$. For $j = -1$, 0, 1, let π_j denote the probability of absorption conditional on $Y_1 = j$. Define

$$\lambda = \frac{\alpha_1(1-\beta_1) + \alpha_2\beta_3}{\gamma_1(1-\beta_1) + \beta_2\gamma_3}, \qquad \rho = (1-\beta_1)(1-\gamma_1) - \beta_2\gamma_3.$$

Show that for $\lambda \leqslant 1$, absorption is certain while for $\lambda > 1$,

$$\pi_j = \frac{x_j}{\rho\lambda^a},$$

where

$$x_1 = \lambda\{\rho\lambda - (1-\beta_1)(\lambda-1)\}, \qquad x_2 = \lambda\rho - \beta_3(\lambda-1), \qquad x_3 = \rho.$$

12. Prove Theorem 2 of Section 3.13 under the extra assumption that $F(s)$, regarded as a function of a complex variable s, is regular at $s = 1$. (Hint: show that $1 - F(s)$ has no zeros in $|s| \leqslant 1$ except at $s = 1$ and hence that $(1-s)F(s)$ is regular in $|s| \leqslant 1 + \delta$, for some $\delta > 0$.)

CHAPTER FOUR

Markov Processes with Discrete States in Continuous Time

4.1. The Poisson process

In Chapters 2 and 3 we have considered processes in discrete time. In the next three chapters, we deal with processes in continuous time. The methods to be used are essentially similar to those already studied, consideration of transitions occurring between times n and $n+1$ being replaced by that of transitions occurring in a short time interval $(t, t+\Delta t)$. The general theory in continuous time is, however, mathematically appreciably more difficult than that of analogous processes in discrete time and therefore we shall proceed largely through the analysis of particular examples.

While the distinction between discrete time and continuous time is mathematically clear-cut, we may in applied work use a discrete time approximation to a continuous time phenomenon and *vice versa*. Broadly, discrete time models are usually easier for numerical analysis, whereas simple analytical solutions are more likely to emerge in continuous time.

In the present chapter we consider Markov processes in continuous time and with a discrete state space. Unless stated otherwise, we label the states 0, 1, 2, ... One of the simplest such processes, and one that is fundamental to the definition and study of more complex processes, is the Poisson process, already defined in Example 1.3. We first discuss the properties of the Poisson process more fully and give some generalizations. Then, in Section 4.3, we introduce some general techniques by discussing a number of important special processes. Section 4.4 deals with the calculation of equilibrium distributions. The remaining sections are more specialized. Section 4.5 is an informal sketch of the general theory of Markov processes with discrete states in continuous time and Section 4.6 discusses some miscellaneous topics.

The defining properties of the Poisson process are given in equations (1.7)–(1.9), which we now repeat. Consider a process of point events occurring on the real axis. Let $N(t, t+\Delta t)$ denote the number of events in $(t, t+\Delta t]$. Suppose that, for some positive constant ρ, as $\Delta t \to 0$,

$$\text{prob}\{N(t, t+\Delta t) = 0\} = 1 - \rho\Delta t + o(\Delta t), \tag{1}$$

$$\text{prob}\{N(t, t+\Delta t) = 1\} = \rho\Delta t + o(\Delta t), \tag{2}$$

so that

$$\text{prob}\{N(t, t+\Delta t) > 1\} = o(\Delta t), \tag{3}$$

and that further $N(t, t+\Delta t)$ is completely independent of occurrences in $(0, t]$. We then say that the stochastic series of events is a Poisson process of rate ρ. The independence condition specifies the 'randomness' of the series. Equation (3) specifies that events occur singly.

We begin by obtaining from first principles some elementary properties of the process.

Suppose that we take a new time origin at t_0, which may be a point at which an event has just occurred, or any fixed point. Any property of the process referring to its behaviour after t_0 is independent of what happens at or before t_0. In particular, if $t_0 + Z$ is the time of the first event after t_0, the random variable Z is independent of whether an event occurs at t_0 and of occurrences before t_0. To calculate the probability distribution of Z, we argue as follows. Let

$$P(x) = \text{prob}(Z > x).$$

Then, for $\Delta x > 0$,

$$
\begin{aligned}
P(x + \Delta x) &= \text{prob}(Z > x + \Delta x) \\
&= \text{prob}\{Z > x \text{ and no event occurs in} \\
&\qquad (t_0 + x, \ t_0 + x + \Delta x]\} \\
&= \text{prob}(Z > x)\,\text{prob}\{\text{no event occurs in} \\
&\qquad (t_0 + x, \ t_0 + x + \Delta x] | Z > x\}. \tag{4}
\end{aligned}
$$

Equation (4) would hold for any process of events. A special feature of the Poisson process, a form of the Markov property, is that the conditional probability in (4) is not affected by the condition $Z > x$, which refers to what happens at or before $t_0 + x$. Hence the unconditional probability (1) can be used.

That is, (4) becomes

$$P(x + \Delta x) = P(x)\{1 - \rho\Delta x + o(\Delta x)\},$$

or

$$P'(x) = -\rho P(x). \tag{5}$$

Thus $P(x) = P(0)e^{-\rho x}$, or, since $P(0) = \text{prob}(Z > 0) = 1$, we have that $P(x) = e^{-\rho x}$. The distribution function of Z is therefore $1 - e^{-\rho x}$, and the probability density function (p.d.f.) is

$$\rho e^{-\rho x} \quad (x > 0), \tag{6}$$

the exponential distribution with parameter ρ.

An alternative proof is to show by an argument very similar to (4) that $P(x + y) = P(x)P(y)$, from which the exponential form of $P(x)$ follows.

The identification of the constant ρ is obtained from the behaviour of $P(x)$ near $x = 0$.

Now consider the process starting from time zero. Let events occur at times $\{Z_1, Z_1 + Z_2, Z_1 + Z_2 + Z_3, \ldots\}$, i.e. let Z_n be the time interval between the $(n-1)$th and the nth events. On taking $t_0 = 0$ in the above argument, we have that Z_1 has the p.d.f. (6). We can now repeat the argument, taking $t_0 = Z_1$, to show that Z_2 has the p.d.f. (6) and is independent of Z_1, and so on. That is, the sequence $\{Z_1, Z_2, \ldots\}$ are independently and identically distributed random variables with the p.d.f. (6). This property characterizes the Poisson process.

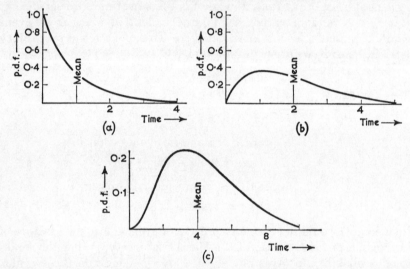

Fig. 4.1. Distribution of time up to rth event in a Poisson process of unit rate. (a) $r = 1$; (b) $r = 2$; (c) $r = 4$.

Now the moment generating function (m.g.f.) of (6) is

$$E(e^{-\theta Z}) = \int_0^\infty \rho e^{-\rho x} e^{-\theta x} dx = \frac{\rho}{\rho + \theta}. \tag{7}$$

Hence the m.g.f. of the random variable $Z_1 + \ldots + Z_r$, the time from the origin to the rth subsequent event, is

$$E[\exp\{-\theta(Z_1 + \ldots + Z_r)\}] = \{E(e^{-\theta Z})\}^r = \frac{\rho^r}{(\rho + \theta)^r}, \tag{8}$$

whence the p.d.f. is

$$\frac{\rho(\rho x)^{r-1} e^{-\rho x}}{(r-1)!} \quad (x > 0), \tag{9}$$

a gamma distribution.

Now it follows from (6), or from the m.g.f. (7), that if Z has the p.d.f. (6),

$$E(Z) = 1/\rho, \qquad V(Z) = 1/\rho^2. \tag{10}$$

Thus, since successive intervals are independent, the mean and variance of $Z_1 + \ldots + Z_r$ are respectively r/ρ and r/ρ^2. Further, by the central limit theorem, $Z_1 + \ldots + Z_r$ is, as $r \to \infty$, asymptotically normally distributed. Figure 4.1 shows the form of the distribution for small values of r.

We now illustrate a general technique for dealing with processes in continuous time and having discrete state space.

Let $N(t)$ be the number of events in an interval of fixed length, say $(0,t]$. Note that the numbers of events in non-overlapping intervals are mutually independent, because of the defining properties of the Poisson process. We write

$$p_i(t) = \text{prob}\{N(t) = i\}$$

and proceed to build up differential equations for the sequence $\{p_i(t)\}$ by a method analogous to that leading to (5). We have for $\Delta t > 0$ that

$$
\begin{aligned}
p_i(t + \Delta t) &= \text{prob}\{N(t + \Delta t) = i\} \\
&= \text{prob}\{N(t) = i \quad \text{and} \quad N(t, t + \Delta t) = 0 \\
&\quad \text{or} \quad N(t) = i - 1 \quad \text{and} \quad N(t, t + \Delta t) = 1 \\
&\quad \text{or} \quad N(t) = i - l \quad \text{and} \quad N(t, t + \Delta t) = l; l = 2, 3, \ldots\} \\
&= \text{prob}\{N(t) = i\}\,\text{prob}\{N(t, t + \Delta t) = 0 | N(t) = i\} \\
&\quad + \text{prob}\{N(t) = i - 1\}\,\text{prob}\{N(t, t + \Delta t) \\
&\quad\quad\quad\quad = 1 | N(t) = i - 1\} \\
&\quad + \sum_{l \geqslant 2} \text{prob}\{N(t) = i - l\}\,\text{prob}\{N(t, t + \Delta t) \\
&\quad\quad\quad\quad = l | N(t) = i - l\}. \tag{11}
\end{aligned}
$$

In this, probabilities corresponding to negative states are, by definition, zero. Equation (11) holds for any process in which the transitions are never downwards.

The special feature of the Markov processes to be considered in this chapter is that the conditional probabilities in (11) are, as $\Delta t \to 0$, known as fairly simple functions of the state occupied at t. In particular, for the Poisson process, the conditional probabilities are constant and are given by (1)–(3). Thus on the right-hand side of (11), the first factor in each term is one of the $p_i(t)$ and the second factor is known from (1)–(3). Hence, as $\Delta t \to 0$, (11) becomes

$$p_i(t + \Delta t) = p_i(t)(1 - \rho \Delta t) + p_{i-1}(t)\rho \Delta t + o(\Delta t),$$

which leads to the differential equation

$$p_i'(t) = -\rho p_i(t) + \rho p_{i-1}(t) \quad (i = 0, 1, \ldots). \tag{12}$$

It is understood that when $i = 0$ in (12), $p_{-1}(t)$ is identically zero.

The differential equations (12) have to be taken in conjunction with the initial condition

$$p_0(0) = 1, \qquad p_i(0) = 0 \quad (i = 1, 2, \ldots). \tag{13}$$

There are several methods for the solution of equations of this type. The simplest method, which is however only applicable in very special cases such as the present problem, is to solve recursively. Thus (12) and (13) with $i = 0$ lead to $p_0(t) = e^{-\rho t}$ and substitution of this into the equation for $i = 1$ leads to $p_1(t) = \rho t e^{-\rho t}$, and so on.

Usually the best method to deal with simultaneous equations such as (12) is to convert them, if possible, into a single equation for a generating function. Let

$$G(z, t) = \sum_{i=0}^{\infty} p_i(t) z^i. \tag{14}$$

For any fixed t, this is the probability generating function of $N(t)$. Multiply the equations (12) by $(z^0, z^1, z^2, \ldots)$ and add. We obtain the equation

$$\frac{\partial G(z, t)}{\partial t} = -\rho G(z, t) + \rho z G(z, t), \tag{15}$$

whereas, from the initial condition (13),

$$G(z, 0) = 1. \tag{16}$$

For any fixed z, (15) is an ordinary linear differential equation in t with the solution

$$G(z, t) = A(z) e^{-\rho t + \rho t z},$$

where the arbitrary 'constant', $A(z)$, is a function of z. By (16), $A(z) = 1$. Therefore

$$G(z, t) = e^{-\rho t + \rho t z}. \tag{17}$$

This, as a function of z, is the probability generating function of the Poisson distribution of mean ρt, as can be seen by isolating the coefficient of z^i to give

$$\text{prob}\{N(t) = i\} \equiv p_i(t) = e^{-\rho t} \frac{(\rho t)^i}{i!} \quad (i = 0, 1, 2, \ldots). \tag{18}$$

The Poisson distribution (18) is one of the standard elementary distributions of probability theory; its main properties are that the mean and variance are both equal to ρt and that as $\rho t \to \infty$ the distribution tends to normality.

The specification of (a) intervals $\{Z_1, Z_2, \ldots\}$ between successive events and (b) numbers of events occurring in fixed intervals must in a general way be equivalent. One illustration of this is that $Z_1 + \ldots + Z_i$ is the time from the origin up to the ith event. Further, the events $N(t) < i$ and

$Z_1 + \ldots + Z_i > t$ are equivalent. If we calculate the probabilities of these respectively from (18) and (9), we have that

$$\sum_{j=0}^{i-1} e^{-\rho t} \frac{(\rho t)^j}{j!} = \int_t^\infty \rho \frac{(\rho x)^{i-1} e^{-\rho x}}{(i-1)!} \, dx, \tag{19}$$

a well-known identity that can be proved directly by integrating by parts.

If we think of $\{N(t)\}$ as a stochastic process, i.e. as a random function of t, each possible realization is a step function starting at 0 and with a jump one upwards each time an event occurs. In view of the independence of the occurrences in non-overlapping time periods, the process $\{N(t)\}$ can be regarded as a random walk in continuous time, the jump in any small interval having high probability of being zero and a small probability of being plus one. Many of the results in the theory of random walk in discrete time, for example Wald's identity (Section 2.4), could be adapted to the Poisson process, but the very special nature of the Poisson process makes it easier to proceed from first principles.

The probabilistic argument (11), leading to the differential equations (12), will be used repeatedly with minor variations in subsequent sections, and a general statement of the method is set out in Section 4.5(iii). It is useful here to discuss briefly the relation between (11) and the general Chapman–Kolmogorov equations (Section 1.3), and to obtain a second alternative set of differential equations for the process.

First, a more elaborate notation is useful for general discussion. The probability $p_i(t)$, and the whole discussion up to now, depends on starting from the initial condition $N(0) = 0$. We can stress this by writing

$$\text{prob}\{N(t) = i | N(0) = 0\} = p_{0i}(0, t).$$

In general, if $t < u$, we write

$$\text{prob}\{N(u) = i | N(t) = j\} = p_{ji}(t, u). \tag{20}$$

Now the general Chapman–Kolmogorov equation (1.19) in continuous time relates the states of the process at three time points $t < u < v$. For systems with a discrete state space (1.19) can be rewritten for three times $t < u < v$ as

$$p_{ki}(t, v) = \sum_{j=0}^\infty p_{kj}(t, u) \, p_{ji}(u, v). \tag{21}$$

If now we replace t, u, v by $0, t, t + \Delta t$, we have that

$$p_{ki}(0, t + \Delta t) = \sum_{j=0}^\infty p_{kj}(0, t) \, p_{ji}(t, t + \Delta t). \tag{22}$$

If, further, we start from the zero state, $k = 0$, (22) is exactly (11), the term $p_{ji}(t, t + \Delta t)$ specifying the probabilities of the transitions occurring in $(t, t + \Delta t]$.

We now obtain a new set of differential equations by taking in (21) the times $t < u < v$ to be $-\Delta t < 0 < t$. These new equations are

$$p_{ki}(-\Delta t, t) = \sum_{j=0}^{\infty} p_{kj}(-\Delta t, 0) p_{ji}(0, t). \tag{23}$$

Suppose that we start from the zero state, $k = 0$, and simplify the notation by using the fact that the number of events in an interval depends only on the length of the interval so that, for example, $p_{ki}(-\Delta t, t)$ can be written $p_{ki}(t + \Delta t)$. Then

$$p_{0i}(t + \Delta t) = \sum_{j=0}^{\infty} p_{0j}(\Delta t) p_{ji}(t).$$

Thus, using the key properties (1)–(3) of the Poisson process, we have that, as $\Delta t \to 0$,

$$p_{0i}(t + \Delta t) = (1 - \rho \Delta t) p_{0i}(t) + \rho \Delta t \, p_{1i}(t) + o(\Delta t),$$

leading to

$$p_{0i}'(t) = -\rho p_{0i}(t) + \rho p_{1i}(t) \quad (i = 0, 1, \ldots). \tag{24}$$

These are to be compared with (12), which, in the new notation, can be written

$$p_{0i}'(t) = -\rho p_{0i}(t) + \rho p_{0, i-1}(t) \quad (i = 0, 1, \ldots). \tag{25}$$

In fact, for the Poisson process, (24) and (25) are trivially different, because $p_{1i}(t)$ and $p_{0, i-1}(t)$ are both equal to the probability that $i-1$ events occur in a time period t. In general, however, two different but normally equivalent sets of equations result. We construct these equations by relating the states at three time points. If we take an initial point followed later by two time points close together, we generate the *forward equations*. Equation (23) is an example. If we take two initial points close together followed by a single time point later, we generate the *backward equations*. Equation (24) is an example.

The general characteristic of the forward equations is that in any single equation the same initial state is involved throughout, whereas in a particular backward equation the same final state enters in all terms. Hence the forward equations are most likely to be useful if there is a single initial state of particular interest and we want the probabilities at time t for various final states. On the other hand, the backward equations would normally be used if there is a single final state of particular interest and we want the probabilities of reaching this at time t for various initial states.

A final general point is that in deriving (11) and (12) we have, for clarity, considered the number of occurrences in $(0, t]$. Now for all the processes considered in this chapter, there is zero probability of a transition occurring at some specified time point. Hence it is immaterial whether we consider open or closed time intervals and from now on we do not normally distinguish them.

4.2. Generalizations of the Poisson process

A few of the many possible generalizations of the Poisson process will now be considered. These are

(a) the Poisson process in several dimensions;
(b) the time-dependent Poisson process;
(c) the combination of several independent Poisson processes;
(d) the Poisson process with multiple occurrences.

First, we sometimes deal with a Poisson process in several dimensions. In three dimensions, for example, we replace (1)–(3) by the requirements that if $N(\Delta v)$ is the number of events in an arbitrary element of volume Δv, then as $\Delta v \to 0$,

$$\text{prob}\{N(\Delta v) = 0\} = 1 - \rho \Delta v + o(\Delta v),$$

$$\text{prob}\{N(\Delta v) = 1\} = \rho \Delta v + o(\Delta v),$$

and that the numbers of events in non-overlapping volumes are mutually independent. It can then be shown that the number of events in a volume v has a Poisson distribution of mean ρv.

The second generalization is to retain the original specification of the Poisson process with the modification that ρ in (1)–(3) is a function of time, $\rho(t)$, say. We call the new process a time-dependent Poisson process. We stress that the number of events in $(t, t + \Delta t]$ is still completely independent of occurrences in $(0, t]$. Equations (11)–(16) are unchanged except for the replacement of ρ by $\rho(t)$, and the solution of (15) and (16) is

$$G(z, t) = \exp \left\{ (z - 1) \int_0^t \rho(u)\, du \right\},$$

so that $N(t)$ has a Poisson distribution of mean

$$\int_0^t \rho(u)\, du. \tag{26}$$

An alternative and instructive proof of this is obtained by introducing a non-linear transformation of the time scale, so that in terms of a suitable new time variable τ, we have a simple Poisson process. In fact, if $\tau(t)$ is a strictly increasing function of t, $\tau(0) = 0$, the probability of an event in $(\tau, \tau + \Delta\tau)$ is approximately

$$\rho(t)\frac{dt}{d\tau}\Delta\tau.$$

If we arrange that

$$\rho(t)\frac{dt}{d\tau} = 1,$$

i.e.

$$\tau(t) = \int_0^t \rho(u)\,du, \tag{27}$$

it is easy to show that in the new τ scale we have a Poisson process of unit rate. Hence the number of events in an interval of length τ on the new scale has a Poisson distribution of mean τ. This proves (26). The device of transforming the time scale is often useful in dealing with more general problems involving time-dependent transition probabilities, but requires that a single transformation of the time scale shall remove all time dependencies.

A number of the more complex processes considered later in this chapter are in effect built up from combinations of Poisson processes. Suppose that there are k independent Poisson processes of rates $\rho_1, \ldots, \rho_k$. Let $N(t, t + \Delta t)$ denote the total number of events of any type in $(t, t + \Delta t)$, $N^{(i)}(t, t + \Delta t)$ being the number in the ith process. Then

$$\text{prob}\{N(t, t + \Delta t) = 0\} = \prod_{i=1}^{k} \text{prob}\{N^{(i)}(t, t + \Delta t) = 0\}$$

$$= \prod_{i=1}^{k} \{1 - \rho_i \Delta t + o(\Delta t)\}$$

$$= 1 - \rho\Delta t + o(\Delta t), \tag{28}$$

where $\rho = \sum \rho_i$, since the component processes are independent. Similarly

$$\text{prob}\{N(t, t + \Delta t) = 1\} = \sum_{i=1}^{k} \text{prob}\{N^{(i)}(t, t + \Delta t) = 1\}$$

$$\text{prob}\{N^{(j)}(t, t + \Delta t) = 0; \quad j \neq i\}$$

$$= \rho\Delta t + o(\Delta t). \tag{29}$$

Finally, the number of events in $(t, t + \Delta t)$ is independent of occurrences in all processes before or at t. Thus the combined series of events is a Poisson process of rate $\rho = \sum \rho_i$. In particular, if the k separate processes have the same rate ρ_1, the combined process has rate $\rho = k\rho_1$.

Now the time, Z, from the origin to the first event in the combined process has p.d.f. $\rho e^{-\rho x}$. Let $Z_1, \ldots, Z_k$ be the corresponding times for the separate processes, so that $Z = \min(Z_1, \ldots, Z_k)$ and $Z_1, \ldots, Z_k$ are independently distributed with p.d.f.'s $\rho_i e^{-\rho_i x}$ $(i = 1, \ldots, k)$. We now consider the joint distribution of Z and of the types of event to occur first. In fact, given that $Z = x$, the probability that the event is from the first process is

$$\text{prob}\{Z_1 = x, \ Z_i > x \ (i = 2, \ldots, k)|Z = x\}$$
$$= \rho_1 e^{-\rho_1 x} . e^{-\rho_2 x} \ldots e^{-\rho_k x}/(\rho e^{-\rho x}) = \rho_1/\rho. \tag{30}$$

This does not involve x. It follows that we can describe the set of independent Poisson processes as follows. Intervals between successive events are independently distributed with the p.d.f. $\rho e^{-\rho x}$. Events are then assigned to individual types randomly and with constant probabilities ρ_1/ρ, etc.

To give one concrete interpretation of these results, consider components, say valves, which have two types of failure, and suppose that immediately on failure, a component is replaced by a new one. Then the following two systems are equivalent:

(a) failures of the two types occur in independent Poisson processes of rates ρ_1, ρ_2;
(b) failures occur in a Poisson process of rate $\rho = \rho_1 + \rho_2$, and the probability that a particular failure is of Type 1 is ρ_1/ρ, independently for all failures.

A fourth generalization of the Poisson process is obtained by allowing simultaneous occurrences. Suppose that points are distributed in a Poisson process of rate ρ and that at the ith point there are V_i events, where $\{V_i\}$ are independently and identically distributed with

$$\text{prob}(V_i = j) = \varpi_j \quad (j = 1, 2, \ldots), \tag{31}$$

the probability generating function being $\Pi(z) = \sum \varpi_j z^j$. In time t, let $M(t)$ be the number of points and $N(t)$ the number of events. Given $M(t) = m$, the distribution of $N(t)$ is the m-fold convolution of (31) and therefore has probability generating function $\{\Pi(z)\}^m$. Hence, allowing for the Poisson distribution of $M(t)$, we have for the probability generating function of $N(t)$

$$\sum_{m=0}^{\infty} \{\Pi(z)\}^m \frac{e^{-\rho t}(\rho t)^m}{m!} = \exp\{\rho t\Pi(z) - \rho t\}. \tag{32}$$

From this, the properties of $N(t)$ can be obtained.

Other generalizations of the Poisson process, including a very important one, the renewal process, will be considered in Chapter 9, on point processes. In the renewal process the intervals between successive events are independently and identically distributed with an arbitrary distribution, instead of with an exponential distribution. In the general theory of point processes, we consider more or less arbitrary distributions of points in time. The Poisson process will recur frequently in the remainder of the book.

4.3. Some simple processes of the birth–death type

In this section we illustrate by examples an important group of stochastic processes which can be solved by arguments similar to those used in Section 4.1.

Example 4.1. *Linear birth process.* Consider a collection of individuals reproducing according to the following rules. An individual present at

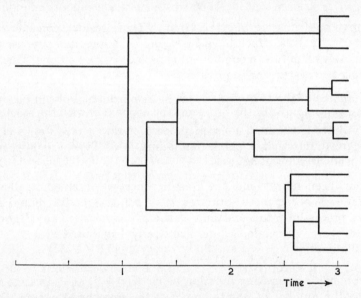

Fig. 4.2. Realization of linear birth process.

time t has probability $\lambda \Delta t + o(\Delta t)$ of splitting into two, i.e. of giving birth to one new individual, in time $(t, t + \Delta t)$. Further, this probability is the same for all individuals present and is independent of the age of the individual; events referring to different individuals are independent. Thus, in view of the relation with a Poisson process, the times between successive births to the same individual are independently exponentially distributed. The parameter λ is the birth rate per individual. We call the process the linear birth process, or sometimes the Yule–Furry process.

Figure 4.2 shows a realization of the system with $\lambda = 1$, starting at $t = 0$ with one individual; at each birth the relevant line in the diagram divides into two. The construction of the realization is straightforward using the table of exponentially distributed numbers in Appendix 1. If only the population size is required, the procedure can be simplified; see Exercise 12.

In this process, decreases in the number of individuals cannot occur and the number of individuals in the system will, with probability one, increase indefinitely. The process is a reasonable approximate model for the growth of bacteria, the fission of neutrons, etc., under conditions such that 'death' is relatively unimportant. Later we shall consider generalizations which allow for death.

Before analysing the system as a stochastic process, it is useful to make a simpler analysis ignoring statistical fluctuations in the number of individuals. Suppose then that there are $n(t)$ individuals at time t, where $n(t)$ is so large that it can be treated as a continuous function of t. Then $\lambda n(t)\Delta t$ births occur in $(t, t + \Delta t)$, so that

$$n'(t) = \lambda n(t), \tag{33}$$

i.e.
$$n(t) = n_0 e^{\lambda t},$$

where n_0 is the initial number of individuals. We call this approach the *deterministic analysis* of the problem; the connexion with the stochastic analysis will appear shortly.

Suppose now that $N(t)$, a random variable, is the number of individuals at time t and that $\text{prob}\{N(t) = i\}$ is denoted by $p_i(t)$. By (29) of Section 4.2, the probability of a birth in $(t, t + \Delta t)$, given that $N(t) = i$, is $i\lambda\Delta t + o(\Delta t)$, since we have a superposition of i Poisson processes each of rate λ. Thus, exactly as in (11) for the Poisson process, we have the forward equations that

$$
\begin{aligned}
p_i(t + \Delta t) &= \text{prob}\{N(t) = i \text{ and no birth occurs in } (t, t + \Delta t)\} \\
&\quad + \text{prob}\{N(t) = i - 1 \text{ and one birth occurs in } (t, t + \Delta t)\} \\
&\quad + o(\Delta t) \\
&= p_i(t)\,(1 - i\lambda\Delta t) + p_{i-1}(t)\,(i - 1)\,\lambda\Delta t + o(\Delta t). \tag{34}
\end{aligned}
$$

The reason that $i - 1$ occurs in the coefficient of $p_{i-1}(t)$ is that the probability of a birth has to be taken conditionally on $N(t) = i - 1$. Equation (34) leads to

$$p_i'(t) = -i\lambda p_i(t) + (i - 1)\lambda p_{i-1}(t) \quad (i = n_0, n_0 + 1, \ldots), \tag{35}$$

where in the equation for $i = n_0$, the term $p_{n_0-1}(t)$ is identically zero. Equation (35) is to be taken with the initial condition

$$p_i(0) = \delta_{in_0}, \tag{36}$$

where δ_{ij} is the Kronecker delta symbol.

We introduce as before a generating function

$$G(z, t) = \sum_{i=0}^{\infty} p_i(t) z^i.$$

Note that

$$\frac{\partial G(z, t)}{\partial z} = \sum_{i=0}^{\infty} i p_i(t) z^{i-1}.$$

On multiplying (35) by z^i and summing, we have that

$$\frac{\partial G(z, t)}{\partial t} + \lambda z(1 - z) \frac{\partial G(z, t)}{\partial z} = 0, \tag{37}$$

with

$$G(z, 0) = z^{n_0}. \tag{38}$$

This partial differential equation can be solved in several ways, for example by applying a Laplace transformation. It is easier, however, to treat (37) as a Lagrange equation; see, for example, Piaggio (1942, Chapter 12). The auxiliary equations are

$$\frac{dt}{1} = \frac{dz}{\lambda z(1 - z)} = \frac{dG}{0}.$$

Two independent solutions are $G = a$, $z e^{-\lambda t}/(1-z) = b$, where a, b are arbitrary constants. The general solution is thus

$$G(z, t) = f\left(\frac{z}{1 - z} e^{-\lambda t}\right).$$

Inserting the initial condition (38), we have that

$$z^{n_0} = f\left(\frac{z}{1 - z}\right),$$

so that $f(\xi) = \{\xi/(1 + \xi)\}^{n_0}$. Thus

$$G(z, t) = \left(\frac{z e^{-\lambda t}}{1 - z + z e^{-\lambda t}}\right)^{n_0}. \tag{39}$$

Thus the distribution of $N(t)$ is of the negative binomial form, with first term at n_0. The moments of $N(t)$ can be derived by differentiating (39) with respect to z. Alternatively it is instructive to obtain differential equations directly for the moments. The method is to pass from the probability generating function to the moment generating function by writing $z = e^{-\theta}$. This replaces $G(z, t)$ by a new function

$$H(\theta, t) = E\{e^{-\theta N(t)}\}.$$

Since, in operator form

$$\frac{\partial}{\partial z} = -e^{\theta} \frac{\partial}{\partial \theta},$$

the differential equation (37) becomes

$$\frac{\partial H(\theta,t)}{\partial t} - \lambda(1-e^{-\theta})\frac{\partial H(\theta,t)}{\partial \theta} = 0. \tag{40}$$

The usual procedure would now be to expand (40) in powers of θ, noting that the coefficient of $(-\theta)^r/r!$ in $H(\theta,t)$ is the rth moment about the origin of $N(t)$. In fact, if the moments about the mean are required, it is slightly easier to use the cumulant generating function

$$K(\theta,t) = \log H(\theta,t),$$

which clearly also satisfies (40). If $\kappa_r(t)$ is the rth cumulant of $N(t)$, the coefficient of $(-\theta)^r/r!$ in $K(\theta,t)$, we have that

$$\kappa_1'(t) = \lambda\kappa_1(t), \qquad \kappa_2'(t) = 2\lambda\kappa_2(t) + \lambda\kappa_1(t), \dots,$$

with initial conditions

$$\kappa_1(0) = n_0, \qquad \kappa_r(0) = 0 \quad (r = 2, 3, \dots).$$

Thus

$$\kappa_1(t) = n_0 e^{\lambda t}, \qquad \kappa_2(t) = n_0(e^{2\lambda t} - e^{\lambda t}), \dots \tag{41}$$

These results can be obtained also directly from (35) or from (39). The point of the present approach is that we obtain directly ordinary differential equations for the cumulants, and it may, in more complex cases, be easier to solve for the first few cumulants than to obtain a complete solution for the probability distribution.

The coefficient of variation of $N(t)$ is, from (41),

$$\frac{(1-e^{-\lambda t})^{\frac{1}{2}}}{\sqrt{n_0}} \sim \frac{1}{\sqrt{n_0}}$$

as $t \to \infty$. It is easy to show that as $t \to \infty$ the limiting distribution of $N(t)$ is a gamma distribution of index n_0. One proof depends on examining the limiting cumulant generating function of $N(t)/\kappa_1(t)$.

Equations (33) and (41) show that the deterministic theory gives exactly the stochastic mean. In later examples, this agreement will not always occur. A sufficient condition for equality is that for an arbitrary positive integer r, the structure of the process starting from rn_0 individuals is identical with that of the sum of r separate systems each starting individually from initial state n_0. Note that even when deterministic theory and stochastic mean do agree, they do not necessarily give a good idea of the behaviour of individual realizations.

First passage time problems for the process are in principle straightforward because $N(t)$ cannot decrease. Therefore if T_a is the time at which state a is first occupied ($a > n_0$), we have that

$$\text{prob}(T_a \leqslant t) = \text{prob}\{N(t) \geqslant a\}. \tag{42}$$

Explicit results can be obtained from (39) and (42), especially when n_0 is small. An alternative approach is to represent T_a as the sum of independent random variables, as follows. Let Z_i be the length of time spent in state i, being exponentially distributed with parameter $i\lambda$. Then different Z's are independent and

$$T_a = Z_{n_0} + \ldots + Z_{a-1}. \tag{43}$$

Thus

$$E(T_a) = \frac{1}{\lambda} \sum_{j=n_0}^{a-1} \frac{1}{j} \sim \frac{1}{\lambda} \log\left(\frac{a}{n_0}\right),$$

if a/n_0 and n_0 are large. Note the agreement between this limit and the deterministic solution from (33). A better approximation for large a and small n_0 is that

$$E(T_a) - \frac{\gamma}{\lambda} + \frac{1}{\lambda} \sum_{j=1}^{n_0-1} \frac{1}{j} \sim \frac{1}{\lambda} \log(a-1),$$

where $\gamma \simeq 0.5772$ is Euler's constant. Further

$$V(T_a) = \frac{1}{\lambda^2} \sum_{j=n_0}^{a-1} \frac{1}{j^2}$$

$$\sim \frac{\pi^2}{6\lambda^2} - \frac{1}{\lambda^2} \sum_{j=1}^{n_0-1} \frac{1}{j^2},$$

as a increases. An interesting point here is the finiteness of $V(T_a)$ as a increases.

It follows from (43) that

$$E(e^{-\theta T_a}) = \prod_{j=n_0}^{a-1} \left(\frac{j\lambda}{\theta + j\lambda}\right).$$

This can be expressed in partial fractions and hence inverted in simple terms. To examine the limiting distribution as $a \to \infty$, we suppose for simplicity that $n_0 = 1$ and introduce the standardized random variable

$$U_a = \lambda T_a - \log a.$$

Then

$$E(e^{-\theta U_a}) \sim a^\theta \prod_{j=1}^{a-1} \left(\frac{j}{\theta + j}\right).$$

If we use Gauss's product form of the gamma function, namely

$$\Gamma(1+\theta) = \lim_{a \to \infty} \frac{(a-1)!\, a^\theta}{(\theta+1)(\theta+2)\ldots(\theta+a-1)},$$

we have that

$$E(e^{-\theta U_a}) \sim \Gamma(1+\theta).$$

The inverse of this, considered as a (two-sided) Laplace transform is the 'extreme value' distribution

$$\exp\left(-x-e^{-x}\right) \quad (-\infty < x < \infty),$$

which is therefore the limiting p.d.f. of U_a.

We have dealt with the properties of the linear birth process at length in order to illustrate important techniques. We now give further examples treated more briefly.

Example 4.2. *Generalized Markov birth process.* There are several generalizations of Example 4.1 which retain the Markov property and are pure birth processes in that $N(t)$ is non-decreasing.

First, we can have a time-dependent process in which the conditions of Example 4.1 are changed only by allowing λ to be a function of t, say $\lambda(t)$. This process is easily reduced to Example 4.1 by transforming the time scale, as in (27). The time-dependent process must not be confused with an age-dependent process, in which the probability of a birth for a particular individual depends on the age of that individual. That is a much more complex process.

Second, we can have a non-linear process in which the probability of a birth in $(t, t+\Delta t)$, given $N(t) = i$, is say $\lambda_i \Delta t + o(\Delta t)$. Previously $\lambda_i = i\lambda$. By suitable choice of λ_i, we can represent such effects as a reduction in the birth rate per individual due to crowding. The differential equation (35) becomes

$$p_i'(t) = -\lambda_i p_i(t) + \lambda_{i-1} p_{i-1}(t) \quad (i = n_0, n_0+1, \ldots). \tag{44}$$

This can be solved, in principle, for particular sequences $\{\lambda_i\}$. Example 4.3 deals with an important special case.

Third, we can deal with multiple births. For example, suppose that for each individual there are in each interval $(t, t+\Delta t)$ probabilities $\lambda' \Delta t + o(\Delta t)$ and $\lambda'' \Delta t + o(\Delta t)$ of one and two new individuals being formed. Then (35) is replaced by

$$p_i'(t) = -i(\lambda' + \lambda'') p_i(t) + (i-1)\lambda' p_{i-1}(t) + (i-2)\lambda'' p_{i-2}(t).$$

The fourth generalization is to have two or more types of individual in the system.

The non-linear process (44) can be used to illustrate a phenomenon, called dishonesty, that is important in the general mathematical theory of Markov processes in continuous time. The time spent in the state i is exponentially distributed with parameter λ_i and the Laplace transform of the distribution is $\lambda_i/(s + \lambda_i)$. Further, since the process goes in increasing order through the successive states and the times spent in different states are independent, the Laplace transform of the distribution of the time taken for the population to increase from one individual to a individuals is

$$\prod_{i=1}^{a} \frac{1}{\left(1 + \dfrac{s}{\lambda_i}\right)}.$$

Now if and only if $\sum 1/\lambda_i$ is convergent this infinite product tends as $a \to \infty$ to a finite limit (Titchmarsh, 1939, p. 13) which is, in fact, the Laplace transform of a well-defined distribution of a finite random variable. If $\sum 1/\lambda_i$ is divergent, the product diverges to zero for all non-zero s. That is, if $\sum 1/\lambda_i$ converges, the population size becomes infinite in a finite time.

A very similar conclusion is reached from a deterministic analysis, based on the differential equation

$$\frac{dn(t)}{dt} = \lambda\{n(t)\},$$

where on the right-hand side the birth rate at t is a function of the population size, $n(t)$. Thus the time for the population to grow from size n_0 to size n is

$$\int_{n_0}^{n} \frac{dx}{\lambda(x)}$$

and the condition for explosion to infinite size in a finite time is the convergence of

$$\int^{\infty} \frac{dx}{\lambda(x)}.$$

Since $\sum 1/i$ diverges, escape to infinity, i.e. explosion, in a finite time does not occur with the linear birth process of Example 4.1; it would, however, arise if $\lambda_i \propto i^{1+\delta}$ ($\delta > 0$).

These facts have an important consequence for the solution of the differential equations of the process. For

$$\sum_{i=1}^{\infty} p_i(t) = \lim_{a \to \infty} \sum_{i=1}^{a} p_i(t)$$

is the probability that at time t the population size is finite. Hence the above argument suggests, what is in fact true, that the solution of (44) is such that for all t,

$$\sum_{i=1}^{\infty} p_i(t) = 1 \quad \text{if } \sum 1/\lambda_i \text{ is divergent;} \tag{45}$$

whereas, for some t,

$$\sum_{i=1}^{\infty} p_i(t) < 1 \quad \text{if } \sum 1/\lambda_i \text{ is convergent.} \tag{46}$$

A process satisfying (45) is called *honest*, and one satisfying (46) *dishonest*. In the latter case, the missing probability is the probability of having escaped to infinity at or before t.

The simplest special case of a dishonest process is $\lambda_i = i^2$. Then the Laplace transform of the *distribution function* of the time to reach infinite population size is

$$L^*(s) = \frac{1}{s} \prod_{i=1}^{\infty} \frac{1}{\left(1 + \dfrac{s}{i^2}\right)}$$

$$= \frac{\alpha_0}{s} + \sum_{i=1}^{\infty} \frac{\alpha_i}{s + i^2}.$$

Hence

$$\alpha_0 = 1$$

$$\alpha_i = \lim_{s \to -i^2} \{(s + i^2) L^*(s)\}$$

$$= 2(-1)^i \quad (i = 1, 2, \ldots),$$

after some calculation. Thus

$$L^*(s) = \frac{1}{s} + 2 \sum_{i=1}^{\infty} \frac{(-1)^i}{s + i^2},$$

leading on inversion to

$$L(t) = 1 + 2 \sum_{i=1}^{\infty} (-1)^i e^{-t i^2}.$$

This is a theta function. Table 4.1 gives a few numerical values.

Table 4.1. *Dishonest birth process. Probability, $L(t)$, that infinite population size is reached at or before t*

t	$L(t)$	t	$L(t)$
$\frac{1}{2}$	0·0360	3	0·9004
1	0·3006	6	0·9950
2	0·7300	10	1·0000

The mean of the distribution is $\sum 1/i^2 = \pi^2/6 \simeq 1{\cdot}6$.

The remaining special processes we shall consider will all be honest, but the above special case shows that quite ordinary processes can be dishonest.

Example 4.3. *Simple stochastic epidemic.* An interesting example of a non-linear birth process is the following very idealized model of an epidemic. Consider a population of a individuals consisting at time $t = 0$ of one 'infective' and $a-1$ 'susceptibles'. The process changes only by susceptibles becoming infective. That is, in this very simple model an individual once infective remains so indefinitely; an obvious generalization is to allow for the removal of infectives from the system. The book by Bailey (1957) contains a thorough discussion of mathematical models connected with epidemics.

Suppose that the individuals are at all times homogeneously mixed. If at some time there are i infectives, one natural assumption is to postulate for each susceptible a chance $i\eta \varDelta t + o(\varDelta t)$ of becoming infective in $(t, t + \varDelta t)$, independently for all susceptibles. Since there are $a-i$ susceptibles, the chance of a 'birth' in $(t, t + \varDelta t)$ is $(a-i)i\eta \varDelta t + o(\varDelta t)$, that is in the notation of Example 4.2,

$$\lambda_i = \begin{cases} (a-i)i\eta & (i = 1, \ldots, a), \\ 0 & \text{(otherwise)}. \end{cases}$$

Equations (44) hold. The general solution of them is quite complicated, considering how simple the process is, but the quantity of main interest, the time T to complete the epidemic, can be analysed easily by an immediate generalization of (43).

For the epidemic is complete when the absorbing state is reached in which all individuals are infective. If Z_i is the time spent in state i, exponentially distributed with parameter $\lambda_i = (a-i)i\eta$, and Laplace transform $\lambda_i/(s+\lambda_i)$, we have

$$T = Z_1 + \ldots + Z_{a-1},$$

so that

$$E(e^{-\theta T}) = \prod_{i=1}^{a-1} \frac{(a-i)i\eta}{\theta + (a-i)i\eta}.$$

The cumulant generating function is the logarithm of this, and hence the rth cumulant of T is

$$(r-1)! \sum_{i=1}^{a-1} \left\{ \frac{1}{\eta i(a-i)} \right\}^r.$$

In particular

$$E(T) \sim \frac{2\log a}{\eta a}$$

as $a \to \infty$ with η fixed. The limiting p.d.f. of $\eta aT - 2\log a$ is given in Exercise 13.

Example 4.4. *Linear birth–death process.* Suppose now that the population size changes by births and deaths. The simplest model, which is a reasonable first approximation to the behaviour of some biological organisms, is to assume that in $(t, t + \Delta t)$ each individual alive at t has a chance $\lambda \Delta t + o(\Delta t)$ of giving birth to a new individual and a chance $\mu \Delta t + o(\Delta t)$ of dying, these probabilities being independent of the age of the individual. Events referring to different individuals alive at t are assumed independent.

By the discussion of Section 4.2, this model is equivalent to having intervals between events exponentially distributed with parameter $\lambda + \mu$; the probability that the event is a birth is $\lambda/(\lambda + \mu)$ and that it is a death $\mu/(\lambda + \mu)$. Further, if $N(t)$, the number of individuals at t, equals i, the conditional probabilities of a birth and of a death in $(t, t + \Delta t)$ are $i\lambda \Delta t + o(\Delta t)$ and $i\mu \Delta t + o(\Delta t)$.

In the deterministic theory, the population size at t is $n_0 e^{(\lambda - \mu)t}$. For the stochastic theory, let

$$p_i(t) = \text{prob}\{N(t) = i\},$$

and suppose $N(0) = n_0$. Then, by the argument used previously, we have that for $i = 0, 1, \ldots$,

$$p_i'(t) = -i(\lambda + \mu) p_i(t) + (i - 1) \lambda p_{i-1}(t) + (i + 1) \mu p_{i+1}(t), \qquad (47)$$

$$p_i(0) = \delta_{i n_0}.$$

In (47), with $i = 0$, $p_{-1}(t)$ is identically zero. It is most important to understand why the factors $i, i - 1, i + 1$ enter the right-hand side of (47); for instance, the coefficient of $p_{i-1}(t)$ arises from considering the conditional probability of a birth given that the population size is $i - 1$. If

$$G(z, t) = \sum_{i=0}^{\infty} z^i p_i(t),$$

we have from (47) that

$$\frac{\partial G(z, t)}{\partial t} = (\lambda z - \mu)(z - 1) \frac{\partial G(z, t)}{\partial z}.$$

The auxiliary equations are

$$\frac{dt}{1} = \frac{-dz}{(\lambda z - \mu)(z - 1)} = \frac{dG}{0},$$

from which it follows that the solution satisfying the initial condition is

$$G(z, t) = \left\{ \frac{\mu(1 - z) - (\mu - \lambda z) e^{-(\lambda - \mu)t}}{\lambda(1 - z) - (\mu - \lambda z) e^{-(\lambda - \mu)t}} \right\}^{n_0}. \qquad (48)$$

The zero state, corresponding to extinction, is an absorbing state. The probability that extinction has occurred at or before t is the coefficient of z^0 in (48), i.e. is, when $\lambda \neq \mu$,

$$\left\{\frac{\mu - \mu\, e^{-(\lambda-\mu)t}}{\lambda - \mu\, e^{-(\lambda-\mu)t}}\right\}^{n_0} \tag{49}$$

When $\lambda = \mu$, this is replaced by $\{\lambda t/(1 + \lambda t)\}^{n_0}$. The value of (49) for $n_0 = 1$ is the probability that a single individual and all his progeny are extinct at or before t.

The probability of ultimate extinction of the system is obtained by letting t tend to infinity in (49) and is

$$
\begin{array}{ll}
1 & (\lambda \leqslant \mu), \\
(\mu/\lambda)^{n_0} & (\lambda > \mu).
\end{array} \tag{50}
$$

Note particularly that the probability of ultimate extinction is unity if birth and death rates are exactly equal. If the birth rate exceeds the death rate, the probability of extinction is less than unity but may still be appreciable, particularly when n_0 is small or when μ/λ is near one.

Further properties of the process can be obtained from the probability generating function (48). In particular,

$$E\{N(t)\} = n_0\, e^{(\lambda-\mu)t}, \tag{51}$$

agreeing with the deterministic solution. However, when $\lambda = \mu$, (51) does not give a good idea of the behaviour of individual realizations, extinction being certain.

An interesting alternative approach to (50) is by considering the imbedded process (Example 1.4). Consider the process at $t = 0$ and at the instants just after the first, second, etc., changes of state. We now have a process in discrete time which is a simple random walk starting at n_0, with zero as an absorbing barrier, and with respective probabilities $\lambda/(\lambda + \mu)$ and $\mu/(\lambda + \mu)$ for a single step to be one upwards or one downwards. Making the necessary changes of notation, we have the probability of ultimate absorption from (2.38), in agreement with (50).

Further, if N is the number of steps up to absorption in this random walk in discrete time, the path of the system up to extinction must consist of $\frac{1}{2}(N - n_0)$ births and $\frac{1}{2}(N + n_0)$ deaths. Hence the discussion of equations (2.39) and (2.40) gives the properties of the number of new individuals born into the system before absorption occurs.

Yet another derivation of (50) can be obtained from the theory of the branching process of Section 3.7. In other applications we may be concerned with the attainment of an upper level of population size, say at a. The analysis of the imbedded random walk gives, by Section 2.2, the probability that a is reached before extinction. If, however, we are interested in the distribution of the time T_a at which level a is reached, a more complex analysis is necessary.

To investigate T_a, consider a new process in which the transition probabilities are modified by making a an absorbing state. Equation (47) is unaltered for $i = 0, 1, \ldots, a-2$, and for $i = a-1, a$ the equation is replaced by

$$p'_{a-1}(t) = -(a-1)(\lambda+\mu)\,p_{a-1}(t) + (a-2)\,\lambda p_{a-2}(t),$$
$$p'_a(t) = (a-1)\,\lambda p_{a-1}(t). \tag{52}$$

In fact, however, this is a situation where the backward equations are rather more convenient than the forward equations.

To obtain the backward equations, we introduce the more elaborate notation that

$$p_{jk}(t) = \text{prob}\{N(t) = k | N(0) = j\}, \tag{53}$$

so that $p_i(t)$ in the previous notation now becomes $p_{n_0 i}(t)$. The forward equations were obtained by fixing $N(0) = n_0$ and relating the states of the process at times $0, t, t+\Delta t$. The backward equations are obtained by fixing $N(t) = a$ and relating the states of the process at times $-\Delta t, 0, t$. For the process with an absorbing state at a, we have that

$$p_{ia}(t+\Delta t) = p_{ia}(t)\{1 - i(\lambda+\mu)\,\Delta t\} + p_{i-1,a}(t)\,i\mu\Delta t$$
$$+ p_{i+1,a}(t)\,i\lambda\Delta t + o(\Delta t),$$

leading to

$$p'_{ia}(t) = -i(\lambda+\mu)\,p_{ia}(t) + i\mu p_{i-1,a}(t) + i\lambda p_{i+1,a}(t) \quad (i = 0, 1, \ldots, a-1),$$

with $p_{aa}(t) = 1$, $p_{ia}(0) = 0$ $(i = 0, 1, \ldots, a-1)$. The function $p_{ia}(t)$ is the distribution function of first passage time from state i to the absorbing barrier at a. The solution of these equations is simple for small values of a but the general solution is complicated (Saaty, 1961a).

The linear birth–death process can be generalized in many ways. One which can be handled without essential change is to have a special form of time-dependent birth and death rates of the form $\lambda\rho(t)$ and $\mu\rho(t)$.

An important class of processes, generalized birth–death processes, can be defined by having conditional transition probabilities $\lambda_i\,\Delta t + o(\Delta t)$ and $\mu_i\,\Delta t + o(\Delta t)$ for a birth and for a death in $(t, t+\Delta t)$ given that $N(t) = i$, the Markov property being satisfied. In this general family of processes, successive transitions are always one step up or one step downwards. Some general results are available for these processes.

Four important special forms for the birth and death rates are as follows:

if λ_i is a constant α, we have *immigration* at rate α;

if $\lambda_i = i\lambda$, we have (simple linear) *birth* at rate λ per individual;

if $\mu_i = \beta$ $(i = 1, 2, \ldots)$, $\mu_0 = 0$, we have *emigration* at rate β;

if $\mu_i = i\mu$, we have (simple linear) *death* at rate μ per individual.

The reason for these names should be clear. For example, the distinction between immigration and birth is that the former occurs at a total rate independent of the state of the process, whereas the latter occurs at a certain rate per individual already in the system; there is a similar distinction between emigration and death.

In this classification, the Poisson process (Section 4.1) is a pure immigration process. Example 4.1 is a pure birth process and Example 4.4 a birth–death process. We now consider a few of the further possibilities.

Example 4.5. Immigration–death process. This is often a reasonable model for describing variations in such quantities as the number of dust particles contained in a given small volume, the number of pedestrians in a given section of footpath, and so on. It is assumed that new individuals enter the system at random at a constant rate α, and that each individual currently in the system leaves it at random, at a constant rate μ per individual; thus when i individuals are present the total death rate is $i\mu$. These assumptions would be unreasonable if there was, for example, appreciable 'crowding' in the volume.

If $N(t)$ is the number of individuals present at time t and if $p_i(t)$ denotes prob$\{N(t) = i\}$, we have, exactly as before, the forward equations

$$p_i'(t) = -(\alpha + i\mu)\, p_i(t) + (i+1)\,\mu p_{i+1}(t) + \alpha p_{i-1}(t) \quad (i = 0, 1, \ldots); \quad (54)$$

in the equation for $i = 0$, the function $p_{-1}(t)$ is identically zero.

This can be converted into the following partial differential equation for the probability generating function, $G(z, t)$, of $N(t)$;

$$\frac{\partial G(z, t)}{\partial t} + \mu(z-1)\frac{\partial G(z, t)}{\partial z} = \alpha(z-1)\, G. \quad (55)$$

The auxiliary equations are

$$\frac{dt}{1} = \frac{dz}{\mu(z-1)} = \frac{dG}{\alpha(z-1)\, G}$$

and the general solution is

$$G(z, t) = e^{\alpha z/\mu}\, f\{e^{-\mu t}(z-1)\}, \quad (56)$$

where $f(x)$ is an arbitrary function to be determined from the initial conditions.

If there are initially n_0 individuals present, the solution (56) becomes

$$G(z, t) = \exp\left\{\frac{\alpha}{\mu}(z-1)(1 - e^{-\mu t})\right\}\{1 + (z-1)\, e^{-\mu t}\}^{n_0}. \quad (57)$$

A very important aspect of (57) is that, independently of n_0,

$$\lim_{t \to \infty} G(z, t) = \exp\left\{\frac{\alpha}{\mu}(z-1)\right\}, \quad (58)$$

the probability generating function of a Poisson distribution of mean α/μ. That is, $N(t)$ has a limiting equilibrium distribution, independent of the initial conditions. The general properties of equilibrium distributions will be discussed more fully in the next section.

It is worth noting the interpretation of the stochastic mean α/μ in terms of deterministic theory. Suppose the parameters are such that there are always a large number n of individuals present. New individuals enter at rate α and individuals leave at rate $n\mu$. There is thus equilibrium if and only if $n = \alpha/\mu$. Further if we start with $n_0 \neq \alpha/\mu$ individuals, deterministic theory leads to

$$n(t) = \left(n_0 - \frac{\alpha}{\mu}\right) e^{-\mu t} + \frac{\alpha}{\mu},$$

an exponential approach to equilibrium.

If we assume that in the stochastic model an equilibrium distribution exists, it can be calculated without solving the differential equations. For in (54) we can replace $p_i(t)$ by p_i, independent of t, which must satisfy the equilibrium equations

$$0 = -(a + i\mu)\, p_i + (i + 1)\mu\, p_{i+1} + a p_{i-1} \quad (i = 0, 1, \ldots). \quad (59)$$

Equations (59) assert, in effect, that the rate $(\alpha + i\mu)p_i$ at which probability leaves state i is balanced by the rate at which probability enters state i.

Equations (59) can be solved recursively to express p_1 in terms of p_0, then p_2 in terms of p_0, and so on. The normalizing condition $\sum p_i = 1$ then determines the distribution, which is the Poisson distribution (58). Alternatively (59) can be converted into an ordinary differential equation for the probability generating function, $G(z)$, which can be solved subject to $G(1) = 1$.

One practical application of the non-equilibrium solution (57) is to the estimation of μ from observations on the stochastic process $\{N(t)\}$. It is intuitively clear, for example from the deterministic theory, that if μ is large the values of $N(t)$ at time points quite close together will be nearly independent, for individuals present at one time will have been replaced by a completely new set a little later, the average time an individual spends in the system being $1/\mu$. In general, the values of $N(t)$ and $N(t + h)$ will be positively correlated for values of h of the order of $1/\mu$.

Example 4.6. *Immigration–emigration process.* In the simple immigration–emigration process, individuals arrive in a Poisson process of rate α and leave in a Poisson process of rate β, so long as some individuals are present. This is exactly the simple queueing process of Example 1.4 in which there is a single server, the random arrival rate is α, and the p.d.f. of customer's service-time is exponential with parameter β.

If $p_i(t)$ is the probability that there are i customers present at time t, we obtain as before the forward equations

$$p_0'(t) = -\alpha p_0(t) + \beta p_1(t), \tag{60}$$

$$p_i'(t) = -(\alpha+\beta)\,p_i(t) + \beta p_{i+1}(t) + \alpha p_{i-1}(t) \quad (i = 1, 2, \ldots). \tag{61}$$

The anomalous form of (60) arises because the constant emigration rate β falls to zero when the zero state is occupied. The anomaly makes the solution of (60) and (61) more difficult than the previous processes. We defer this until Section 4.6(iii), where some more advanced methods are discussed.

It is very plausible that an equilibrium distribution for $N(t)$ will exist provided that the immigration rate is less than the emigration rate, $\alpha < \beta$. Thus, writing $p_i(t) = p_i$, independent of t, in (60) and (61), we have that

$$0 = -\alpha p_0 + \beta p_1,$$

$$0 = -(\alpha+\beta)\,p_i + \beta p_{i+1} + \alpha p_{i-1} \quad (i = 1, 2, \ldots).$$

The solution of these equations, subject to the normalizing condition $\sum p_i = 1$, is

$$p_i = (1-\rho)\rho^i, \tag{62}$$

where $\rho = \alpha/\beta$ is in queueing problems called the *traffic intensity*. Thus the distribution of the number in the system is geometric with mean $\rho/(1-\rho)$. Equation (62) defines a probability distribution if and only if $0 \leqslant \rho < 1$; thus an equilibrium distribution is impossible if $\rho \geqslant 1$. It can be shown rigorously that when $0 \leqslant \rho < 1$ (62) is in fact the unique equilibrium distribution.

A point of special interest is that an equilibrium distribution does not exist when $\rho = 1$, i.e. when there is exact balance between the immigration and emigration rates. This result is in line with results in the theory of random walk (Section 2.2(vi)), because the imbedded process obtained by considering the state immediately after a transition, has a probability $\alpha/(\alpha+\beta)$ for a jump one up, and a probability $\beta/(\alpha+\beta)$ for a jump one down, except that when the zero state is occupied the next step must be upwards. The imbedded process is thus a simple random walk with a reflecting barrier at the origin. When $\rho = 1$, i.e. $\alpha = \beta$, the steps are equally likely to be up or down and it is known (Section 3.8(iii)) that there is then no equilibrium distribution in the random walk. The relation with the imbedded process will be used again later.

Further properties of the process can be deduced from the equilibrium distribution (62). For example, consider the interpretation of the process as a queue. Let customers be served in order of arrival (first come, first served) and let W be the total time spent in the system by an individual arriving a long time after the start of the process; i.e. W is the sum of the service-time and the time, if any, spent queueing. Since arrivals are random, the probability that the customer arrives to find i customers ahead of him is given by (62). Conditionally on i, the total time W is the sum of $i+1$ random variables independently exponentially distributed with parameter β. Thus the conditional Laplace transform of the p.d.f. of W is $\{\beta/(\beta+\theta)\}^{i+1}$. Therefore the Laplace transform of the marginal p.d.f. of W is

$$\sum_{i=0}^{\infty} \left(\frac{\beta}{\beta+\theta}\right)^{i+1} \rho^i(1-\rho) = \frac{\beta-\alpha}{\beta-\alpha+\theta}.$$

It follows that the p.d.f. of W is exponential with parameter $\beta-\alpha$.

To sum up this section, we have considered a number of important special processes with the Markov property. An essential feature is that the time spent in a particular stay in any state is exponentially distributed. In a natural sense, the processes are generalizations of the Poisson process. We have seen that it is possible, in a routine way, to write down two sets of simultaneous differential equations, the forward equations and the backward equations, each determining the probability distribution of the state occupied at an arbitrary time t. Solution of the equations may, of course, be mathematically difficult, but in suitably simple cases we can determine from the equations the main properties of the stochastic process. In more complex cases, numerical solution of the differential equations may be the best method of analysis.

Further important ideas that have been illustrated in this section are the existence and determination of equilibrium distributions, the use of an imbedded process in discrete time, and the determination of first passage time distributions.

In the remaining sections of this chapter, we develop further the ideas introduced in the present section.

4.4. Equilibrium distributions

In the previous section we gave examples of equilibrium probability distributions. The important properties of these equilibrium distributions correspond to those of equilibrium distributions in discrete time (Section 3.6) and are as follows:

(a) for all initial conditions, the probability $p_i(t)$ that state i is occupied at time t, tends as $t \to \infty$ to a limit p_i and $\{p_i\}$ is a probability distribution;

(b) if the initial state has the probability distribution $\{p_i\}$, then $p_i(t) = p_i$ for all t;

(c) over a long time t_0, the proportion of time spent in any state i converges in probability to p_i as $t_0 \to \infty$;

(d) the equilibrium probabilities, if they exist, satisfy a system of ordinary simultaneous linear equations.

In order to discuss these properties more thoroughly it is useful to consider a very simple process with just two states, in fact the analogue in continuous time of the Markov chain of Section 3.2.

Example 4.7. *The two-state process.* Suppose that two states are labelled 0, 1 and that, given that state 0 is occupied at t there is a probability $\alpha \Delta t + o(\Delta t)$ of a transition to state 1 in $(t, t + \Delta t)$, independently of all occurrences before t; if state 1 is occupied we assume similarly a constant rate β for transitions to state 0. An alternative specification of the process is in terms of two sequences of mutually independent random variables $\{U_1, U_2, \ldots\}$ and $\{U_1', U_2', \ldots\}$, exponentially distributed with parameters equal respectively to α and β. If, for example, the process starts in state 0, there is a transition to state 1 at time U_1, a transition back to 0 after a further time U_1', and so on; see Fig. 4.3.

One concrete interpretation of the process is in terms of a single telephone line which is either free (state 0) or busy (state 1), it being assumed that calls arising when the line is busy are lost and do not queue. If calls arise at random at rate α and the duration of calls is exponentially distributed with parameter β, the above process is obtained.

Let $p_i(t)$ $(i = 0, 1)$ give the probability distribution at time t and let $p_0(0)$ and $p_1(0) = 1 - p_0(0)$ be specified. The (forward) equations for the process are

$$p_0'(t) = -\alpha p_0(t) + \beta p_1(t),$$
$$p_1'(t) = -\beta p_1(t) + \alpha p_0(t). \tag{63}$$

The solution satisfying the initial conditions is

$$p_0(t) = \frac{\beta}{\alpha + \beta} + \left\{ p_0(0) - \frac{\beta}{\alpha + \beta} \right\} e^{-(\alpha + \beta)t},$$
$$p_1(t) = \frac{\alpha}{\alpha + \beta} + \left\{ p_1(0) - \frac{\alpha}{\alpha + \beta} \right\} e^{-(\alpha + \beta)t}. \tag{64}$$

From this simple explicit solution, we can now discuss the general properties (a)–(d) of equilibrium distributions. First, as $t \to \infty$, it is clear that, independently of the initial conditions, the probability distribution tends to

$$p_0 = \beta/(\alpha + \beta), \qquad p_1 = \alpha/(\alpha + \beta). \tag{65}$$

Thus if we have two processes, one starting from 0 and one from 1, then after a time t large compared with $1/(\alpha + \beta)$, both processes have the same probability, namely $\beta/(\alpha+\beta)$, of being in state 0. The independence of the initial condition is extremely plausible; so many transitions occur over a long time interval that the states occupied at distant times become independent.

For the second property, note that in (64), the time-dependent terms vanish if and only if $p_i(0) = p_i$. That is (65) is the unique stationary distribution, in the sense that if the process starts with this distribution, then at all future times the distribution is the same.

Property (a) gives us one empirical interpretation of the equilibrium distribution. For this we consider a large number of independent realizations and count the proportion which, at some fixed t remote from

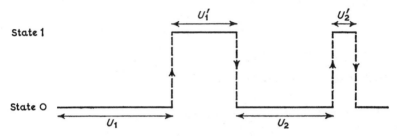

Fig. 4.3. The two-state process.

the time origin, are in state 1. This proportion will be nearly p_1, independently of the initial conditions. This interpretation refers to an average over a large number of realizations taken at a single time. An alternative, and often more important, interpretation is as follows. For a single long realization the limiting proportion of time spent in state 1 is, with probability one, p_1: This is of especial importance when we wish to examine in practice the behaviour of a system over a long time period.

We sketch two proofs of this result. First, suppose for simplicity that the system starts in state 0 and consider the process after $2m$ transitions occur. It will then have spent a time $U_1 + \ldots + U_m$ in state 0 and a time $U_1' + \ldots + U_m'$ in state 1. Thus the proportion of time spent in state 1 up to the $2m$th transition is

$$R_1^{(m)} = \frac{\overline{U}'}{\overline{U} + \overline{U}'},$$

where $\overline{U}$, $\overline{U}'$ are sample means each of m independent random variables. By the strong law of large numbers (Feller, 1957, p. 243; Parzen, 1960, p. 417)

$$\text{prob}\{\lim_{m \to \infty} \overline{U} = 1/\alpha\} = \text{prob}\{\lim_{m \to \infty} \overline{U}' = 1/\beta\} = 1$$

and this implies that

$$\text{prob}\left\{\lim_{m \to \infty} R_1^{(m)} = \frac{\alpha}{\alpha+\beta} = p_1\right\} = 1. \tag{66}$$

This proves that p_1 has, almost certainly, an interpretation as the limit in a single realization of the proportion of time spent in state 1, this limit being evaluated over that particular set of time points in the history of the process at which transitions occur. It is not difficult to show that the same limit holds taken over all time points.

The second proof requires more calculation but is in a sense more elementary, since it does not use the strong law of large numbers; also it enables the rapidity of approach to the limit to be assessed. It gives a result in terms of convergence in probability rather than almost certain convergence. Let $X(t)$ denote the state occupied at t. The proportion of the interval $(0, t_0)$ spent in state 1 is

$$R_1(t_0) = \frac{1}{t_0} \int_0^{t_0} X(t)\, dt, \tag{67}$$

where the right-hand side can be regarded as the limit as $\varDelta \to 0$ of a time average taken at the points $(0, \varDelta, 2\varDelta, \ldots)$. Now

$$E\{R_1(t_0)\} = \frac{1}{t_0} \int_0^{t_0} E\{X(t)\}\, dt$$

$$= \frac{1}{t_0} \int_0^{t_0} \mathrm{prob}\{X(t) = 1\}\, dt$$

$$= \frac{\alpha}{\alpha + \beta} + \left\{ p_1(0) - \frac{\alpha}{\alpha + \beta} \right\} \left\{ \frac{1 - e^{-(\alpha + \beta)t_0}}{(\alpha + \beta)\, t_0} \right\}, \tag{68}$$

from (64). This tends to $p_1 = \alpha/(\alpha + \beta)$ as $t_0 \to \infty$. Incidentally, note that the expected total 'excess' time in state 1 associated with the departure of the initial condition from the equilibrium distribution is

$$\frac{1}{(\alpha + \beta)} \left\{ p_1(0) - \frac{\alpha}{\alpha + \beta} \right\}.$$

To prove that $R_1(t_0)$ converges in probability to p_1 as t_0 tends to infinity, it is therefore sufficient to show that $V\{R_1(t_0)\}$ tends to zero as t_0 increases. Now

$$V\{R_1(t_0)\} = \frac{1}{t_0^2} \int_0^{t_0} \int_0^{t_0} C\{X(t), X(u)\}\, dt\, du. \tag{69}$$

Further if $t \leqslant u$

$$C\{X(t), X(u)\} = \mathrm{prob}\{X(t) = X(u) = 1\} - p_1(t) p_1(u)$$

$$= \mathrm{prob}\{X(u) = 1 | X(t) = 1\} p_1(t) - p_1(t) p_1(u). \tag{70}$$

The conditional probability in (70) is obtained from (64) on taking

$p_1(0) = 1$, and replacing t by $u-t$. It is enough for the present purpose to note that, for $t \leqslant u$,

$$C\{X(t), X(u)\} = a\,e^{-(\alpha+\beta)t} + b\,e^{-(\alpha+\beta)u} + c\,e^{-(\alpha+\beta)(u-t)}, \tag{71}$$

where a, b, c are constants. If the distribution at $t = 0$ is the equilibrium distribution, $a = b = 0$. On substituting (71) into (69) it follows that

$$V\{R_1(t_0)\} = O(t_0^{-1})$$

and hence can be made arbitrarily small by choice of t_0. An application of Chebychev's inequality completes the proof that $R_1(t_0)$ converges in probability to p_1. In fact $R_1(t_0)$ is asymptotically normally distributed around (68) so that, if required, the magnitude of the deviations of $R_1(t_0)$ from p_1 that would be likely to occur can be assessed.

The final point in the discussion of this example is that if the equilibrium distribution is assumed to exist, it must give a solution of the two differential equations (63) independent of t. That is

$$0 = -\alpha p_0 + \beta p_1, \qquad 0 = -\beta p_1 + \alpha p_0. \tag{72}$$

The two equations are, of course, trivially equivalent; this would not happen in general. Both equations express a balance between the rate at which probability leaves and enters a particular state. If we add the normalizing condition

$$1 = p_0 + p_1 \tag{73}$$

we obtain immediately the equilibrium distribution (65). In dealing with more complex problems, like Example 4.6, it is most helpful to be able to calculate the equilibrium distribution without having to obtain the full time-dependent solution first.

We return later to the general theory of equilibrium distributions; the essential ideas have, however, all appeared in Example 4.7. We give now some more examples of their calculation.

Example 4.8. *Generalized birth–death process.* Consider the process of Section 4.3 in which the only transitions are one up or one down, with respective transition rates λ_i, μ_i from the ith state. If an equilibrium distribution $\{p_i\}$ exists, it must satisfy the equations

$$0 = -(\lambda_i + \mu_i)\,p_i + \lambda_{i-1}\,p_{i-1} + \mu_{i+1}\,p_{i+1} \quad (i = 0, 1, \ldots). \tag{74}$$

Equation (74) has the same physical interpretation as other equilibrium equations. We add the normalizing condition

$$1 = \sum_{i=0}^{\infty} p_i.$$

Because negative states cannot occur in a birth–death process, $\mu_0 = 0$, $p_{-1} = 0$, and therefore (74) with $i = 0$ gives

$$0 = -\lambda_0 p_0 + \mu_1 p_1,$$

that is $p_1 = (\lambda_0/\mu_1)\, p_0$. The equation with $i = 1$ gives

$$\mu_2 p_2 = (\lambda_1 + \mu_1) p_1 - \lambda_0 p_0$$

$$= \frac{\lambda_1 \lambda_0}{\mu_1} p_0,$$

that is $p_2 = \{(\lambda_1 \lambda_0)/(\mu_2 \mu_1)\}\, p_0$. In general

$$p_i = \frac{\lambda_{i-1} \dots \lambda_0}{\mu_i \dots \mu_1} p_0. \tag{75}$$

The normalizing condition now gives

$$p_0 = 1/S, \tag{76}$$

where

$$S = \sum_{i=0}^{\infty} \frac{\lambda_{i-1} \dots \lambda_0}{\mu_i \dots \mu_1}, \tag{77}$$

the term in S corresponding to $i = 0$ being one.

Equations (75)–(77) determine the equilibrium distribution; they define a distribution if and only if S is finite. Thus if (77) is divergent, an equilibrium distribution cannot exist. Of course, if only a finite number of states can be occupied, i.e. if $\lambda_a = 0$ for some a greater than the initial level, S is necessarily finite.

There are a number of special cases of interest.

First if $\lambda_i = \alpha$, $\mu_i = i\mu$ we have the immigration–death process of Example 4.5. The series S is always convergent and the solution (75) is easily shown to be a Poisson distribution of mean α/μ.

Secondly if $\lambda_i = \alpha$, $\mu_i = \beta$ $(i = 1, 2, \dots)$, we have the immigration–emigration process of Example 4.6. The series S is a geometric progression with common ratio $\rho = \alpha/\beta$, so that the condition for the existence of an equilibrium distribution is $\rho < 1$ and the solution (62) is easily recovered.

Next suppose that we have the same conditions as in Example 4.6, except that there are k servers. With arrivals forming a Poisson process of rate α, and an exponential distribution of service-time of parameter σ, the process has $\lambda_i = \alpha$, $\mu_0 = 0$, $\mu_1 = \sigma$, $\mu_2 = 2\sigma$, ..., $\mu_k = k\sigma$, $\mu_i = k\sigma$ $(i = k+1, \dots)$. The point here is that so long as the number of customers present does not exceed k, all can be served and we have effectively linear death $(\mu_i = i\sigma)$; when there are k or more customers present only k can be served at a time and we have emigration, with a constant μ_i.

It is natural to define the traffic intensity ρ as the ratio of the arrival rate to the maximum service rate $k\sigma$, that is $\rho = \alpha/(k\sigma)$. Then

$$S = 1 + k\rho + \frac{(k\rho)^2}{2!} + \ldots + \frac{(k\rho)^{k-1}}{(k-1)!} + \sum_{i=k}^{\infty} \frac{k^{k-1}}{(k-1)!} \rho^i.$$

Thus the condition for convergence is $\rho < 1$; in this case

$$S = 1 + k\rho + \frac{(k\rho)^2}{2!} + \ldots + \frac{(k\rho)^{k-1}}{(k-1)!} + \frac{(k\rho)^k}{k!(1-\rho)}$$

and

$$p_i = S^{-1} \frac{(k\rho)^i}{i!} \quad (i < k),$$

$$p_i = S^{-1} \frac{k^k}{k!} \rho^i \quad (i \geqslant k).$$

For numerical work, this is most easily expressed in terms of the Poisson distribution of mean $k\rho$. When $k = 2$ the mean queue size is $2\rho/(1 - \rho^2)$.

As a final example, consider a telephone exchange with k lines in which calls arriving when all lines are occupied are lost. That is, suppose that there are k servers and no facilities for customers to queue. Then consider the same distributional assumptions as before, $\lambda_i = \alpha$ ($i = 0, \ldots, k$), $\lambda_i = 0$ ($i = k+1, \ldots$), $\mu_i = i\sigma$ ($i = 0, \ldots, k$). The series S, being finite, is always convergent, and the solution (75) can be written

$$p_i = \frac{e^{-\alpha/\sigma} \dfrac{(\alpha/\sigma)^i}{i!}}{\displaystyle\sum_{i=0}^{k} e^{-\alpha/\sigma} \dfrac{(\alpha/\sigma)^j}{j!}} \quad (i = 0, \ldots, k). \tag{78}$$

Thus the number of lines occupied has a truncated Poisson distribution. The probability that all lines are occupied is p_k and, since arrivals are random, this is the probability that an arriving call finds the system full. That is, p_k is the probability that a call is lost. Equation (78) with $i = k$ is often called Erlang's loss formula.

In all the examples so far, transitions are one step at a time. Although this makes the calculations simpler the general argument does not depend on this restriction. The following is an example where multiple steps upwards are possible.

Example 4.9. *Queue with bulk arrivals.* Consider again the simple single-server queueing process of Example 4.6, but suppose that arrivals occur in batches of fixed size b. That is, at each occurrence in a Poisson process of rate α, b customers arrive simultaneously. Of course a more general process would have batch size as a random variable.

If p_i is the equilibrium probability that there are i customers in the system, we have by the previous arguments that

$$0 = -\alpha p_0 + \beta p_1,$$
$$0 = -(\alpha+\beta)\,p_i + \beta p_{i+1} + \alpha p_{i-b} \quad (i = 1, 2, \ldots), \tag{79}$$

with the understanding that any p_i with a negative suffix is zero. To solve, put

$$G(z) = \sum_{i=0}^{\infty} z^i p_i,$$

multiply (79) by z^0, z^1, ... and add. Then we have that

$$G(z) = \frac{\beta p_0(1-z)}{\beta + \alpha z^{b+1} - (\alpha+\beta)\,z}. \tag{80}$$

This in effect expresses all the unknown probabilities in terms of p_0. We now use the normalizing condition $G(1) = 1$, evaluating the right-hand side of (80) by the method of indeterminate forms. The result is that $p_0 = 1 - \rho$, where $\rho = b\alpha/\beta$ is the traffic intensity. Thus, finally,

$$G(z) = \frac{(1-\rho)\,(1-z)}{\left(1 + \dfrac{\rho}{b}\,z^{b+1}\right) - \left(1 + \dfrac{\rho}{b}\right)z}. \tag{81}$$

This gives the equilibrium properties of the number of customers in the system. If customers are served in order of arrival, the argument of Example 4.6 can be used to find the distribution of the total time spent in the system, although the calculation is rather more complicated (see Exercise 20).

4.5. General formulation

(i) INTRODUCTION

In the previous sections of this chapter we have illustrated important methods for dealing with Markov processes with discrete states in continuous time. The discussion has been almost entirely in terms of particular processes. The development of a general theory parallel to that for Markov chains given in Chapter 3 is beyond the scope of this book. This is primarily because in continuous time there are a number of types of pathological behaviour that cannot arise in discrete time. Therefore we shall in this section only outline the general theory. Chung (1960, part II) has given a very thorough account of the theory.

(ii) INSTANTANEOUS TRANSITION RATES

Let $X(t)$ denote the state occupied at time t. A Markov process with discrete states in continuous time is in principle defined when we have a set of functions $p_{ij}(t)$ with the interpretation

$$p_{ij}(t) = \text{prob}\{X(t+u) = j \mid X(u) = i\},$$

where we restrict attention for the moment to time-homogeneous processes for which the probability is independent of u. The functions must satisfy

$$0 \leqslant p_{ij}(t) \leqslant 1, \tag{82}$$

$$p_{ik}(t) = \sum_j p_{ij}(v)\, p_{jk}(t-v) \quad (t > v), \tag{83}$$

$$\sum_j p_{ij}(t) \leqslant 1. \tag{84}$$

Equation (83) is the Chapman–Kolmogorov equation for a time-homogeneous Markov process; see Section 1.3. If in equation (84) we have equality for all i and t, the process is called *honest*, whereas if there is inequality for some i and t, the process is *dishonest* and has a non-zero probability of 'escaping to infinity'. In Example 4.2 we have discussed a particular case of a dishonest process.

In matrix form (83) can be written

$$\mathbf{P}(t) = \mathbf{P}(v)\,\mathbf{P}(t-v),$$

or more neatly

$$\mathbf{P}(x+y) = \mathbf{P}(x)\,\mathbf{P}(y) \quad (x, y > 0),$$

with

$$\mathbf{P}(0) = \mathbf{I}.$$

This means that the family of matrices $\mathbf{P}(x)$, defined for $x \geqslant 0$, forms a semi-group. Although we shall not use this fact here, it can be made the basis of a general treatment; see, for example, Reuter (1957).

In applications, we normally start from a knowledge of the transition rates over short time intervals. We therefore now assume that as $\Delta t \to 0$,

$$
\begin{aligned}
p_{ij}(\Delta t) &= q_{ij}\,\Delta t + o(\Delta t) \quad (i \neq j), \\
p_{ii}(\Delta t) &= 1 + q_{ii}\,\Delta t + o(\Delta t),
\end{aligned}
\tag{85}
$$

where the q_{ij}'s are constants. The conditions under which these limits exist are examined in the general theory, but will not be discussed here.

If $q_{ii} = 0$, the state i is called *absorbing*. The first major awkward possibility is that there are processes for which $q_{ii} = -\infty$; the state i is then called *instantaneous*. There are even processes for which all states are instantaneous. However, these possibilities, while of much mathematical interest, do not arise in applications and henceforth we suppose that all q_{ii}'s are finite.

Now, given that state i is occupied at t, suppose that one of the states must be occupied at $t + \Delta t$. Then we have

$$q_{ii} + \sum_{j \neq i} q_{ij} = 0. \tag{86}$$

However, it is possible to have Markov processes which can 'disappear instantaneously to infinity' and for these equality in (86) is replaced by less than or equals. Here, however, we shall consider only processes for which (86) is true for all i; such processes are called *conservative*.

In applications we usually start from the matrix $\mathbf{Q}$ of q_{ij}'s and the question naturally arises as to whether a given $\mathbf{Q}$ can correspond to a Markov process and, if it can, whether the process is uniquely defined. Even if we restrict attention to matrices satisfying (86), the answer is complicated, and will not be given here. For honest processes, the q_{ij}'s do, however, define the process uniquely.

The matrix $\mathbf{Q}$ is, of course, finite or infinite depending on the number of states. Thus for the linear birth–death process of Example 4.4, $\mathbf{Q}$ is the infinite matrix

$$\begin{bmatrix} 0 & 0 & 0 & 0 & . \\ \mu & -(\mu+\lambda) & \lambda & 0 & . \\ 0 & 2\mu & -2(\mu+\lambda) & 2\lambda & . \\ . & . & . & . & . \\ . & . & . & . & . \end{bmatrix}.$$

This is a conservative process, the row sums (86) all vanishing.

(*iii*) THE FORWARD AND BACKWARD EQUATIONS

Suppose now that initially state i is occupied, $X(0) = i$, and let

$$p_{ij}(t) = \text{prob}\{X(t) = j | X(0) = i\}.$$

The forward equations are now obtained by following the argument that has been used repeatedly in special cases. Thus, for $\Delta t > 0$,

$$p_{ik}(t + \Delta t) = p_{ik}(t)\{1 + q_{kk}\Delta t\} + \sum_{j \neq k} p_{ij}(t) q_{jk} \Delta t + o(\Delta t),$$

leading to

$$p'_{ik}(t) = \sum_j p_{ij}(t) q_{jk}.$$

If $\mathbf{p}_i(t)$ denotes the *row* vector of probabilities at time t corresponding to the initial state i, it follows that

$$\mathbf{p}'_i(t) = \mathbf{p}_i(t)\mathbf{Q}.$$

If, more generally, we define a matrix $P(t)$, having $p_{ij}(t)$ as its (i,j)th element, then

$$\mathbf{P}'(t) = \mathbf{P}(t)\mathbf{Q}. \tag{87}$$

The initial condition for (87) is

$$\mathbf{P}(0) = \mathbf{I},$$

where $\mathbf{I}$ is the unit diagonal matrix.

If the initial distribution at $t = 0$ is given by the row vector $\mathbf{p}^0$ the distribution at t is $\mathbf{p}^0 \mathbf{P}(t)$.

Now consider the backward equations. A consequence of the constancy in time of $\mathbf{Q}$ is that

$$\text{prob}\{X(t) = j | X(-\varDelta t) = i\}$$
$$= \text{prob}\{X(t+\varDelta t) = j | X(0) = i\}.$$

Now

$$\text{prob}\{X(t) = j | X(-\varDelta t) = i\}$$
$$= \text{prob}\{X(t) = j | X(0) = i\}\text{prob}\{X(0) = i | X(-\varDelta t) = i\}$$
$$+ \sum_{k \neq i} \text{prob}\{X(t) = j | X(0) = k\}\text{prob}\{X(0) = k | X(-\varDelta t) = i\}.$$

Hence

$$p_{ij}(t+\varDelta t) = p_{ij}(t)(1+q_{ii}\,\varDelta t) + \sum_{k \neq i} p_{kj}(t)\,q_{ik}\,\varDelta t + o(\varDelta t).$$

Thus

$$p'_{ij}(t) = \sum_k q_{ik}p_{kj}(t). \tag{88}$$

The matrix version of (88), analogous to (87) is

$$\mathbf{P}'(t) = \mathbf{Q}\mathbf{P}(t). \tag{89}$$

We have derived (87) and (89) for processes for which the matrix $\mathbf{Q}$ is constant. There is no difficulty in extending the argument to processes for which the transition rates are functions of time, that is for which q_{ij}, now denoted $q_{ij}(t)$, and hence $\mathbf{Q}$, now denoted $\mathbf{Q}(t)$, are functions of t. We have a family of matrices $\mathbf{P}(t,u)$, for $u > t$, with elements

$$\text{prob}\{X(u) = j | X(t) = i\}.$$

We can no longer write $\mathbf{P}(t,u) = \mathbf{P}(u-t)$. The forward equations become

$$\frac{\partial \mathbf{P}(t,u)}{\partial u} = \mathbf{P}(t,u)\,\mathbf{Q}(u) \tag{90}$$

and the backward equations become

$$-\frac{\partial \mathbf{P}(t,u)}{\partial t} = \mathbf{Q}(t)\,\mathbf{P}(t,u). \tag{91}$$

Equations (90) and (91) are called the Kolmogorov differential equations.

In the rest of this section we shall, however, deal only with processes with constant $\mathbf{Q}$. The equations

$$\mathbf{P}'(t) = \mathbf{P}(t)\,\mathbf{Q}, \qquad \mathbf{P}'(t) = \mathbf{Q}\mathbf{P}(t),$$

with initial condition $\mathbf{P}(0) = \mathbf{I}$ have the formal solution

$$\mathbf{P}(t) = \exp(\mathbf{Q}t) \tag{92}$$

$$= \sum_{r=0}^{\infty} \mathbf{Q}^r \frac{t^r}{r!}, \tag{93}$$

where (93), with $\mathbf{Q}^0 = \mathbf{I}$, is to be regarded as the definition of $\exp(\mathbf{Q}t)$.

When $\mathbf{Q}$ is a finite matrix, that is when the number of states of the process is finite, the series (93) is convergent and (92) is the unique solution of both forward and backward equations. When $\mathbf{Q}$ is an infinite matrix, however, (93) may not converge and a careful treatment of processes with an infinite number of states is necessary.

(*iv*) SOLUTION OF THE DIFFERENTIAL EQUATIONS

We can now consider the forward and backward systems of differential equations taken with the initial condition

$$p_{ij}(0) = \delta_{ij},$$

where δ_{ij} is the Kronecker delta symbol and can ask

- (a) whether the forward equations separately and the backward equations separately have unique solutions;
- (b) what is the connexion between the solutions of the two sets of equations;
- (c) whether the solution, if unique, defines the probability distribution of a Markov process, i.e. satisfies (82)–(84).

If the number of states is finite and the matrix $\mathbf{Q}$ conservative, there are no difficulties. Both sets of equations have the same unique solution which defines an honest Markov process. The theory when the number of states is infinite and the matrix $\mathbf{Q}$ conservative is set out very clearly in elementary terms by Reuter and Ledermann (1953). They make a careful examination of the limiting behaviour of the solution for a process with a finite number r of states as $r \to \infty$.

We refer to Reuter and Ledermann's paper for the detailed answers to (a)–(c). Briefly, however, there is a unique so-called *minimal* solution $\{f_{ij}(t)\}$ of both sets of differential equations, such that any other solution $\{p_{ij}(t)\}$ satisfies for all i, j and t,

$$p_{ij}(t) \geqslant f_{ij}(t) \geqslant 0.$$

If the minimal solution is such that for all i and t,

$$\sum_j f_{ij}(t) = 1,$$

and so corresponds to an honest process, this is the unique solution of both sets of equations. The other possibility is that for some i and t

$$\sum_j f_{ij}(t) < 1,$$

corresponding to a dishonest process. Then the situation is more complicated. In certain simple cases, however, as for example for the generalized birth process of Example 4.2, the forward equations have the unique solution $\{f_{ij}(t)\}$ representing a dishonest process, but the backward equations have many solutions.

A final general point is that a dishonest process can be converted into an honest process by arranging that when infinity is reached, the process returns to a finite state, say to state i with probability h_i and starts afresh from there. Each distribution $\{h_i\}$ defines a new Markov process with the same instantaneous transition rates $\mathbf{Q}$, showing that if the solution $\{f_{ij}(t)\}$ is dishonest there are infinitely many Markov processes with the same $\mathbf{Q}$. All satisfy the same backward equations but not the same forward equations. Of course this reconstruction changes the physical character of the process.

(v) FINITE NUMBER OF STATES

We now sketch a general treatment in continuous time for processes with a finite number r of states. For a fixed initial state i, the probabilities $p_{ij}(t)$ $(j = 1,\ldots,r)$ satisfy a set of r simultaneous linear differential equations with constant coefficients, determined by $\mathbf{Q}$. It is well known (Bellman, 1960, p. 192) that the solution of such equations is given by linear combinations of exponential terms, the coefficients of t in the exponents being the eigenvalues of $\mathbf{Q}$. There are minor complications if $\mathbf{Q}$ has multiple eigenvalues. We know already that $\mathbf{Q}$ has a zero eigenvalue and it is extremely plausible that the remaining eigenvalues have negative real parts. We then have an equilibrium distribution determined, as a limit as t tends to infinity, from the contribution of the zero eigenvalue plus a transient part dying away exponentially as t increases.

More formally, we introduce the spectral resolution of $\mathbf{Q}$, namely

$$\mathbf{Q} = \mathbf{B}\,\mathrm{diag}\,(\lambda_1,\ldots,\lambda_r)\,\mathbf{C}', \tag{94}$$

where $\lambda_1, \ldots, \lambda_r$ are the eigenvalues of $\mathbf{Q}$ and $\lambda_1 = 0$, and

$$\mathbf{BC}' = \mathbf{I}.$$

The matrices $\mathbf{B}$ and $\mathbf{C}'$ are formed respectively from the right and left eigenvalues of $\mathbf{Q}$. The representation (94) requires that the eigenvectors are distinct, which we assume for simplicity. Then, from (93), we have

$$\mathbf{P}(t) = \mathbf{B} \operatorname{diag}(e^{\lambda_1 t}, \ldots, e^{\lambda_r t}) \mathbf{C}'.$$ (95)

We assume that all states inter-communicate. Thus $\mathbf{Q} - \operatorname{diag}(q_{ii})$ is irreducible, for if we consider the imbedded Markov chain obtained by observing the process immediately after each change of state, the matrix of this chain is irreducible and has non-zero elements in the same off-diagonal positions as those of $\mathbf{Q}$.

To study the eigenvalues we convert $\mathbf{Q}$ into a matrix with non-negative elements by adding to all diagonal elements $m = -\min(q_{ii}) > 0$. Then the matrix $\mathbf{Q} + m\mathbf{I}$ is *a fortiori* irreducible and has non-negative elements. The theorem of Perron and Frobenius, Section 3.10(ii), therefore shows that the equation in μ,

$$|\mathbf{Q} + m\mathbf{I} - \mu\mathbf{I}| = 0,$$ (96)

has a maximal positive solution μ_1 with an associated eigenvector of positive elements. If $\mu_1, \ldots, \mu_r$ are the roots then the eigenvalues of $\mathbf{Q}$ are $\lambda_i = \mu_i - m$. It is easily shown that the zero eigenvalue of $\mathbf{Q}$, $\lambda_1 = 0$, corresponds to the maximal root $\mu_1 = m$ and that all other μ_i have real parts less than m. That is λ_i $(i = 2, \ldots, r)$ have negative real parts.

Thus in (95) the diagonal matrix tends to $\operatorname{diag}(1, 0, \ldots, 0)$, the rate of approach to the limit being exponential at a rate governed asymptotically by the largest non-zero eigenvalue of $\mathbf{Q}$. Further the limiting form of $P(t)$ is

$$\mathbf{bc}' = \begin{pmatrix} b_{11} \\ \vdots \\ b_{r1} \end{pmatrix} (c_{11} \ldots c_{1r}),$$ (97)

where $\mathbf{b}$, the first column of $\mathbf{B}$, and $\mathbf{c}'$, the first row of $\mathbf{C}'$, are respectively right and left eigenvectors of $\mathbf{Q}$ corresponding to $\lambda_1 = 0$. Thus $\mathbf{b}$ satisfies $\mathbf{Qb} = 0$ and we may therefore take $\mathbf{b}' = (1, \ldots, 1)$ since the row sums of $\mathbf{Q}$ vanish.

The row $\mathbf{c}'$ is a suitably normalized solution of the equation in $\mathbf{p}$,

$$\mathbf{pQ} = 0,$$ (98)

and since $\mathbf{C}'\mathbf{B} = \mathbf{I}$, we have in fact $\sum c_{1i} = 1$. Further, as noted above, the theorem of Perron and Frobenius ensures that the c_{1i} are all positive and unique.

Corresponding to an initial probability row vector $\mathbf{p}(0)$ there is a limiting distribution

$$\mathbf{p}(0) \begin{pmatrix} 1 \\ \vdots \\ 1 \end{pmatrix} (c_{11} \ldots c_{1r}) = \left\{ \sum_i p_i(0) \right\} (c_{11} \ldots c_{1r}) = (c_{11} \ldots c_{1r}).$$

We have thus proved property (a) of equilibrium distributions stated in Section 4.4. For all processes with a finite number of states, mutually inter-communicating, the probability distribution tends for all initial conditions to the unique normalized solution of the equilibrium equations (98). The further properties (b) and (c) of Section 4.4 follow easily. Thus if the initial distribution satisfies (98), we have $\mathbf{pQ}^m = 0$ ($m = 1, 2, \ldots$) and by (93), the same distribution holds for all t.

Finally, we can prove the very important property relating the equilibrium probability p_j to the proportion of a long realization spent in state j. The argument follows very closely that used for Example 4.7, the two-state process. We define

$$X_j(t) = \begin{array}{ll} 1 & (\text{if } X(t) = j), \\ 0 & (\text{otherwise}), \end{array} \tag{99}$$

so that

$$R_j(t_0) = \frac{1}{t_0} \int_0^{t_0} X_j(t)\, dt$$

is the proportion of the time t_0 spent in state j. We then show that

$$E\{R_j(t_0)\} \sim p_j, \qquad V\{R_j(t_0)\} = O(1/t_0).$$

The limit of $E\{R_j(t_0)\}$ follows, as in (68), from the fact that prob$\{X_j(t) = 1\}$ differs from p_j by expressions exponentially small in t. The behaviour of the variance is investigated as in (69)–(71); the only modification is to (71) which has to be generalized to allow for contributions from all non-zero eigenvalues.

To sum up, the rather detailed investigation of the two-state process, Example 4.7, applies with relatively minor changes to a general process with a finite number of states in which all states inter-communicate.

(*vi*) SPECTRAL THEORY FOR INFINITE PROCESSES

The spectral representation used for finite processes in (*v*) is useful both in discussing general theory and in analysing particular processes. For if it is possible to find analytically or numerically the largest eigenvalues and corresponding eigenvectors the behaviour for large t can be approximated.

It would be desirable therefore to develop the spectral theory for processes with an infinite number of states. In general it is difficult to do this in a form that is useful for analysing particular processes. However, Karlin and McGregor (1957a, b; 1958a, b) have given a very elegant account of the spectral theory for the generalized birth–death process of

Example 4.8, i.e. of a process for which, if the possible states are all the non-negative integers,

$$q_{i,i+1} = \lambda_i > 0, \qquad q_{i,i-1} = \mu_i,$$

with $\mu_0 = 0$, $\mu_i > 0 (i = 1, 2, \ldots)$ and for $j \neq i-1, i, i+1$, the q_{ij} are zero. Then Karlin and McGregor introduce a system of polynomials by the recurrence equations

$$Q_0(x) = 1, \qquad -xQ_n(x) = \mu_n Q_{n-1}(x) + \lambda_n Q_{n+1}(x) - (\lambda_n + \mu_n) Q_n(x).$$

They then show that there is a distribution function $\phi(x)$ on $[0, \infty)$ with respect to which the $Q_n(x)$'s are orthogonal polynomials, in fact with

$$\int_0^\infty Q_i(x) Q_j(x) \, d\phi(x) = \frac{\delta_{ij}}{\varpi_j},$$

where

$$\varpi_0 = 1, \qquad \varpi_j = \frac{\lambda_0 \lambda_1 \ldots \lambda_{j-1}}{\mu_1 \mu_2 \ldots \mu_j}.$$

Karlin and McGregor's spectral representation is that

$$p_{ij}(t) = \varpi_j \int_0^\infty e^{-xt} Q_i(x) Q_j(x) \, d\phi(x). \tag{100}$$

The essential simplification here is that the eigenvalues are real. The main difficulty in applying the results is in finding $\phi(x)$.

For processes with an equilibrium distribution, $\phi(x)$ has a discontinuity at $x = 0$ of amount ϕ_0, say, and as $t \to \infty$, $p_{ij}(t)$ tends to $\varpi_j \phi_0$, since $Q_n(0) = 1$ for all n. In fact, $\phi_0 = 1/\sum \varpi_j$. The rate of approach to the equilibrium distribution is determined by the smallest non-zero x making a contribution to (100).

4.6. Some miscellaneous topics

In this section, we deal with a number of unrelated topics arising in applying the theory set out earlier in the chapter. The subjects dealt with are

  (i) a special device for dealing with some problems containing non-exponential distributions;
 (ii) some processes with multidimensional state variables;
(iii) certain function-theoretic problems sometimes arising in solving the differential equations of birth–death and allied processes;
(iv) the calculation of first passage and recurrence time distributions.

(i) THE METHOD OF STAGES

All the processes considered so far involve exponentially distributed random variables, in that the individual times spent in a particular state are independently exponentially distributed. Further, the underlying quantities specifying the process, such as service-times and arrival intervals in a queueing process, are exponentially distributed.

The central difficulty in processes involving non-exponential distributions is that the probability of an event, death, etc., in $(t, t + \Delta t)$ is a function of the time since the start of the life-time in question. This makes the process in a simple form non-Markovian. In Chapter 6 we deal more systematically with non-Markovian processes. Now we describe a special device that often enables the methods of the present chapter to be applied straightforwardly.

Consider, for example, the service-time of a customer in a queueing problem. Suppose that service consists of k stages, the lengths of which are independently exponentially distributed with parameter β. That is, the service-time, Y, has the form $Y = Z_1 + \ldots + Z_k$, where the Z's are independently exponentially distributed. The Laplace transform of the distribution of a Z is $\beta/(\beta + s)$, that for Y is $\beta^k/(\beta + s)^k$, and hence the p.d.f. of Y is of the gamma form with mean k/β, index k, and coefficient of variation $1/\sqrt{k}$.

Now let us argue in reverse. Suppose that the distribution of service-time is of the gamma form with index an integer k and with mean μ, say. Let $\beta = k/\mu$. Then we can proceed as if there are k stages of service, each characterized by an exponential distribution; these states need have no physical significance. If now we include in the specification of the state of the system a description of the stage of service reached, all distributions associated with the process are exponential and the methods of the present chapter are applicable.

This can best be explained by examples.

Example 4.10. *Simple queue with gamma distribution of service-time.* Suppose that we have exactly the model of Example 4.6, except that the exponential distribution of service-time is replaced by a gamma distribution with index k and mean $\mu = k/\beta$. In Example 3.10 we have analysed a more general problem, with an arbitrary distribution of service-time, using an imbedded Markov chain. It is instructive, however, to see how the method of stages can be applied to the problem.

Define the possible states by the number of customers in the system and, if that is non-zero, also by the stage of service of the customer being served. That is, the states are $\{0; (1, 1), \ldots, (1, k); (2, 1), \ldots, (2, k); \ldots\}$. In state (i, j), there are i customers in the queue and the one being served is in the jth stage of service $(i = 1, 2, \ldots; j = 1, 2, \ldots, k)$.

We consider only the equilibrium theory. Let p_0, p_{ij} be the equilibrium probabilities. Then

$$\alpha p_0 = \beta p_{1k},$$

$$(\alpha+\beta)\,p_{11} = \alpha p_0 + \beta p_{2k}, (\alpha+\beta)\,p_{12} = \beta p_{11}, \dots, (\alpha+\beta)\,p_{1k} = \beta p_{1,\,k-1},$$

$$(\alpha+\beta)\,p_{21} = \alpha p_{11} + \beta p_{3k}, (\alpha+\beta)\,p_{22} = \alpha p_{12} + \beta p_{21}, \dots,$$

$$(\alpha+\beta)\,p_{2k} = \alpha p_{1k} + \beta_{2,\,k-1};\dots$$

These equations are most easily dealt with by denoting p_{ij} by p_{ki-j+1}, i.e. by using the number of stages of service still to be dealt with as a one-dimensional state variable.

The equations are then identical with (79) of Example 4.9, the queue with bulk arrivals, with batch size being replaced by k. The equivalence of the two problems can be seen directly on identifying customers in Example 4.9 with stages of service in Example 4.10. The equilibrium distribution is given by (81) with $b = k$; the probability that there is, for example, one customer in the system is the sum of the coefficients of $z, \dots, z^k$.

We can obtain the equilibrium distribution of the total time, W, spent in the system, under the queue discipline first-come first-served, by following the argument of Example 4.6. Conditionally on the customer's arriving to find i stages of service waiting to be completed, W will be the sum of $i+k$ exponentially distributed random variables. The Laplace transform of the distribution of W can thereby be obtained; the result is equivalent to a special case of a formula to be proved by a different method in Section 5.12.

More particularly, the conditional mean of W is $(i+k)/\beta$. Further, the mean value of the number of stages of service, i, is on differentiating (81) once at $z = 1$, $\frac{1}{2}(k+1)\rho/(1-\rho)$. We have that

$$
\begin{aligned}
E(W) &= \frac{1}{\beta}\left\{k + \frac{(k+1)\rho}{2(1-\rho)}\right\} \\
&= \mu\left\{1 + \frac{\rho(1+1/k)}{2(1-\rho)}\right\},
\end{aligned}
\tag{101}
$$

where $\mu = k/\beta$ is the mean service-time, and $\rho = k\alpha/\beta$ is the traffic intensity. It will be shown in (5.171) that if $1/k$ is replaced by the square of the coefficient of variation of service-time, (101) is true for arbitrary distributions of service-time. One important special case, constant service-time, is obtained by letting k tend to infinity. For this

$$E(W) = \frac{\mu(1-\frac{1}{2}\rho)}{(1-\rho)},$$

compared with

$$E(W) = \frac{\mu}{1-\rho}$$

for an exponential distribution of service-time.

We have illustrated above the simplest and most useful application of the method of stages, the representation of a gamma distribution with integral index. An obvious generalization is to allow the exponential distributions associated with the different stages to have different parameters. Another possibility, important in representing distributions with more dispersion than the exponential, is to have two or more stages 'in parallel'. That is, there are exponential distributions with parameters $\beta_1, \ldots, \beta_k$, and a probability distribution $\{\varpi_1, \ldots, \varpi_k\}$. Then the random variable of interest, for example the service-time, has with probability ϖ_i an exponential distribution with parameter β_i. The random variable thus has the p.d.f.

$$\sum_{i=1}^{k} \varpi_i \beta_i e^{-\beta_i x}.$$

A more general discussion of the method of stages will be given in Section 6.2.

(ii) MULTIDIMENSIONAL PROCESSES

No essentially new principles are involved in dealing with a Markov process in which the random variable defining the state is multidimensional. In sufficiently simple cases it will be possible to form a differential equation for the joint probability generating function of the process. The method is illustrated by the following simple process.

Example 4.11. *Birth–death process with two sexes.* The following is a very simplified model for the growth of a population consisting of individuals of two sexes. It is assumed that

(i) any female alive at t has probabilities $\lambda_1 \, \Delta t + o(\Delta t)$, or $\lambda_2 \, \Delta t + o(\Delta t)$ of producing respectively a female or male offspring in $(t, t + \Delta t)$;

(ii) there are probabilities of death in $(t, t + \Delta t)$ per individual of $\mu_1 \, \Delta t + o(\Delta t)$ for females and $\mu_2 \, \Delta t + o(\Delta t)$ for males;

(iii) the sub-populations generated by different individuals are mutually independent, and all probabilities referring to $(t, t + \Delta t)$ depend only on the numbers of females and males alive at t.

It follows directly from the assumptions (i)–(iii) that the number of females, considered as a one-dimensional process, is a linear birth–death process with birth and death rates λ_1 and μ_1. Extinction of the females, which is in this model a necessary and sufficient condition for the extinction of the population as a whole, thus occurs with probability

$$1 \quad (\mu_1 \geqslant \lambda_1),$$

$$(\mu_1/\lambda_1)^{n_1} \quad (\mu_1 < \lambda_1),$$

where n_1 is the initial number of females. The number of males considered separately is not a simple process. The assumption (i) that the birth rates per female are independent of the number of males present is, of course, likely to be a reasonable approximation only in rather special circumstances.

If $p_{ij}(t)$ is the probability that there are i females and j males alive at t, we have the forward equations

$$\begin{aligned}
p'_{ij}(t) = &-(i\mu_1+j\mu_2+i\lambda_1+i\lambda_2)p_{ij}(t) \\
&+(i+1)\mu_1 p_{i+1,j}(t) + (j+1)\mu_2 p_{i,j+1}(t) \\
&+(i-1)\lambda_1 p_{i-1,j}(t) + i\lambda_2 p_{i,j-1}(t).
\end{aligned} \tag{102}$$

It is understood that in (102) any p_{ij} with one or both suffices negative is identically zero. Note especially that the two suffixes define the state occupied at time t. If we wished to introduce the initial state explicitly, as in Section 4.5, two further suffixes would be required.

Equation (102) is easily converted into a partial differential equation for the probability generating function

$$G(z_1,z_2;t) = \sum z_1^i z_2^j p_{ij}(t).$$

We have that

$$\frac{\partial G}{\partial t} = \{-\mu_1 z_1 - (\lambda_1+\lambda_2)z_1 + \mu_1 + \lambda_1 z_1^2 + \lambda_2 z_1 z_2\}\frac{\partial G}{\partial z_1} + \mu_2(1-z_2)\frac{\partial G}{\partial z_2}. \tag{103}$$

To obtain from (103) the means and variances of the numbers of females and males alive at t, we follow the device used in (40) and (41) and write $z_1 = e^{-\theta_1}$, $z_2 = e^{-\theta_2}$,

$$K(\theta_1,\theta_2;t) = \log G(e^{-\theta_1}, e^{-\theta_2};t),$$

so that $K(\theta_1,\theta_2;t)$ is the joint cumulant generating function. Since

$$\frac{\partial}{\partial z_i} = -e^{\theta_i}\frac{\partial}{\partial \theta_i},$$

we have that

$$\frac{\partial K}{\partial t} = (\mu_1+\lambda_1+\lambda_2-\mu_1 e^{\theta_1}-\lambda_1 e^{-\theta_1}-\lambda_2 e^{-\theta_2})\frac{\partial K}{\partial \theta_1} - \mu_2(e^{\theta_2}-1)\frac{\partial K}{\partial \theta_2}.$$

We now expand in powers of θ_1, θ_2 and equate coefficients of corresponding powers. Let $\kappa_{ij}(t)$ denote the mixed cumulant of order (i,j), so that, for instance, $\kappa_{10}(t)$ is the mean number of females, $\kappa_{11}(t)$ is the covariance of the numbers of males and females, etc. Then

$$\kappa'_{10}(t) = (\lambda_1 - \mu_1)\kappa_{10}(t),$$

$$\kappa'_{01}(t) = \lambda_2 \kappa_{10}(t) - \mu_2 \kappa_{01}(t),$$

$$\kappa'_{20}(t) = 2(\lambda_1 - \mu_1)\kappa_{20}(t) + (\lambda_1 - \mu_1)\kappa_{10}(t),$$

$$\kappa'_{11}(t) = \lambda_2 \kappa_{20}(t) + (\lambda_1 - \mu_1 - \mu_2)\kappa_{11}(t),$$

$$\kappa'_{02}(t) = 2\lambda_2 \kappa_{11}(t) - 2\mu_2 \kappa_{02}(t) + \lambda_2 \kappa_{10}(t) - \mu_2 \kappa_{01}(t).$$

These equations are easily solved recursively; the equations referring to the marginal distribution of the number of females are, as noted above, those for a simple birth–death process. The solutions for the means are, if (n_1, n_2) denote the initial numbers,

$$\kappa_{10}(t) = n_1 e^{(\lambda_1 - \mu_1)t},$$

$$\kappa_{01}(t) = \frac{\lambda_2 n_1}{\lambda_1 - \mu_1 + \mu_2} e^{(\lambda_1 - \mu_1)t} + \left(n_2 - \frac{\lambda_2 n_1}{\lambda_1 - \mu_1 + \mu_2}\right) e^{-\mu_2 t}.$$

Thus if $\lambda_1 > \mu_1$, the limiting ratio of the expected numbers of females and males is $(\lambda_1 - \mu_1 + \mu_2)/\lambda_2$.

For further discussion of this problem, see Goodman (1953a) and Joshi (1954).

Example 4.12. *Simple model of bacterial multiplication.* The following illustrates a rather different method for handling an essentially bivariate process, using the backward equations. Suppose that individuals can be in two states I, II. An individual starts in I, and changes to II at a probability rate λ; an individual of type II divides into two new individuals of type I at rate ν. The usual assumptions are made about the independence of different individuals and the Markov character of the process.

Let $G_1(z_1, z_2; t)$ and $G_2(z_1, z_2; t)$ be the joint probability generating functions of the numbers of individuals of the two types alive at t, given that there is initially one individual of, respectively, type I or type II. Now we could form the forward equations; the advantage, however, of the backward equations is that near the time origin the number of individuals is small. To obtain $G_1(z_1, z_2; t + \Delta t)$, we consider a process starting with one type I individual at time $-\Delta t$. This leads

(i) with probability $1 - \lambda \Delta t + o(\Delta t)$, to the process developing from one type I individual at $t = 0$ and hence having generating function $G_1(z_1, z_2; t)$;

(ii) with probability $\lambda \Delta t + o(\Delta t)$, to the process developing from one type II individual at $t = 0$.

Thus

$$G_1(z_1, z_2; t + \Delta t) = G_1(z_1, z_2; t)(1 - \lambda \Delta t) + G_2(z_1, z_2; t) \lambda \Delta t + o(\Delta t).$$

Therefore

$$\frac{\partial G_1(z_1, z_2; t)}{\partial t} = \lambda \{G_2(z_1, z_2; t) - G_1(z_1, z_2; t)\}.$$

Similarly, if we start with one type II individual at $-\Delta t$, then this leads

(a) with probability $1 - \nu \Delta t + o(\Delta t)$, to the process developing from one type II individual at $t = 0$ and hence having generating function $G_2(z_1, z_2; t)$;

(b) with probability $\nu \Delta t + o(\Delta t)$, to the process developing from *two* type I individuals at $t = 0$ and hence having generating function $\{G_1(z_1, z_2; t)\}^2$.

Thus

$$\frac{\partial G_2(z_1, z_2; t)}{\partial t} = \nu [\{G_1(z_1, z_2; t)\}^2 - G_2(z_1, z_2; t)].$$

We shall not deal with the solution of these equations. For example, identical equations hold for the moment generating functions and these may be expanded in series to give equations for the moments.

Multivariate processes arise in essentially three ways:

(a) There may be, as in Example 4.11, individuals of two or more types with statistically separate properties and interacting in some way.

(b) The inclusion of a multidimensional set of states may be required to make the system a Markov process with known transition rates. Thus if in Example 4.11 we were interested in the males only, it would still be necessary to include the females in order to set up the equations of the process.

(c) Even if the basic process is a one-dimensional Markov process, it may be necessary to include further dimensions in the state space in order to study derived properties of the process. For example, in the simple linear birth–death process, Example 4.4, we may be interested not only in the number of individuals alive at t, but also in the total number of individuals born into the system up to t. We would then have a bivariate Markov process.

(iii) SOME FUNCTION-THEORETIC PROBLEMS

We now turn to a mathematical point connected with the solution of the differential equations of the process rather than with the formulation of the equations. It quite often happens that the equations for the process fall into a simple general pattern except for some anomalous equations near to a boundary state. The type of difficulty that this produces can be illustrated by Example 4.6, the immigration–emigration process, or simple queue. The general equation (61) is for $i = 1, 2, \ldots$

$$p_i'(t) = -(\alpha+\beta)\,p_i(t)+\beta p_{i+1}(t)+\alpha p_{i-1}(t), \tag{104}$$

but for $i = 0$ there is the anomalous equation

$$p_0'(t) = -\alpha p_0(t)+\beta p_1(t). \tag{105}$$

Note that (105) is not the special case of (104) obtained by regarding $p_{-1}(t)$ as identically zero. The reason for the anomaly is that emigration occurs at rate β from all states except the zero state.

If now we introduce the generating function

$$G(z,t) = \sum_{i=0}^{\infty} z^i p_i(t),$$

and form a partial differential equation from (104) and (105), we have that

$$\frac{\partial G(z,t)}{\partial t} = \left(-\alpha-\beta+\alpha z+\frac{\beta}{z}\right)G(z,t)+\beta\left(1-\frac{1}{z}\right)p_0(t). \tag{106}$$

The essential difficulty is that (106) involves $p_0(t) = G(0,t)$, which is unknown, so that the solution of (106) has the general character of a boundary value problem.

If we apply a Laplace transformation with respect to t, writing

$$G^*(z,s) = \int_0^\infty G(z,t)\,e^{-st}dt, \qquad p_0^*(s) = \int_0^\infty p_0(t)\,e^{-st}dt,$$

and if the initial condition is that state i_0 is occupied at $t = 0$, $G(z,0) = z^{i_0}$, we have that

$$sG^*(z,s)-z^{i_0} = \left(-\alpha-\beta+\alpha z+\frac{\beta}{z}\right)G^*(z,s)+\beta\left(1-\frac{1}{z}\right)p_0^*(s).$$

Thus

$$G^*(z,s) = \frac{z^{i_0+1}+\beta(z-1)\,p_0^*(s)}{z(s+\alpha+\beta)-\alpha z^2-\beta}. \tag{107}$$

Equation (107) says in effect that if we can find $p_0(t)$, and hence $p_0^*(s)$, the whole generating function is determined. Now one might hope that if we were to put $z = 0$ in (107), and then to use $G^*(0,s) = p_0^*(s)$, the unknown function $p_0^*(s)$ would be determined. But this does not work. We need an essentially new argument to find $p_0^*(s)$.

Such an argument is obtained by examining the analytic nature of (107) (Bailey, 1954). Now since $\sum p_i(t) = 1$, $G(1,t) = 1$, so that $G(z,t)$

must converge in and on the unit circle $|z| = 1$, for all t. Hence, provided $\mathcal{R}(s) \geqslant 0$, $G^*(z,s)$ must converge in and on $|z| = 1$. Therefore if the denominator of (107), considered as a function of z, has zeros inside $|z| = 1$, these must be zeros of the numerator also. The zeros of the denominator are at $z = z_0(s)$, $z_1(s)$, where

$$z_0(s), z_1(s) = \frac{\alpha + \beta + s \mp \{(\alpha + \beta + s)^2 - 4\alpha\beta\}^{\frac{1}{2}}}{2\alpha}, \tag{108}$$

and it is easy to show that $z_0(s)$, taking the minus sign, together with the positive value of the square root, is the single root inside $|z| = 1$. Thus the numerator vanishes for $z = z_0(s)$, i.e.

$$p_0^*(s) = \frac{\{z_0(s)\}^{i_0+1}}{\beta\{1 - z_0(s)\}}. \tag{109}$$

The Laplace transform of the probability generating function follows from (107).

The properties of the process are now in principle uniquely determined, although there is, of course, the non-trivial problem of inverting the Laplace transform. Since we shall in (116) obtain the solution by a different method, the full inversion of $G^*(z,s)$ will not be examined. Note, however, that the Bessel function of imaginary argument, $I_n(at)$, defined by

$$I_n(at) = \sum_{r=0}^{\infty} \frac{(\frac{1}{2}at)^{n+2r}}{r!(n+r)!}$$

has as its Laplace transform

$$\left\{\frac{a}{s + \sqrt{(s^2 - a^2)}}\right\}^n \frac{1}{\sqrt{(s^2 - a^2)}}.$$

It is reasonable to expect the solution to involve combinations of Bessel functions.

To illustrate some simple consequences of these formulae, suppose that initially the system is empty and consider (109) with $i_0 = 0$. Then

$$p_0^*(s) = \frac{(\alpha + \beta + s) - \{(\alpha + \beta + s)^2 - 4\alpha\beta\}^{\frac{1}{2}}}{\beta[\alpha - \beta - s + \{(\alpha + \beta + s)^2 - 4\alpha\beta\}^{\frac{1}{2}}]}.$$

Now as s tends to zero the quantity inside the square root tends to $(\alpha - \beta)^2$ and so the square root tends to $|\alpha - \beta|$. Thus if $\alpha > \beta$,

$$p_0^*(0) = \int_0^{\infty} p_0(t)\, dt = \frac{1}{\alpha - \beta}. \tag{110}$$

When $\alpha > \beta$, the arrival rate exceeds the service rate and the interpretation of (110) is that the expected total time spent in the zero state, given that the zero state is occupied initially is $1/(\alpha - \beta)$. When $\alpha = \beta$,

$$p_0^*(s) \sim \frac{1}{\sqrt{(\alpha s)}} \quad \text{as } s \to 0$$

so that as $t \to \infty$

$$p_0(t) \sim \frac{1}{\sqrt{(\pi \alpha t)}}. \tag{111}$$

Finally if $\alpha < \beta$, it is easily shown that $p_0(t)$ tends to its equilibrium value $1 - \alpha/\beta$.

An alternative approach that can sometimes be used is to regard the general equation, in this case (104), as applying for $i = 0, \pm 1, \pm 2, \ldots$ and then to find a special solution satisfying the anomalous equation. More explicitly, we consider

$$p_i'(t) = -(\alpha + \beta) p_i(t) + \beta p_{i+1}(t) + \alpha p_{i-1}(t) \quad (i = 0, \pm 1, \pm 2, \ldots), \tag{112}$$

with

$$0 = -\beta p_0(t) + \alpha p_{-1}(t). \tag{113}$$

Equation (113) ensures that the solution of (112) satisfies also (105). The initial conditions are

$$p_i(0) = \delta_{ii_0} \quad (i = 0, 1, \ldots),$$

$$p_i(0) = q_i \quad (i = -1, -2, \ldots), \tag{114}$$

where the q_i are arbitrary. Now in terms of the generating function

$$H(z, t) = \sum_{i=-\infty}^{\infty} z^i p_i(t),$$

(112) becomes

$$\frac{\partial H(z, t)}{\partial t} = -(\alpha + \beta) H(z, t) + \frac{\beta}{z} H(z, t) + \alpha z H(z, t),$$

so that

$$H(z, t) = \left(z^{i_0} + \sum_{i=1}^{\infty} q_i z^{-i} \right) \exp \left\{ t \left(-\alpha - \beta + \frac{\beta}{z} + \alpha z \right) \right\}. \tag{115}$$

We must now extract the coefficients of z^0 and z^{-1} from (115) so that the solution can be forced to satisfy (113). For this, note that

$$\exp \left(\frac{\beta t}{z} + \alpha z t \right) = \sum_{n=-\infty}^{\infty} I_n \{ 2t \sqrt{(\alpha \beta)} \} \left(z \sqrt{\frac{\alpha}{\beta}} \right)^n,$$

where $I_n(x)$ is a Bessel function of imaginary argument. Thus, abbreviating $I_n\{2t\sqrt{(\alpha\beta)}\}$ to I_n, writing $\rho = \alpha/\beta$, and recalling that $I_{-n} = I_n$, we have from (115) that

$$-\beta p_0(t) + \alpha p_{-1}(t) = \beta e^{-(\alpha+\beta)t}\left[-\left\{ I_{i_0}\rho^{-\frac{1}{2}i_0} + \sum_{n=1}^{\infty} q_n I_n \rho^{\frac{1}{2}n} \right\} \right.$$

$$\left. + \left\{ I_{i_0+1}\rho^{-\frac{1}{2}i_0+\frac{1}{2}} + \sum_{n=1}^{\infty} q_n I_{n-1}\rho^{\frac{1}{2}n+\frac{1}{2}} \right\} \right].$$

The coefficients q_n are determined uniquely by equating to zero the coefficients of each order of Bessel function and the final solution for $H(z,t)$ becomes

$$H(z,t) = \left\{ z^{i_0} + \rho^{-i_0-1}z^{-i_0-1} + (1-\rho)\sum_{j=i_0+2}^{\infty}(\rho z)^{-j} \right\}$$

$$\times \left\{ \sum_{n=-\infty}^{\infty} I_n\left(z\sqrt{\frac{\alpha}{\beta}}\right)^n \right\} e^{-t(\alpha+\beta)}; \tag{116}$$

from this any required $p_i(t)$ can be expressed as a finite series of Bessel functions.

This form can be obtained also by inverting the Laplace transforms (107) and (109). Appreciable further analysis is necessary to get from (116) insight into the limiting behaviour as t tends to infinity. One approach to this is to obtain a spectral representation (Karlin and McGregor, 1958b).

(iv) FIRST PASSAGE TIME AND RECURRENCE TIME
DISTRIBUTIONS

The main general ideas for dealing with first passage time problems have been dealt with in Chapter 3 in the context of Markov chains and also in connexion with Examples 4.1, 4.3 and 4.4 in the present chapter.

It is possible to develop a qualitative classification, corresponding to that used for Markov chains, depending on whether return to say the ith state has probability one or less than one, and, in the former case, on whether the expected time to return is finite or infinite. However, a separate discussion is not always necessary, since the theory of Chapter 3 can be applied to the imbedded chain defined at the instants that transitions occur.

In order to calculate explicitly the distribution of first passage time to a state a, we consider a new process in which a is made an absorbing state. If T_a is the first passage time in the original process and $r_a(t)$ is the probability in the new process that state a is occupied at t, then

$$\text{prob}(T_a \leqslant t) = r_a(t) \tag{117}$$

and in particular

$$E(e^{-sT_a}) = sr_a^*(s) - r_a(0). \tag{118}$$

Of course if, as in Example 4.1, all the transitions are in the same direction the calculation of the distribution of T_a is much simplified.

To examine the distribution of the time from leaving a certain state a to first returning to a, a combination of first passage time distributions from i to a is formed, appropriately weighted over the other states, i.

The methods can probably be illustrated best by considering a very simplified example.

Example 4.13. *An artificial example with three states.* Suppose that the states are labelled 1, 2, 3 and that the matrix defining the transition rates is

$$\mathbf{Q} = \begin{bmatrix} -2 & 1 & 1 \\ 1 & -3 & 2 \\ 3 & 1 & -4 \end{bmatrix}. \tag{119}$$

For example, given that state 2 is occupied at t, there is a probability $2\Delta t + o(\Delta t)$ of a transition to state 3 in $(t, t+\Delta t)$. Suppose that we are interested in the first passage times T_{13} and T_{23} to state 3 starting from states 1 and 2, respectively.

Consider a new process in which state 3 is absorbing; the new process has matrix

$$\begin{bmatrix} -2 & 1 & 1 \\ 1 & -3 & 2 \\ 0 & 0 & 0 \end{bmatrix} \tag{120}$$

Let $r_{ij}(t)$ be the probability that in the new process state j is occupied at t given that state i is occupied initially. Since we are interested in these probabilities only for $j = 3$, it is natural to consider the backward equations, which are, since $r_{33}(t) = 1$,

$$r'_{13}(t) = -2r_{13}(t) + r_{23}(t) + 1,$$

$$r'_{23}(t) = r_{13}(t) - 3r_{23}(t) + 2.$$

Now $r_{ij}(0) = \delta_{ij}$, so that if we apply a Laplace transformation to these equations, we have that

$$(s+2)r^*_{13}(s) - r^*_{23}(s) = 1/s, \qquad -r^*_{13}(s) + (s+3)r^*_{23}(s) = 2/s.$$

Thus

$$s(s^2 + 5s + 5) \begin{bmatrix} r^*_{13}(s) \\ r^*_{23}(s) \end{bmatrix} = \begin{bmatrix} s+5 \\ 2s+5 \end{bmatrix}. \tag{121}$$

Therefore, by (118), the Laplace transforms of the p.d.f.'s of T_{13} and T_{23}, being $sr_{13}^*(s)$ and $sr_{23}^*(s)$, are

$$\frac{s+5}{s^2+5s+5} \quad \text{and} \quad \frac{2s+5}{s^2+5s+5}. \tag{122}$$

The corresponding p.d.f.'s are obtained, on inversion, as mixtures of exponentials with parameters $-\tfrac{1}{2}(5\pm\sqrt{5})$. Moments of T_{13} and T_{23} are most easily obtained by converting (122) into cumulant generating functions and then expanding in powers of s.

Suppose now that we want for the original process (119) the distribution of the time from first leaving state 3 to first returning to that state. Now, from the last row of (119), it follows that on leaving state 3 the process goes with probability $\tfrac{3}{4}$ to state 1 and with probability $\tfrac{1}{4}$ to state 2. Hence the Laplace transform of the required distribution is

$$\tfrac{3}{4}r_{13}^*(s) + \tfrac{1}{4}r_{23}^*(s). \tag{123}$$

If we are interested in the time from an arbitrary time in which state 3 is occupied to the next return to state 3, we must add an independent random variable exponentially distributed with parameter 4, i.e. we must multiply (123) by $4/(s+4)$.

The methods used in Example 4.13 apply quite generally although, of course, it may be difficult to obtain an explicit solution in a useful form.

Bibliographic Notes

A very careful discussion of the assumptions underlying the Poisson process is given by Khintchine (1960, part I). The pioneer work of Erlang on congestion applications in telephone exchanges is reviewed from a modern point of view by Jensen (1948). Chung (1960, part II) develops the general theory of Markov processes with discrete states in continuous time. Bharucha-Reid (1960) has summarized the theory and discussed a large number of very interesting applications from many fields of study. Karlin and McGregor (1957b) have given criteria for the generalized birth–death process classifying the states as transient, recurrent null or ergodic. Kendall (1959) has given an extension of the Karlin–McGregor integral representation to processes reversible in time. An important series of papers by Keilson on Green's function representations for Markov processes in discrete and continuous time is best approached through his monograph (Keilson, 1965). Whittle (1957) and Daniels (1960) have considered methods for approximating to the solution of processes of the type considered in this chapter.

Exercises

1. Derive the Poisson distribution (18) for $N(t)$, the number of events occurring in $(0,t)$ in a Poisson process, by dividing $(0,t)$ into a large number of small intervals and applying the binomial distribution.

2. An approximate (one-dimensional) model of a textile yarn is that fibre left-ends are distributed along the yarn axis in a Poisson process of rate ρ. Each fibre is of length a and is parallel to the yarn axis. The yarn thickness at any point is the number of fibres crossing that point. Prove that the correlation coefficient between the thicknesses at points h apart is $1 - h/a$ $(h \leqslant a)$, 0 $(h > a)$.

3. Prove that in the Poisson process of rate ρ in three dimensions, the distance R from the origin to the nearest point of the process has p.d.f. $4\pi\rho r^2 \exp\left(-\tfrac{4}{3}\pi\rho r^3\right)$.

4. In the process of (31) and (32), the Poisson process with multiple occurrences, let $M(t)$ be the number of points and $N(t)$ the number of events, that is counting multiple occurrences. Find the mean and variance of $N(t)$ by finding first the values conditionally on $M(t) = m$. Check the answers from (32).

5. Suppose that with the ith event in a Poisson process of rate ρ is associated a random variable W_i, where the W_i are independently exponentially distributed with parameter γ. Let $Z(t)$ be the sum of the W_i over all events occurring before t. Prove that the probability that $Z(t)$ is zero is $e^{-\rho t}$ and that the p.d.f. for $x > 0$ is

$$\sqrt{\left(\frac{\rho t \gamma}{x}\right)} e^{-\rho t - \gamma x} I_1\{2\sqrt{(\rho t \gamma x)}\},$$

where $I_1(y)$ is a Bessel function with imaginary argument.

6. Suppose that the conditions for a Poisson process hold, except that following each event there is a 'dead period' of length τ in which no event can occur. That is the intervals between successive events are independently distributed with the p.d.f. $\rho e^{-\rho(x-\tau)}$ $(x \geqslant \tau)$. Prove that if $N(t)$ is the number of events in time t, then

$$\text{prob}\{N(t) < r\} = \sum_{k=0}^{r-1} e^{-\rho(t-r\tau)} \frac{\{\rho(t - r\tau)\}^k}{k!}.$$

(Consider first the distribution of the time up to the rth event.)

7. A simplified model of accident 'contagion' is obtained by supposing that the probability of an accident in $(t, t + \Delta t)$ is $\lambda_0 \Delta t + o(\Delta t)$ if no previous accidents have occurred, and is $\lambda_1 \Delta t + o(\Delta t)$ otherwise, independently of the actual number of accidents. Obtain the probability

generating function, mean and variance of the number of accidents occurring in time t.

8. In place of the assumptions of Exercise 7, suppose that the probability of an accident in $(t, t + \Delta t)$ given r previous accidents, is $(\lambda + \nu r)\, \Delta t + o(\Delta t)$. Prove that the number of accidents in $(0, t)$ has a negative binomial distribution.

9. A population of individuals subject to accidents is such that for any one individual accidents occur in a Poisson process of rate R, say. The rate R varies from individual to individual in a gamma distribution. Prove that for a randomly selected individual, the number of accidents in $(0, t)$ has a negative binomial distribution.

10. Discuss briefly the types of observation that you would make to distinguish empirically between the models of Exercises 7, 8, 9.

11. Obtain a form of Wald's identity for random walks in continuous time by passing formally to the limit in the identity for discrete time (Section 2.3). Use the identity to obtain the p.d.f. (9) of the time in a Poisson process up to the rth event.

12. Develop a simulation procedure for the linear birth process, Example 4.1, which enables the evolution of population size to be followed, but not the details of the progeny of particular individuals. (For this note that when the population is of size i, the distribution of the time to the next birth is exponential with parameter $i\lambda$.) Construct several independent realizations of the process starting from one individual and examine the distribution of the time taken for the population to grow to size 10.

13. For the simple epidemic model (Example 4.3), prove that the limiting cumulant generating function of $U = \eta a T - 2\log a$ is $-2\eta\theta + 2\log \Gamma(1 - \theta)$ and hence that the limiting p.d.f. of U is $2e^{-u} K_0(2e^{-\frac{1}{2}u})$, where $K_0(x)$ is a modified Bessel function of the second kind.

14. In the linear birth–death process with $\lambda = \mu$, starting from n_0 individuals, let the time to extinction be T. Prove that as $n_0 \to \infty$, the limiting distribution function of T/n_0 is $\exp(-1/x)$. Comment on the connexion of this result with the theory of extreme values.

15. Suppose that in the linear birth–death process the initial number of individuals has a Poisson distribution of mean m_0. Prove that the probability generating function of the number of individuals, N_t, at time t is

$$\exp\left\{\frac{m_0(\mu - \lambda)(1 - z)}{\lambda(1 - z) - (\mu - \lambda z)\, e^{-(\lambda - \mu)t}}\right\}.$$

16. In the linear birth–death process, let $\tau = e^{(\lambda-\mu)t}$ and

$$U_t = N_t/E(N_t) = N_t/(n_0\tau).$$

Obtain the moment generating function of U_t starting from one individual and hence examine the limiting behaviour of N_t as $t \to \infty$, for $\lambda > \mu$, by letting $\tau \to \infty$. Show that the limiting distribution has probability μ/λ at zero and p.d.f. $(1-\mu/\lambda)^2 \exp\{-x(1-\mu/\lambda)\}$ $(x > 0)$. Obtain the corresponding distribution when the initial condition is that of Exercise 15.

17. In the two-state process of Example 4.7, let the initial probabilities be the equilibrium probabilities (p_0, p_1). Let $X(t)$ be the state occupied at t. Prove that the correlation coefficient between $X(t)$ and $X(t+h)$ is $\exp\{-|h|(\alpha+\beta)\}$.

18. Suppose that customers arrive randomly at rate α at a single-server queue, the distribution of service-time being exponential with parameter β. Suppose further that there is waiting room for only m customers, including the one being served. Customers arriving to find the waiting-room full leave without service and do not return. Prove that the equilibrium probability that there are i customers in the system is

$$\frac{(\alpha/\beta)^i (1-\alpha/\beta)}{1-(\alpha/\beta)^{m+1}} \quad (i = 0, \ldots, m).$$

19. Develop the equilibrium theory of the single-server queue with random arrivals, exponential distribution of service-time and a finite population of c customers. (That is, any customer who is not in the queue at time t, has a constant chance of arriving in $(t, t+\Delta t)$.)

20. In the bulk queueing model of Example 4.9, prove that the Laplace transform of the distribution of the time spent by a customer from arrival to the start of the service of his group is obtained by replacing z in (81) by $\beta/(\beta+s)$. Prove further that for a randomly chosen customer C the time from the start of service of his group to the completion of C's service has a p.d.f. with Laplace transform

$$\frac{\beta\{(\beta+s)^b - \beta^b\}}{b(\beta+s)^b s}.$$

Hence obtain the Laplace transform, and in particular the mean, of the distribution of the total time spent in the system. Compare the mean with that for a process with random arrivals and the same traffic intensity and mean service-time.

21. Generalize the forward and backward equations (87) and (89) to Markov processes with time-dependent transition probabilities. That is, let the matrix $\mathbf{Q}(t)$ depend on time, let $p_{ij}(t, u)$ be prob$\{X(u) = j | X(t) = i\}$

and let $\mathbf{P}(t, u)$ be the corresponding matrix. Prove that the forward and backward equations are respectively

$$\frac{\partial \mathbf{P}(t, u)}{\partial u} = \mathbf{P}(t, u)\,\mathbf{Q}(u), \qquad \frac{\partial \mathbf{P}(t, u)}{\partial t} = -\mathbf{Q}(t)\,\mathbf{P}(t, u).$$

Obtain (86) and (88) by noting that when $\mathbf{Q}$ is constant, $\mathbf{P}(t, u)$ is a function of $u - t$.

22. Verify that the formal solution (94), $\mathbf{P}(t) = \exp(\mathbf{Q}t)$, gives the correct probabilities for the two-state process, Example 4.7.

23. Apply the method of stages (Section 4.6(i)) to discuss the simple single-server queue, with random arrivals and with p.d.f. of service-time $\varpi\beta_1 e^{-\beta_1 x} + (1-\varpi)\beta_2 e^{-\beta_2 x}$.

24. A population of organisms consists initially of n_1 male and n_2 females. In time Δt any particular male is equally likely to mate with any particular female with probability $\lambda\Delta t + o(\Delta t)$, where λ is constant. Each mating immediately produces one offspring, equally likely to be male or female. Obtain a differential equation for the joint probability generating function of the numbers $N_1(t)$ and $N_2(t)$ of males and females at time t. Prove further that

$$E\{N_1(t) - N_2(t)\} = n_1 - n_2.$$

(Cambridge Diploma, 1956)

25. Formulate and solve the forward differential equations for an immigration–birth–death process and show that the distribution of population size has the negative binomial form. Show that an equilibrium distribution exists if and only if the birth rate is less than the death rate. Examine the limiting form of the equilibrium distribution when the immigration rate is very small compared with the birth rate.

26. Use the results of Example 3.18 to classify the imbedded process in discrete time corresponding to the generalized birth–death process with transition rates λ_i, μ_i.

Markov Processes in Continuous Time with Continuous State Space

5.1. Introduction

The main characteristic of the processes studied in Chapter 4 is that in a small time interval there is either no change of state or a radical change of state. Therefore in a finite interval there is either no change of state or a finite or possibly denumerably infinite number of discontinuous changes. Realizations of such processes are step functions. Physical situations from which such processes arise are typically those whose state is characterized by an integral number of particles or biological individuals in a system and the changes of state represent the addition or subtraction of individuals to or from the system by various means, such as birth, emigration, etc.

In the present chapter we start by examining processes whose state space is the continuum of real numbers and in which changes of state are occurring all the time. In a small time interval such a process can only undergo a small displacement or change of state. We may expect realizations of such processes to be continuous functions. A typical physical situation is that of particles suspended in a fluid, and moving under the rapid, successive, random impacts of neighbouring particles. If for such a particle the displacement in a given direction were plotted against time we would expect to obtain a continuous if somewhat erratic graph which would in fact be a realization of a stochastic process in continuous time with continuous state space. The physical phenomenon is known as Brownian motion after the botanist Robert Brown who first noticed it in 1827. It was not until 1905 that Einstein first advanced a satisfactory theory and in 1923 Wiener developed a rigorous theory. In physics, the theory of diffusion and the kinetic theory of matter are concerned with the aggregate motion of collections of molecules and these physical topics are intimately connected with the Markov processes discussed in this chapter. For this reason Markov processes in which only continuous changes of state occur are called *diffusion processes*.

In the two preceding chapters, especially in Chapter 3, we were able to carry out in reasonably elementary terms a fairly full mathematical discussion of Markov processes on a discrete state space. A comparably full mathematical discussion of diffusion processes is possible but beyond the scope of the present book. We shall proceed therefore by examining various physical situations which give rise to diffusion processes and also by considering diffusion processes as limiting cases of discrete processes. For most of this chapter we shall be concerned with one-dimensional processes.

As an approximation to the motion of a Brownian particle (in one dimension) we may divide the time axis and the displacement axis each into a very large number of small intervals and consider a particle undergoing a simple random walk, moving one small step up or down in each small time interval. Such an approach gives considerable insight into the continuous process and in many cases we can obtain a complete probabilistic description of the continuous process by proceeding mathematically to the limit of zero time and displacement intervals. Further we may expect the probability functions of interest to satisfy partial differential equations in the state and time variables since both these variables are continuous. The corresponding equations for the processes studied in Chapter 4 are differential equations in the (continuous) time variable but difference equations in the (discrete) state variable.

In the present chapter we shall deal with transition probability densities $p(x,t;y,u)$ giving the p.d.f. of $X(u)$ (as a function of y) conditional on $X(t) = x$. Thus

$$\text{prob}(a < X(u) < b | X(t) = x) = \int_a^b p(x,t;y,u)\,dy \quad (t < u). \quad (1)$$

For example, for a Brownian particle $p(x,t;y,u)dy$ would be the probability of finding the particle at time u in the interval $(y,y+dy)$ when it is known that the particle was at x at an earlier time t. For processes whose transition mechanism does not change with time, i.e. for time-homogeneous processes, the transition probability density depends only on the time interval $u-t$ and can be written $p(x,y;t)$ giving the p.d.f. of $X(t_0+t)$ (as a function of y) conditional on $X(t_0) = x$, for any t_0. The function $p(x,y;t)$ corresponds to the transition probabilities $p_{jk}^{(n)}$ (discrete time) and $p_{jk}(t)$ (continuous time) used in Chapters 3 and 4.

The most general kind of Markov process in continuous time is one in which both continuous and discontinuous changes of state may occur and for such processes we shall in general have to deal with the transition probability distribution function,

$$P(x,t;y,u) = \text{prob}(X(u) < y | X(t) = x) \quad (t < u). \quad (2)$$

The variables x and t are the backward variables since they refer to the earlier time and for corresponding reasons y and u called the forward variables. Because of the Markov property the function (2) will satisfy the Chapman–Kolmogorov equation which connects the distribution function of the process at time u with that at an earlier time t through that at an intermediate time v,

$$P(x,t;y,u) = \int_{-\infty}^{\infty} P(z,v;y,u)\,d_z P(x,t;z,v) \quad (t < v < u). \qquad (3)$$

If the probability densities exist then (3) takes the form

$$p(x,t;y,u) = \int_{-\infty}^{\infty} p(x,t;z,v)\,p(z,v;y,u)\,dz \quad (t < v < u). \qquad (4)$$

The Chapman–Kolmogorov equation is merely a statement of the formula of total probability modified by the Markov property: if A_t, B_v, C_u denote events concerning the process at times t, v, u respectively $(t < v < u)$ then

$$\text{prob}(C_u | A_t) = \sum_v \text{prob}(C_u | B_v)\,\text{prob}(B_v | A_t),$$

the summation being over all possible events B_v at time v. That the conditional probability, $\text{prob}(C_u | B_v)$ for example, depends only on u and v $(v < u)$, and not on events at times earlier than v, is, of course, the Markov property.

5.2. Continuous limit of the simple random walk: the Wiener process

The simple random walk has the property that one-step transitions are permitted only to the nearest neighbouring states. Such local changes of state may be regarded as the analogue for discrete states of the phenomenon of continuous changes for continuous states. Thus if we imagine small steps of magnitude Δ taking place at small time intervals of length τ, then in the limit as Δ and τ approach zero we may expect to obtain a process whose realizations are continuous functions of the time coordinate.

Consider a particle starting at the origin. In each time interval τ it takes a step Z where

$$\text{prob}(Z = +\Delta) = p, \qquad \text{prob}(Z = -\Delta) = q = 1-p, \qquad (5)$$

and all steps are mutually independent. The m.g.f. of one step is

$$E(e^{-\theta Z}) = p\,e^{-\theta \Delta} + q\,e^{\theta \Delta}. \qquad (6)$$

In time t there will be $n = t/\tau$ steps and the total displacement $X(t)$ is the sum of n independent random variables each with the m.g.f. (6). Hence

$$E\{e^{-\theta X(t)}\} = (p\,e^{-\theta \Delta} + q\,e^{\theta \Delta})^n = (p\,e^{-\theta \Delta} + q\,e^{\theta \Delta})^{t/\tau}. \qquad (7)$$

The mean and variance of $X(t)$ are

$$E\{X(t)\} = (t/\tau)\,(p-q)\,\Delta, \qquad V\{X(t)\} = (t/\tau)\,4pq\Delta^2. \qquad (8)$$

We wish to let $\Delta \to 0$ and $\tau \to 0$ in such a way as to obtain a sensible result. In particular, suppose we require the limiting process to have mean μ and variance σ^2 in unit time, i.e. we require Δ and τ to tend to zero in such a way that

$$(p-q)\Delta/\tau \to \mu \quad \text{and} \quad 4pq\Delta^2/\tau \to \sigma^2. \tag{9}$$

This will be satisfied if we take

$$\Delta = \sigma\sqrt{\tau}, \qquad p = \frac{1}{2}\left(1+\frac{\mu\sqrt{\tau}}{\sigma}\right), \qquad q = \frac{1}{2}\left(1-\frac{\mu\sqrt{\tau}}{\sigma}\right). \tag{10}$$

The important points about the relations (10) are that for small τ, $\Delta = O(\tau^{\frac{1}{2}})$, and that p and q are each $\frac{1}{2}+O(\tau^{\frac{1}{2}})$, so that a displacement Δ must be of a considerably larger order of magnitude than the small time interval τ in which it occurs and p, q must not be too different from $\frac{1}{2}$ if we are to avoid degeneracies.

If we now substitute (10) in (7) we obtain

$$E\{e^{-\theta X(t)}\} = \left[\frac{1}{2}\left(1+\frac{\mu\sqrt{\tau}}{\sigma}\right)e^{-\theta\sigma\sqrt{\tau}}+\frac{1}{2}\left(1-\frac{\mu\sqrt{\tau}}{\sigma}\right)e^{\theta\sigma\sqrt{\tau}}\right]^{t/\tau}. \tag{11}$$

If we expand the expression in square brackets in powers of τ, take logarithms and let $\tau \to 0$ we easily find that $K(\theta,t)$, the cumulant generating function of $X(t)$, becomes

$$K(\theta;t) = (-\mu\theta+\tfrac{1}{2}\sigma^2\theta^2)t. \tag{12}$$

This is the cumulant generating function of a normal distribution with mean μt and variance $\sigma^2 t$. Further, it is clear from the properties of the random walk from which we started that for any two time instants t, u ($u > t > 0$), the increments $X(t) - X(0)$ and $X(u) - X(t)$ are independent, the latter being normally distributed with mean $\mu(u-t)$ and variance $\sigma^2(u-t)$. In fact to obtain (12) we have simply carried out a central limiting operation, since we have summed a large number of independent random variables and appropriately scaled the sum so as to give a proper limiting distribution.

For $\mu = 0$ the limiting process $X(t)$ we have obtained is called the *Wiener process* or *Brownian motion process*. In this case the p.d.f. of $X(t)$ is symmetrical about the origin for all t. For $\mu \neq 0$ we shall call the process a Wiener process with *drift* μ and variance parameter σ^2. For such a process the increment $\Delta X(t) = X(t+\Delta t) - X(t)$ in a small time interval Δt is independent of $X(t)$ and has mean and variance proportional to Δt. The process is a Markov process, since, if we are given that $X(t) = a$, the distribution of $X(u)$ ($u > t$) is fixed, namely it is that of the quantity $a+\mu(u-t)+\sigma(u-t)^{\frac{1}{2}}Y$, where Y is a standard normal variate.

As a model for the motion of a Brownian particle the Wiener process has some defects. The change in position $\Delta X(t)$ in a small time interval

Δt is of the order of magnitude $(\Delta t)^{\frac{1}{2}}$. Hence the velocity has the order of magnitude $(\Delta t)^{\frac{1}{2}}/\Delta t$ which becomes infinite as $\Delta t \to 0$. However, for values of t large compared with intervals between successive collisions, the Wiener process has been found to be a very good representation of Brownian motion. We shall examine another model later which is also satisfactory for small t. As a stochastic process, however, the Wiener process is, of course, perfectly proper and of considerable importance, although it may be difficult to imagine the appearance of a realization of the process. But, for example, by imagining a very long realization of a simple symmetrical random walk plotted on a graph with very small time intervals, and making the displacement per time interval have the order of magnitude of the square root of the time interval, we are led to expect that the realization of the Wiener process, although continuous, has an infinite number of small spikes in any finite interval and is therefore non-differentiable.

A useful representation of the process $X(t)$ is in terms of a stochastic differential equation. This is an analogue in continuous time of the difference equation defining the simple random walk. If we start with a sequence Z_1, Z_2, ... of independent random variables each having the distribution $\text{prob}(Z_n = 1) = \text{prob}(Z_n = -1) = \frac{1}{2}$ then we can define a simple symmetrical random walk X_n by the *stochastic difference equation*

$$X_n = X_{n-1} + Z_n \quad (X_0 = x_0, \quad n = 1, 2, \ldots). \tag{13}$$

The appropriate analogue in continuous time of a sequence of independent identically distributed random variables is a purely random process $Z(t)$ (also called pure noise and discussed in more detail in Section 7.4) having the properties that $Z(t)$ has the same distribution for all t and that for any finite set of distinct time points $\{t_j\}$ the random variables $\{Z(t_j)\}$ are mutually independent. This is mathematically a perfectly well defined process although it is impossible to depict its sample functions. A purely random Gaussian process is obtained if, for each t, $Z(t)$ has a normal distribution. Starting with a purely random Gaussian process $Z(t)$ with zero mean and unit variance we define, in analogy with (13), a process $X(t)$ by the equation

$$X(t + \Delta t) = X(t) + Z(t) \sqrt{\Delta t},$$

or

$$\Delta X(t) = Z(t) \sqrt{\Delta t},$$

where Δt is a small time interval. Thus the increment in the process during the interval Δt is a normally distributed random variable with mean zero and variance Δt and is independent of the increments in any other small time intervals. To obtain a process with drift μ and variance σ^2 per unit time we write

$$\Delta X(t) = \mu \Delta t + \sigma Z(t) \sqrt{\Delta t}.$$

We may formally write this as a stochastic differential equation defining a Wiener process $X(t)$ with drift μ,

$$dX(t) = \mu\,dt + \sigma Z(t)\sqrt{dt}. \tag{14}$$

We interpret this equation as follows: the change in $X(t)$ in a small time interval dt is a normal variate with mean $\mu\,dt$ and variance $\sigma^2 dt$ and is independent of $X(t)$ and of the change in any other small time interval. Thus

$$E\{dX(t)\} = \mu\,dt, \qquad V\{dX(t)\} = \sigma^2\,dt, \qquad C\{dX(t), dX(u)\} = 0$$
$$(t \neq u).$$

It is possible to treat equations such as (14) rigorously but we shall not go into the matter here; see Itô (1951).

5.3. The diffusion equations for the Wiener process

We may also apply the limiting procedure of the previous section to the equations governing the transition probabilities of the simple random walk with steps $Z_1, Z_2, \ldots$ each having the distribution

$$\text{prob}(Z_n = 1) = p, \qquad \text{prob}(Z_n = -1) = q = 1-p. \tag{15}$$

The forward equation is

$$p_{jk}^{(n)} = p_{j,k-1}^{(n-1)}\,p + p_{j,k+1}^{(n-1)}\,q. \tag{16}$$

For fixed j, this is a difference equation of the first order in the time coordinate n and of the second order in the state coordinate k.

As in the previous section we consider the above process as a particle taking small steps of magnitude Δx in small time intervals of length Δt. Let $p(x_0, x; t)\,\Delta x$ be the conditional probability that the particle is at x at time t given that it starts at x_0 at time $t = 0$. Then we have $x_0 = j\Delta x$, $x = k\Delta x$, $t = n\Delta t$, and, in terms of this new scale, equation (16) becomes

$$p(x_0, x; t) = p\,p(x_0, x - \Delta x; t - \Delta t) + q\,p(x_0, x + \Delta x; t - \Delta t), \tag{17}$$

the factor Δx cancelling throughout. Suppose that $p(x_0, x; t)$ can be differentiated a suitable number of times and expanded in a Taylor series, for example

$$p(x_0, x + \Delta x; t - \Delta t) = p(x_0, x; t) - \Delta t\frac{\partial p}{\partial t} + \Delta x\frac{\partial p}{\partial x} + \tfrac{1}{2}(\Delta x)^2\frac{\partial^2 p}{\partial x^2} + \ldots \tag{18}$$

If we expand (17) in this way, writing p, q and Δx in the form (10) with $\Delta = \Delta x$ and $\tau = \Delta t$ and finally let $\Delta t \to 0$, then we obtain the forward equation

$$\tfrac{1}{2}\sigma^2\frac{\partial^2}{\partial x^2}p(x_0, x; t) - \mu\frac{\partial}{\partial x}p(x_0, x; t) = \frac{\partial}{\partial t}p(x_0, x; t), \tag{19}$$

a partial differential equation of the second order in x and the first order in t. It is an equation in the state variable x at time t for a given initial state x_0.

The equation (19) is familiar to physicists in two guises. Firstly it is the one-dimensional equation for diffusion with an external field of force (e.g. gravity), in which $p(x;t) = p(x_0, x;t)$ is the density of particles at the point x at time t, σ^2 is a constant associated with the medium and μ a constant associated with the external field. Secondly, with $\mu = 0$ it is the one-dimensional equation for heat conduction, in which $p(x;t)$ represents temperature and $k = \frac{1}{2}\sigma^2$ the temperature conductivity.

The backward version of (16) is

$$p_{jk}^{(n-n_0+1)} = pp_{j+1,k}^{(n-n_0)} + qp_{j-1,k}^{(n-n_0)}, \tag{20}$$

where n_0 is the initial time and $n(> n_0)$ the final time. Similar limiting operations give the backward equation

$$\tfrac{1}{2}\sigma^2 \frac{\partial^2}{\partial x_0^2} p(x_0, x;t) + \mu \frac{\partial}{\partial x_0} p(x_0, x;t) = \frac{\partial}{\partial t} p(x_0, x;t). \tag{21}$$

In particular problems the solution of these equations will depend on initial and boundary conditions. In many cases the initial condition for the forward equation is the natural one that $p(x_0, x;0)$ must have the character of a delta function representing a discrete probability distribution all concentrated at the point x_0. Other boundary conditions depend on whether we are considering an unrestricted process or a process restricted by boundaries such as absorbing or reflecting barriers.

According to the previous section the solution of (19) and (21) for an unrestricted process $X(t)$, conditional on $X(0) = x_0$, should be the normal density function with mean $x_0 + \mu t$ and variance $\sigma^2 t$ represented by (12), namely

$$p(x_0, x;t) = \frac{1}{\sigma\sqrt{(2\pi t)}} \exp\left\{-\frac{(x - x_0 - \mu t)^2}{2\sigma^2 t}\right\}. \tag{22}$$

It is a simple matter to verify by direct differentiation that (22) satisfies (19) and (21), but as a guide to handling more complicated problems, it is instructive to obtain the result (22) by solving, say, the forward equation (19).

The most common method of solving a linear partial differential equation with constant coefficients such as (19) or (21) is the method of separation of variables. However, if, as in the present case, the process is not restricted by boundaries a more convenient method is that of Fourier transforms or, equivalently, moment generating or characteristic functions.

However, let us first simplify the equation by making a change of variable (suggested, in fact, by the already known solution). Consider the process

$$Y(t) = \{X(t) - x_0 - \mu t\}/\sigma. \tag{23}$$

This involves making a change of variable $y = (x - x_0 - \mu t)/\sigma$ in equation (19), thus giving the equation

$$\frac{1}{2}\frac{\partial^2 p}{\partial y^2} = \frac{\partial p}{\partial t}, \tag{24}$$

where $p = p(y,t)$ is the p.d.f. of $Y(t)$ with the initial condition $Y(0) = 0$.
    Let

$$M(\theta;t) = E(e^{-\theta Y(t)}) = \int_{-\infty}^{\infty} p(y,t)e^{-\theta y}\,dy$$

be the m.g.f. of $Y(t)$. Then according to (24), $M(\theta;t)$ satisfies the equation

$$\tfrac{1}{2}\theta^2 M = \partial M/\partial t. \tag{25}$$

with the initial condition $M(\theta;0) = 1$. Thus

$$M(\theta;t) = e^{\frac{1}{2}\theta^2 t}, \tag{26}$$

which is the m.g.f. of a normal distribution with mean 0 and variance t. It follows that $X(t) = x_0 + \mu t + \sigma Y(t)$ has the normal distribution given by (22). It is clear from (22) that the increment $X(t) - X(0)$ is normally distributed with mean μt and variance $\sigma^2 t$ and is therefore independent of the value of $X(0)$. Thus, in addition to the Markov property, the process $X(t)$ has the stronger property of having independent increments.
    We may therefore formally define the Wiener process with drift μ and variance parameter σ^2 as a process $X(t)$ with the following properties

(a) $X(t)$ has independent increments, i.e. for any non-overlapping time intervals (t_1,t_2), (t_3,t_4) the random variables $X(t_2) - X(t_1)$ and $X(t_4) - X(t_3)$ are independent;
(b) for any time interval (t_1,t_2), $X(t_2) - X(t_1)$ is normally distributed with mean $\mu(t_2 - t_1)$ and variance $\sigma^2(t_2 - t_1)$.

5.4. First passage problems for the Wiener process
Suppose that $X(t)$ is a Wiener process starting at x_0 with drift μ and variance parameter σ^2.
    Let us investigate the first passage time T of $X(t)$ to the point $a > x_0$. Thus T is defined by

$$X(0) = x_0, \qquad X(t) < a \quad (0 < t < T), \qquad X(T) = a. \tag{27}$$

We have seen that for processes with discrete states the use of an absorbing barrier is a convenient means of studying passage times. We suppose therefore that there is an absorbing barrier at a and we let $p(x_0,x;t)$ be the probability density that $X(t) = x$ *and* that the process does not reach the barrier in the time interval $(0,t)$. Although we use the same notation, the function $p(x_0,x;t)$ is not the same as that in Section 5.3; in the present case it depends on a but it still satisfies the forward and backward equations (19) and (21) for $x < a$ and $x_0 < a$. We also have that

$$\text{prob}\{X(\tau) < a \quad \text{for } 0 < \tau < t, \ X(t) \leqslant x | X(0) = x_0\}$$

$$= \int_{-\infty}^{x} p(x_0, y; t)\, dy = P(x_0, x; t), \tag{28}$$

say. The function $P(x_0, x; t)$, like $p(\dot{x}_0, x; t)$, satisfies the backward equation (21), as can be seen by integrating over the forward variable in (21). Thus $P(x_0, a; t)$ is the probability that absorption has not yet occurred by time t, i.e.

$$P(x_0, a; t) = \text{prob}(T \geqslant t). \tag{29}$$

The backward equation is the appropriate one to use for passage times since our object is to determine the passage time distribution as a function of the initial state x_0 for a fixed final state a. Let $g(t|x_0, a)$ be the p.d.f. of T. We admit the possibility that

$$\int_0^{\infty} g(t|x_0, a)\, dt < 1$$

since absorption may be uncertain. It follows from (29) that

$$g(t|x_0, a) = -\frac{\partial}{\partial t} P(x_0, a; t). \tag{30}$$

We define the Laplace transform of g,

$$g^*(s|x_0, a) = \int_0^{\infty} e^{-st} g(t|x_0, a)\, dt \tag{31}$$

and, emphasizing the dependence of g^* on x_0, write

$$\gamma(x_0) = g^*(s|x_0, a). \tag{32}$$

Taking Laplace transforms in (21), we obtain

$$\tfrac{1}{2}\sigma^2 \frac{d^2\gamma}{dx_0^2} + \mu \frac{d\gamma}{dx_0} = s\gamma. \tag{33}$$

This is a second-order linear differential equation with constant coefficients, the general solution of which is

$$\gamma(x_0) = A\, e^{x_0\, \theta_1(s)} + B\, e^{x_0\, \theta_2(s)} \tag{34}$$

where

$$\theta_1(s), \theta_2(s) = \frac{-\mu \mp \sqrt{(\mu^2 + 2s\sigma^2)}}{\sigma^2} \tag{35}$$

are the roots of the quadratic equation

$$\tfrac{1}{2}\sigma^2 \theta^2 + \mu\theta = s.$$

Note that if we take the positive square root in (35), then, for s real,

$$\theta_1(s) < 0 < \theta_2(s) \quad (s > 0), \tag{36}$$

$$\theta_1(0) = -2\mu/\sigma^2, \qquad \theta_2(0) = 0 \quad (\mu \geqslant 0),$$
$$\theta_1(0) = 0, \qquad \theta_2(0) = 2|\mu|/\sigma^2 \quad (\mu \leqslant 0). \tag{37}$$

To determine A and B we note from (31) and (32) that for $s > 0$

$$\gamma(x_0) \leqslant \int_0^\infty g \, dt \leqslant 1,$$

and it follows immediately that $A = 0$ in (34), for otherwise $\gamma(x_0)$ would become unbounded as $x_0 \to -\infty$. Further, when $x_0 = a$ absorption occurs immediately and so $\gamma(a) = 1$. We thus obtain the result

$$\gamma(x_0) = g^*(s|x_0, a) = e^{(x_0-a)\,\theta_2(s)} \quad (x_0 \leqslant a), \tag{38}$$

which is the Laplace transform of the first passage time density from x_0 to a. The inversion of (38) presents no real difficulties but we shall obtain the density $g(t|x_0, a)$ by another method in Example 5.1, namely by solving the forward equation for $p(x_0, x; t)$ in conjunction with appropriate boundary conditions and then using (28) and (30).

However we may note that when $s = 0$, (38) gives the probability $\pi(x_0, a)$ of ever reaching a when starting from $x_0 < a$. Thus, we have from (37) that

$$\pi(x_0, a) = g^*(0|x_0, a) = \begin{cases} 1 & (\mu \geqslant 0), \\ e^{-2(a-x_0)|\mu|/\sigma^2} & (\mu < 0), \end{cases} \tag{39}$$

so that the situation is exactly analogous to that of the simple random walk, namely that when there is a drift towards the barrier ($\mu > 0$) or no drift ($\mu = 0$) the probability of ultimate absorption is unity, while when the drift is away from the barrier ($\mu < 0$) there is a non-zero probability of never reaching the barrier.

Note that for $x_0 = 0$ the result (39) gives the distribution of the maximum displacement Y in the positive direction when the process starts at the origin. For Y is defined by

$$Y = \sup_{t>0} \{X(t)|X(0) = 0\}$$

and from (39) we have

$$\text{prob}(Y \leqslant a) = \begin{cases} 0 & (\mu \geqslant 0), \\ 1 - e^{-2a|\mu|/\sigma^2} & (\mu < 0). \end{cases} \tag{40}$$

Thus for $\mu < 0$, Y has an exponential distribution with mean $\sigma^2/(2|\mu|)$.

Further, equation (39) bears comparison with (2.91) to which, in fact, it would be identical after appropriate changes of notation if (2.91) were an equality rather than an approximation. The result (2.91) can be regarded as a diffusion approximation to the probability of absorption for a random walk with normally distributed steps. For such a random walk the Wiener process is an approximation in the following sense: if we consider a Wiener process $X(t)$, $X(0) = 0$, with drift μ and variance parameter σ^2, then the imbedded process $X(1)$, $X(2)$, ... is the unrestricted random walk Z_1, $Z_1 + Z_2$, ..., where Z_i are independently and normally distributed with mean μ and variance σ^2.

5.5. Continuous limits of more general discrete processes

The Wiener process is a diffusion process having the special property of independent increments and it may be regarded, as we saw in Sections 5.2 and 5.3, as a limiting version of a simple random walk, itself a Markov process with independent increments and, in addition, local changes of state. We now consider a more general homogeneous Markov chain, but in which, again, only local changes of state are permissible. More precisely we suppose that in unit time we can move from state k to states $k+1$, $k-1$ with probabilities θ_k, ϕ_k respectively, while with probability $1 - \theta_k - \phi_k$ there is no change of state. We thus have the random walk with variable transition probabilities discussed in Example 3.18.

Let $p_{jk}^{(n)}$ be the n-step transition probabilities. Then corresponding to (16) we have the forward equation

$$p_{jk}^{(n)} = p_{j,k-1}^{(n-1)} \theta_{k-1} + p_{jk}^{(n-1)}(1 - \theta_k - \phi_k) + p_{j,k+1}^{(n-1)} \phi_{k+1}. \tag{41}$$

Again we consider the process as a particle taking small steps of amount $-\Delta x, 0$ or Δx in small time intervals of length Δt. We assume, to begin with, that Δx and Δt are fixed. We suppose that if the particle is at x at time t then the probabilities that it will be at $x + \Delta x$, x, $x - \Delta x$ at time $t + \Delta t$ are respectively $\theta(x)$, $1 - \theta(x) - \phi(x)$, $\phi(x)$. Let $p(x_0, x; t) \Delta x$ be the conditional probability that the particle is at x at time t given that it starts at x_0 at time $t = 0$. The following is not meant to be a rigorous derivation of the partial differential equation satisfied by the transition p.d.f. $p(x_0, x; t)$. Nevertheless it is a useful approach and it indicates the close connexion between diffusion processes and Markov chains with local changes of state.

In terms of our new scale the equation (41) becomes

$$p(x_0, x; t) \Delta x = p(x_0, x - \Delta x; t - \Delta t) \Delta x . \theta(x - \Delta x)$$
$$+ p(x_0, x, t - \Delta t) \Delta x \{1 - \theta(x) - \phi(x)\}$$
$$+ p(x_0, x + \Delta x; t - \Delta t) \Delta x . \phi(x + \Delta x). \tag{42}$$

Note that the factor Δx may be cancelled throughout. Again we shall need to place some restrictions on $\theta(x)$ and $\phi(x)$ and on the approach of Δx and Δt to zero. In Section 5.2 we required the process increment to have mean μ and variance σ^2 in unit time. The quantities μ and σ^2 may also be interpreted as the instantaneous mean and variance of the change in $X(t)$ per unit time, i.e.

$$\mu = \lim_{\Delta t \to 0} \frac{E\{X(t+\Delta t) - X(t)\}}{\Delta t},$$

$$\sigma^2 = \lim_{\Delta t \to 0} \frac{V\{X(t+\Delta t) - X(t)\}}{\Delta t}.$$

In our present more general process these quantities will depend on the state x at time t and we suppose that the instantaneous mean and variance per unit time of the change in $X(t)$, conditional on $X(t) = x$, are $\beta(x)$ and $\alpha(x)$ respectively. Thus

$$\beta(x) = \lim_{\Delta t \to 0} \frac{E[\{X(t+\Delta t) - X(t)\}|X(t) = x]}{\Delta t}, \tag{43}$$

$$\alpha(x) = \lim_{\Delta t \to 0} \frac{V[\{X(t+\Delta t) - X(t)\}|X(t) = x]}{\Delta t}. \tag{44}$$

If at time t the particle is at x, then since in the next small time interval the change in position is $-\Delta x$, 0 or Δx with probabilities $\phi(x)$, $1 - \phi(x) - \theta(x)$, $\theta(x)$, we have that the mean change in position is

$$\{\theta(x) - \phi(x)\}\, \Delta x$$

and that the variance of the change in position is

$$[\theta(x) + \phi(x) - \{\theta(x) - \phi(x)\}^2]\,(\Delta x)^2.$$

The relations (43) and (44) thus become

$$\lim_{\Delta x, \Delta t \to 0} \{\theta(x) - \phi(x)\}\frac{\Delta x}{\Delta t} = \beta(x), \tag{45}$$

$$\lim_{\Delta x, \Delta t \to 0} [\theta(x) + \phi(x) - \{\theta(x) - \phi(x)\}^2]\frac{(\Delta x)^2}{\Delta t} = \alpha(x). \tag{46}$$

In analogy with (9) these relations will now determine the allowable forms of $\theta(x)$ and $\phi(x)$ and the approach of Δx and Δt to zero. Let us suppose that $\alpha(x)$ is a bounded function satisfying

$$\alpha(x) < A \tag{47}$$

for all x. If we now take

$$(\Delta x)^2 = A\,\Delta t,$$

$$\theta(x) = \frac{1}{2A}\{\alpha(x) + \beta(x)\,\Delta x\}, \tag{48}$$

$$\phi(x) = \frac{1}{2A}\{\alpha(x) - \beta(x)\,\Delta x\},$$

then the relations (45) and (46) will be satisfied. Again we have imposed the condition that $\Delta x = O\{(\Delta t)^{\frac{1}{4}}\}$ and that $\theta(x)$, $\phi(x)$ differ from the same constant by amounts $O\{(\Delta t)^{\frac{1}{4}}\}$.

We now return to the equation (42). Again assuming suitable differentiability conditions for the functions involved we obtain, as for (19), the forward equation

$$\tfrac{1}{2}\alpha(x)\frac{\partial^2 p}{\partial x^2} + \{\alpha'(x)-\beta(x)\}\frac{\partial p}{\partial x} + \{\tfrac{1}{2}\alpha''(x)-\beta'(x)\}p = \frac{\partial p}{\partial t} \qquad (49)$$

or

$$\frac{1}{2}\frac{\partial^2}{\partial x^2}\{\alpha(x)\,p\} - \frac{\partial}{\partial x}\{\beta(x)\,p\} = \frac{\partial p}{\partial t}, \qquad (50)$$

where $p = p(x_0,x;t)$. Similarly, by considering the backward version of (41) we obtain the backward equation

$$\tfrac{1}{2}\alpha(x_0)\frac{\partial^2 p}{\partial x_0^2} + \beta(x_0)\frac{\partial p}{\partial x_0} = \frac{\partial p}{\partial t}. \qquad (51)$$

In the theory of partial differential equations the equations (50) and (51) are called parabolic equations and are adjoints of one another; see, for example, Sommerfeld (1949, p. 44). The solution will, of course, depend on initial and boundary conditions.

5.6. The Kolmogorov equations

The forward and backward equations of Section 5.5 are particular cases of differential equations of a more general type which can be obtained by a similar method if we allow the transition mechanism to depend not only on the state variable x but also on the time t. In this case we are led to define $\beta(x,t)$ and $\alpha(x,t)$, both depending on x and t, with $\alpha(x,t) > 0$, as the instantaneous mean and variance per unit time of the change in $X(t)$, given that $X(t) = x$. The right-hand sides of equations (43) and (44) respectively still serve to define these quantities.

If x_0 denotes the state variable at time t_0 and x that at a later time t then the transition probability density $p(x_0,t_0;x,t)$ satisfies the *Kolmogorov equations*, namely the *forward equation*

$$\frac{1}{2}\frac{\partial^2}{\partial x^2}\{\alpha(x,t)\,p\} - \frac{\partial}{\partial x}\{\beta(x,t)\,p\} = \frac{\partial p}{\partial t}, \qquad (52)$$

in which the backward variables x_0 and t_0 are essentially constant and only enter by way of boundary conditions, and the *backward equation*

$$\tfrac{1}{2}\alpha(x_0,t_0)\frac{\partial^2 p}{\partial x_0^2} + \beta(x_0,t_0)\frac{\partial p}{\partial x_0} = -\frac{\partial p}{\partial t_0}, \qquad (53)$$

in which the forward variables x and t enter only through boundary conditions. The forward equation (52) is also known as the Fokker–Planck diffusion equation. The functions $\beta(x,t)$ and $\alpha(x,t)$ are sometimes

called the *infinitesimal mean and variance* of the process. For a proof of
(52) and (53) and precise conditions under which these results hold, see
Gnedenko (1962, p. 344).

More generally still the backward equation (53) is satisfied by the
transition probability distribution function

$$P(x_0, t_0; x, t) = \text{prob}\{X(t) \leqslant x | X(t_0) = x_0\}$$

$$= \int\limits_{-\infty}^{x} p(x_0, t_0; \xi, t) \, d\xi.$$

This can be shown by proceeding to the limit from a discrete process or
by integrating the backward equation (53) with respect to the forward
variable x. The forward equation, however, is essentially an equation
for the probability density. It is interesting that apart from initial and
boundary conditions, the transition probabilities of a diffusion process
are completely determined by the first and second moments of the
infinitesimal increments of the process.

From a practical point of view both the forward and backward
equations have their uses in appropriate circumstances, as is indeed the
case for all Markov processes. If we are interested in the probability
distribution of $X(t)$ for a given initial value $X(t_0) = x_0$, then clearly the
appropriate approach is via the forward equation. But if we are interested
in, say, the first passage time distribution to a fixed state a as a function
of the initial position x_0, then the backward equation provides the
appropriate method.

When we come to consider boundary conditions it can be seen that
certain types of behaviour on a boundary can lead to a forward equation
which is no longer a differential equation of the type (52). For example,
suppose we have a diffusion process inside a finite interval and when the
diffusing particle reaches a given boundary it is held there for an interval
of time T which has an exponential distribution. It then returns instantaneously to an interior point x having a probability distribution with
p.d.f. $\rho(x)$ and starts its diffusion motion again. Such a process is still a
Markov process, but when we consider formulating the forward equation
we see that during a small time interval a point x can be reached not only
from neighbouring points but also from a point on the boundary. The
transitions are no longer of a local character and this fact destroys the
simple differential character of the forward equation but does not
change the form of the backward equation. The forward and backward
equations are no longer the formal adjoints of one another.

It is often convenient to take Laplace transforms with respect to the
time variable, for this operation changes a partial differential equation
involving the space coordinate x and time coordinate t into an ordinary
differential equation in x. This was in fact done in deriving equation (33).

Thus, for example, in the time-homogeneous case the transition probability

$$P(x_0, x, t) = \int_{-\infty}^{x} p(x_0, \xi, t)\, d\xi$$

satisfies the backward equation (51). Letting

$$P^*(x_0, x; s) = \int_{0}^{\infty} e^{-st} P(x_0, x; t)\, dt,$$

we see that P^* satisfies

$$\tfrac{1}{2}\alpha(x_0)\frac{d^2 P^*}{dx_0^2} + \beta(x_0)\frac{dP^*}{dx_0} = P^* - P(x_0, x; 0) \tag{54}$$

$$= \begin{array}{ll} P^* - 1 & (x_0 \leqslant x), \\[2mm] P^* & (x_0 > x). \end{array} \tag{55}$$

Another approach to the Fokker–Planck (forward) equation is obtained by considering a stochastic differential equation defining the process $X(t)$, which in the case of the Wiener process, is given by (14). In the present more general case we consider a purely random process $Z(t)$ with $E\{Z(t)\} = 0$, $V\{Z(t)\} = 1$ and $E\{|Z(t)|^3\} = O(1)$ (in fact $o(1/\sqrt{dt})$ is sufficient). Generalizing (14), we consider the stochastic differential equation

$$dX(t) = \beta(x, t)\, dt + Z(t)\sqrt{\{\alpha(x, t)\, dt\}}, \tag{56}$$

which holds conditionally on $X(t) = x$. Thus, given that $X(t) = x$, the increment $dX(t)$ in a small time interval dt has mean $\beta(x, t)\, dt$ and variance $\alpha(x, t)\, dt$ and is independent of all previous increments. This may be written in terms of moment generating functions

$$E\{e^{-\theta\, dX(t)}|X(t) = x\} = 1 - \theta\{\beta(x, t)\, dt\} + \frac{\theta^2}{2!}\{\alpha(x, t)\, dt\} + o(dt), \tag{57}$$

where we assume that the term $o(dt)$ is uniform in x.

Consider now the m.g.f. of $X(t)$, namely

$$\phi(\theta; t) = E\{e^{-\theta X(t)}\} = \int_{-\infty}^{\infty} e^{-\theta x} p(x; t)\, dx,$$

where $p(x, t)$ is the p.d.f. of $X(t)$ subject to some given initial conditions. By differentiating under the integral sign we obtain

$$\frac{\partial \phi}{\partial t} = \int_{-\infty}^{\infty} e^{-\theta x}\frac{\partial p}{\partial t}\, dx. \tag{58}$$

Alternatively we can obtain another expression for $\partial\phi/\partial t$ by noting that

$$\frac{\partial\phi}{\partial t} = \lim_{\Delta t \to 0} \frac{1}{\Delta t}\{\phi(\theta;t+\Delta t) - \phi(\theta;t)\}$$

$$= \lim_{\Delta t \to 0} \frac{1}{\Delta t} E[e^{-\theta\{X(t)+\Delta X(t)\}} - e^{-\theta X(t)}]. \tag{59}$$

The expected value on the right can be written

$$E[\{e^{-\theta\Delta X(t)} - 1\}e^{-\theta X(t)}] = \int_{-\infty}^{\infty} E[\{e^{-\theta\Delta X(t)} - 1\}e^{-\theta X(t)}|X(t) = x]\,p(x,t)\,dx$$

$$= \int_{-\infty}^{\infty} E[(e^{-\theta\Delta X(t)} - 1)|X(t) = x]\,e^{-\theta x}p(x;t)\,dx.$$

Now using (57) and letting $\Delta t \to 0$, we find that

$$\frac{\partial\phi}{\partial t} = \int_{-\infty}^{\infty} \{-\theta\beta(x,t) + \tfrac{1}{2}\theta^2\,\alpha(x,t)\}\,e^{-\theta x}p(x,t)\,dx, \tag{60}$$

and, on integration by parts,

$$\frac{\partial\phi}{\partial t} = \int_{-\infty}^{\infty} e^{-\theta x}\left[\frac{1}{2}\frac{\partial^2}{\partial x^2}\{\alpha(x,t)\,p(x,t)\} - \frac{\partial}{\partial x}\{\beta(x,t)\,p(x,t)\}\right]dx. \tag{61}$$

Equating (58) and (61), we obtain the Fokker–Planck equation (52).

A useful equation for the m.g.f. of a Markov process has been given by Bartlett (1955, p. 83). We may use equation (60) to derive Bartlett's equation for diffusion processes. Suppose that $\alpha(x,t)$ and $\beta(x,t)$ may be expanded in power series in x,

$$\alpha(x,t) = \sum_{r=0}^{\infty} \alpha_r(t)\,x^r, \qquad \beta(x,t) = \sum_{r=0}^{\infty} \beta_r(t)\,x^r.$$

Substituting in (60), we obtain

$$\frac{\partial\phi}{\partial t} = -\theta \int_{-\infty}^{\infty} e^{-\theta x}\left\{\sum_{r=0}^{\infty} x^r \beta_r(t)\right\}p(x,t)\,dx$$

$$+ \tfrac{1}{2}\theta^2 \int_{-\infty}^{\infty} e^{-\theta x}\left\{\sum_{r=0}^{\infty} x^r \alpha_r(t)\right\}p(x,t)\,dx.$$

Now

$$\int_{-\infty}^{\infty} e^{-\theta x}\,x^r\,p(x,t)\,dx = \left(-\frac{\partial}{\partial\theta}\right)^r \phi(\theta;t).$$

Hence we have

$$\frac{\partial \phi}{\partial t} = -\theta \left[\sum_{r=0}^{\infty} \beta_r(t) \left(-\frac{\partial}{\partial \theta} \right)^r \right] \phi(\theta;t) + \tfrac{1}{2}\theta^2 \left[\sum_{r=0}^{\infty} \alpha_r(t) \left(-\frac{\partial}{\partial \theta} \right)^r \right] \phi(\theta;t),$$

which, in a suitable formalism, may be written

$$\frac{\partial \phi}{\partial t} = \left\{ -\theta \beta \left(-\frac{\partial}{\partial \theta}, t \right) + \tfrac{1}{2}\theta^2 \alpha \left(-\frac{\partial}{\partial \theta}, t \right) \right\} \phi(\theta;t). \tag{62}$$

This is Bartlett's equation for a diffusion process. To recover it we write down the expression $-\theta\beta(x,t) + \tfrac{1}{2}\theta^2 \alpha(x,t)$ which, apart from a multiple dt, is the m.g.f. of $dX(t)$ conditional on $X(t) = x$; we then form a differential operator by replacing x by $(-\partial/\partial\theta)$ and operate on ϕ to obtain the right-hand side of (62).

We note further that for a diffusion process the probability of a finite increment in a small time interval Δt is $o(\Delta t)$; for, using a generalization of Chebychev's inequality, namely

$$\mathrm{prob}(|X - E(X)| > \epsilon) \leqslant (1/\epsilon^3) E\{|X - E(X)|^3\}$$

for a random variable X with finite third moment, we have

$$\mathrm{prob}\{|\Delta X(t)| > \epsilon | X(t) = x\} \leqslant (1/\epsilon^3) E\{|\Delta X(t)|^3 | X(t) = x\}$$

and in view of (56) the right-hand side is $o(\Delta t)$ for any fixed $\epsilon > 0$. We may recall that the Markov processes in continuous time of Chapter 4, while having the same properties as diffusion processes in that for a small time interval Δt the increment has mean and variance proportional to Δt, have the probability of a change of state during Δt of the form $\alpha \Delta t + o(\Delta t)$.

5.7. Boundary conditions for homogeneous diffusion processes

When a diffusion process is restricted by one or two barriers then the solution of the Kolmogorov equations is subject to boundary conditions. We now consider the conditions to be satisfied by the transition p.d.f. at boundaries of various kinds. We deal only with time-homogeneous processes.

(i) ABSORBING BARRIER

Let $X(t)$ be a time-homogeneous diffusion process representing the position of a particle at time t with the initial condition $X(0) = x_0$. The p.d.f. $p(x;t)$ of $X(t)$ satisfies the Kolmogorov equations (50) and (51). For convenience we suppress x_0 in the notation $p(x_0, x;t)$. Suppose that an absorbing barrier is placed at $a > x_0$ and that the infinitesimal variance $\alpha(x)$ does not vanish when $x = a$. We shall show that the appropriate boundary condition for an absorbing barrier is that $p(x;t)$ must vanish at $x = a$ for all t, i.e.

$$p(a;t) = 0 \quad (t > 0). \tag{63}$$

For suppose to the contrary that in the interval $(a - \Delta x, a)$ we have

$$p(x;t) \geqslant \eta > 0 \quad (t_1 \leqslant t \leqslant t_2).$$

Then for a small time interval $(t, t + \Delta t)$ contained in (t_1, t_2) the probability $g(t)\, \Delta t$ that the particle is absorbed is roughly the probability that the particle is near a at time t and that the increment carries the particle beyond a; certainly we have

$$g(t)\, \Delta t > \eta \Delta x \, \mathrm{prob}\{X(t + \Delta t) - X(t) > \Delta x \,|\, a - \Delta x < X(t) < a\}. \tag{64}$$

Since $\alpha(x) > 0$ near $x = a$ we may conclude that

$$\alpha(x)\, \Delta t = V\{\Delta X(t) \,|\, X(t) = x\} \geqslant \alpha \Delta t$$

near $x = a$ for some $\alpha > 0$ and hence that $\Delta X(t)$ will exceed $\epsilon \sqrt{\Delta t}$ with finite probability, not less than p, say, for some $\epsilon > 0$. Hence if we take $\Delta x = \epsilon \sqrt{\Delta t}$ in (64) we have

$$g(t)\, \Delta t \geqslant \eta \epsilon p \sqrt{\Delta t} \quad (t_1 < t < t_2). \tag{65}$$

This implies that $g(t)$ is infinite for $t_1 < t < t_2$. But $g(t)$ is in fact the probability density of the first passage time to a and so we have a contradiction, i.e. $p(a;t)$ must be zero.

Example 5.1. The Wiener process with absorbing barriers. Consider the Wiener process $X(t)$ with $\alpha(x) = \sigma^2$, $\beta(x) = \mu$ and $X(0) = 0$ and suppose that there is an absorbing barrier at $x = a$. We must therefore solve the forward equation for $p(x;t)$, namely

$$\tfrac{1}{2}\sigma^2 \frac{\partial^2 p}{\partial x^2} - \mu \frac{\partial p}{\partial x} = \frac{\partial p}{\partial t} \quad (x < a), \tag{66}$$

subject to the conditions

$$p(x;0) = \delta(x), \tag{67}$$

$$p(a;t) = 0 \quad (t > 0). \tag{68}$$

Consider the solution to the unrestricted process starting from x_0, namely

$$p_1(x_0, x; t) = \frac{1}{\sigma \sqrt{(2\pi t)}} \exp\left\{ -\frac{(x - x_0 - \mu t)^2}{2\sigma^2 t} \right\}. \tag{69}$$

The function $p_1(0, x; t)$ clearly satisfies the equation (66) and the initial condition (67). Further, for $x_0 > a$, a linear combination

$$p(x, t) = p_1(0, x; t) + A p_1(x_0, x; t) \tag{70}$$

also satisfies (66) and (67) and the question now is whether suitable values of A and x_0 can be found to make the linear combination (70) also satisfy the boundary condition (68). These considerations lead to the method of images used in solving problems of heat conduction and diffusion. We imagine the barrier as a mirror and place an 'image source' at $x = 2a$, the image of the origin in the mirror. This leads to a solution

$$p(x, t) = p_1(0, x; t) + Ap_1(2a; x; t).$$

By simply substituting from (69) it will be seen that if A is given the value $-\exp(2\mu a/\sigma^2)$ then the resulting solution satisfies (66), (67) and (68) and is therefore the required solution, namely

$$p(x, t) = \frac{1}{\sigma\sqrt{(2\pi t)}}\left[\exp\left\{-\frac{(x-\mu t)^2}{2\sigma^2 t}\right\} - \exp\left\{\frac{2\mu a}{\sigma^2} - \frac{(x-2a-\mu t)^2}{2\sigma^2 t}\right\}\right]. \quad (71)$$

We can regard this solution as a superposition of a source of unit strength at the origin and a source of strength $-\exp(2\mu a/\sigma^2)$ (i.e. a 'sink') at $x = 2a$.

We can now reconsider the problem of finding the p.d.f. $g(t)$ of the first passage time T from 0 to a in the case when $\mu \geqslant 0$; then the probability of reaching the barrier is unity. The equation (30) takes the form

$$g(t) = g(t|0, a) = -\frac{d}{dt}\int_{-\infty}^{a} p(x, t)\, dx$$

$$= -\frac{d}{dt}\left\{\Phi\left(\frac{a-\mu t}{\sigma\sqrt{t}}\right) - \exp\left(\frac{2\mu a}{\sigma^2}\right)\Phi\left(\frac{-a-\mu t}{\sigma\sqrt{t}}\right)\right\}, \quad (72)$$

where $\Phi(x)$, the standard normal integral, is given by

$$\Phi(x) = \frac{1}{\sqrt{(2\pi)}}\int_{-\infty}^{x} e^{-\frac{1}{2}y^2}\, dy.$$

After carrying out the differentiation in (72) we obtain

$$g(t) = \frac{a}{\sigma\sqrt{(2\pi t^3)}}\exp\left\{-\frac{(a-\mu t)^2}{2\sigma^2 t}\right\} \quad (73)$$

as the p.d.f. of T. For a process with zero drift, i.e. $\mu = 0$, we have

$$g(t) = \frac{a}{\sigma\sqrt{(2\pi t^3)}}\exp\left\{-\frac{a^2}{2\sigma^2 t}\right\}, \quad (74)$$

which is of order $t^{-3/2}$ as $t \to \infty$, so that T has no finite moments in this case.

Note that (38) gives the m.g.f. of T, namely

$$g^*(s) = \int_0^{\infty} e^{-st}g(t)\, dt = e^{-a\theta_2(s)} = \exp\left[\frac{a}{\sigma^2}\{\mu - (\mu^2 + 2s\sigma^2)^{\frac{1}{2}}\}\right]. \quad (75)$$

Differentiation with respect to s will give the moments of T and we find that for $\mu > 0$,

$$E(T) = \frac{a}{\mu}, \qquad V(T) = \frac{a\sigma^2}{2\mu^3}. \tag{76}$$

Similar methods can be applied to the case of two absorbing barriers, one at $-b$ $(b > 0)$ and the other at $a > 0$. The solution must satisfy (66) for $-b < x < a$ and in addition to the conditions (67) and (68) we must have, for all $t > 0$,

$$p(-b, t) = 0. \tag{77}$$

A doubly infinite system of images is now required and we simply quote the solution. We place sources at the points $x'_n = 2n(a+b)$ $(n = 0, \pm 1, \pm 2, \ldots)$ with strengths $\exp(\mu x'_n/\sigma^2)$ and sources at the points $x''_n = 2a - x'_n$ with strengths $-\exp(\mu x''_n/\sigma^2)$. Then the required solution is the linear superposition of solutions for each source weighted by the corresponding strength, i.e.

$$p(x, t) = \frac{1}{\sigma\sqrt{(2\pi t)}} \sum_{n=-\infty}^{\infty} \left[\exp\left\{ \frac{\mu x'_n}{\sigma^2} - \frac{(x - x'_n - \mu t)^2}{2\sigma^2 t} \right\} \right.$$
$$\left. - \exp\left\{ \frac{\mu x''_n}{\sigma^2} - \frac{(x - x''_n - \mu t)^2}{2\sigma^2 t} \right\} \right]. \tag{78}$$

The series is absolutely convergent and differentiable term by term and since each term satisfies (66) so does the series. All the sources except x'_0, which has unit strength, lie outside the barriers and so the condition (67) is satisfied. When $x = a$ the source at x'_n annuls that at x''_n and when $x = -b$, the source at x'_n annuls that at x''_{n+1}, and so the conditions (68) and (77) are satisfied.

Another approach to the two-barrier problem is to seek a Fourier series solution suggested by the method of separation of variables. Consider a solution of the form

$$e^{\kappa x - \lambda t} \sin\left\{ \frac{n\pi(x+b)}{a+b} \right\}. \tag{79}$$

We choose the sine function in this form since it vanishes when $x = a$ and $x = -b$, thus satisfying the boundary conditions (68) and (77). We find that (79) will satisfy the equation (66) provided that

$$\kappa = \frac{\mu}{\sigma^2}, \qquad \lambda = \lambda_n = \frac{1}{2}\left\{ \frac{\mu^2}{\sigma^2} + \frac{n^2\pi^2\sigma^2}{(a+b)^2} \right\}. \tag{80}$$

A linear superposition of terms of the form (79) gives

$$p(x, t) = \exp\left(\frac{\mu x}{\sigma^2}\right) \sum_{n=1}^{\infty} a_n e^{-\lambda_n t} \sin\left\{ \frac{n\pi(x+b)}{a+b} \right\}, \tag{81}$$

where the coefficients a_n have to be determined from the initial condition (67). We find finally that for

$$a_n = \frac{2}{a+b} \sin\left(\frac{n\pi b}{a+b}\right) \tag{82}$$

and for λ_n given by (80), the function (81) is the required solution. The series (78) and (81) are thus alternative representations of the same function. For an arbitrary initial value x_0 we simply use (78) or (81) but replace a, b and x by $a - x_0$, $b + x_0$ and $x - x_0$ respectively.

(ii) REFLECTING BARRIER

We shall show that the boundary condition for a reflecting barrier at a is

$$\left[\frac{1}{2}\frac{\partial}{\partial x}\{\alpha(x)\,p(x;t)\} - \beta(x)\,p(x;t)\right]_{x=a} = 0. \tag{83}$$

Consider the forward equation for $p(x;t)$,

$$\frac{1}{2}\frac{\partial^2}{\partial x^2}\{\alpha(x)\,p(x;t)\} - \frac{\partial}{\partial x}\{\beta(x)\,p(x;t)\} = \frac{\partial}{\partial t}p(x;t). \tag{84}$$

For a large number of particles all starting at x_0 and all undergoing independently the same diffusion process, $p(x;t)\,\Delta x$ will be approximately the proportion of particles in the interval $(x, x + \Delta x)$ at time t. The existence of a reflecting barrier at $a > x_0$ means that no particle can pass beyond a. We therefore have that

$$\int_{-\infty}^{a} p(x;t)\,dx = 1 \tag{85}$$

and consequently

$$\frac{\partial}{\partial t}\int_{-\infty}^{a} p(x;t)\,dx = \int_{-\infty}^{a} \frac{\partial p}{\partial t}\,dx = 0. \tag{86}$$

Now from (86) and (84) we have that

$$0 = \int_{-\infty}^{a} \frac{\partial p}{\partial t}\,dx = \int_{-\infty}^{a} \frac{\partial}{\partial x}\left[\frac{1}{2}\frac{\partial}{\partial x}\{\alpha(x)\,p(x;t)\} - \beta(x)\,p(x;t)\right]dx$$

$$= \left[\frac{1}{2}\frac{\partial}{\partial x}\{\alpha(x)\,p(x;t)\} - \beta(x)\,p(x;t)\right]_{x=a},$$

and so we obtain the condition (83).

Example 5.2. *The Wiener process with a reflecting barrier.* Consider a Wiener process $X(t)$ with drift μ and variance parameter σ^2 taking place

8*

on the positive half-line where the origin is a reflecting barrier. We suppose that $X(0) = x_0 > 0$. The p.d.f. $p(x_0, x; t)$ of $X(t)$ satisfies the forward equation

$$\tfrac{1}{2}\sigma^2 \frac{\partial^2 p}{\partial x^2} - \mu \frac{\partial p}{\partial x} = \frac{\partial p}{\partial t} \quad (x > 0), \tag{87}$$

For convenience we again write $p(x_0, x; t) = p(x, t)$. We then have the initial condition

$$p(x; 0) = \delta(x - x_0) \tag{88}$$

and the boundary condition

$$\left[\frac{1}{2}\sigma^2 \frac{\partial p}{\partial x} - \mu p \right]_{x=0} = 0. \tag{89}$$

We may again use the method of images. It is known from the theory of partial differential equations (see, for example, Sommerfeld (1949, p. 66)) that the appropriate image system for this problem consists of a point image at $x = -x_0$ and a continuous system of images in the range $x < -x_0$. Accordingly we take

$$p(x; t) = \frac{1}{\sigma\sqrt{(2\pi t)}} \left[\exp\left\{ -\frac{(x - x_0 - \mu t)^2}{2\sigma^2 t} \right\} + A \exp\left\{ -\frac{(x + x_0 - \mu t)^2}{2\sigma^2 t} \right\} \right.$$
$$\left. + \int_{-\infty}^{-x_0} \exp\left\{ -\frac{(x - \xi - \mu t)^2}{2\sigma^2 t} \right\} k(\xi)\, d\xi \right] \tag{90}$$

which satisfies the equation (87) and the initial condition (88). We must determine the constant A and the function $k(\xi)$ so that the boundary condition (89) is satisfied. Using (90), the condition (89) leads to the equations

$$\tfrac{1}{2}\sigma^2 \frac{dk}{d\xi} - \mu k(\xi) = 0$$

$$(x_0 - \mu t) \exp\left(-\frac{x_0 \mu}{\sigma^2} \right) - \left\{ x_0 A + \mu t A + t\sigma^2 k(-x_0) \right\} \exp\left(\frac{x_0 \mu}{\sigma^2} \right) = 0 \quad (t > 0).$$

It follows that

$$A = \exp\left(-\frac{2x_0 \mu}{\sigma^2} \right), \qquad k(\xi) = \frac{2\mu}{\sigma^2} \exp\left(\frac{2\mu \xi}{\sigma^2} \right).$$

We finally obtain the solution

$$p(x; t) = \frac{1}{\sigma\sqrt{(2\pi t)}} \left[\exp - \left\{ \frac{(x - x_0 - \mu t)^2}{2\sigma^2 t} \right\} + \exp\left\{ -\frac{4x_0 a\mu t - (x + x_0 - \mu t)^2}{2\sigma^2 t} \right\} \right.$$
$$\left. + \frac{2\mu}{\sigma^2} \exp\left(\frac{2\mu x}{\sigma^2} \right) \left\{ 1 - \Phi\left(\frac{x + x_0 + \mu t}{\sigma\sqrt{t}} \right) \right\} \right], \tag{91}$$

where $\Phi(x)$ is the standard normal integral.

We may also investigate the possibility of an equilibrium distribution. Letting $t \to \infty$ we find that for, $\mu \geqslant 0$, $p(x;t) \to 0$ as $t \to \infty$, whereas for $\mu < 0$ we obtain the equilibrium distribution

$$p(x;\infty) = \frac{2|\mu|}{\sigma^2} \exp\left(\frac{-2|\mu|x}{\sigma^2}\right) \quad (x > 0,\ \mu < 0), \tag{92}$$

an exponential distribution independent of the initial position x_0.

(*iii*) OTHER BOUNDARY CONDITIONS

We shall state some other possible types of boundary conditions. We have already mentioned in Section 5.6 that boundary conditions involving jumps off the boundary, while preserving the Markov property, can destroy the basic form of the forward equation. Instantaneous return, for example, involves the immediate transition from a boundary point to an interior point x, distributed with a given density $f(x)$. Or we may have a jump transition to the interior after a visit to the boundary, the duration of the visit being exponentially distributed. On the other hand we cannot, after such a visit, have continuous movement off the boundary without destroying the Markovian character of the process. The reason for this will be seen in Example 5.4.

The boundary conditions discussed so far are those which can be imposed on the process. It may happen, however, that zeros or singularities of one or both of the functions $\alpha(x)$ and $\beta(x)$ preclude us from obtaining a solution with given boundary conditions. In such cases the boundary conditions are implicit in the equations. Feller (1952) has given a detailed discussion of such problems based on the theory of semi-groups and has classified boundary points into different categories according to the behaviour of $\alpha(x)$ and $\beta(x)$ at such points. Boundaries which admit the imposition of conditions such as absorption or reflection are called regular.

5.8. The Ornstein–Uhlenbeck process

We remarked in Section 5.2 that the Wiener process, regarded as a model of Brownian motion, fails for small values of t, but has been found to fit physical data for large values of t. We now consider an alternative model of Brownian motion, due to Uhlenbeck and Ornstein (1930), which overcomes this defect.

In this model we consider the velocity $U(t)$ of a Brownian particle at time t rather than the displacement $X(t)$, since the momentum mass $\times U(t)$, rather than the displacement, is affected in a known manner by the phenomenon of impact. In a small time interval there are two factors which affect the change in velocity, firstly the frictional resistance of the surrounding medium, the effect of which is proportional to $U(t)$, and secondly the random impacts of neighbouring particles whose effect in successive small time intervals can be represented by independent random variables with zero mean, which we take to be increments of a Wiener process (without drift). Thus we have

$$dU(t) = -\beta U(t)dt + dY(t), \tag{93}$$

where $Y(t)$ is a Wiener process with variance parameter σ^2. Using the expression (14) for the increment of a Wiener process we may write

$$dU(t) = -\beta U(t)dt + \sigma Z(t)\sqrt{dt}, \tag{94}$$

where $Z(t)$ is a purely random Gaussian process with zero mean and unit variance. Conditional on $U(t) = u$ we may write (94) as

$$dU(t) = -\beta u\,dt + \sigma Z(t)\sqrt{dt} \tag{95}$$

and comparison with the stochastic differential equation (56) for a general diffusion process shows that $U(t)$ is a diffusion process with infinitesimal mean and variance $\beta(u,t) = -\beta u$ and $\alpha(u,t) = \sigma^2$ respectively. The forward Kolmogorov equation is accordingly

$$\tfrac{1}{2}\sigma^2 \frac{\partial^2 p}{\partial u^2} + \beta \frac{\partial}{\partial u}(up) = \frac{\partial p}{\partial t}, \tag{96}$$

where $p(u;t)$ is the p.d.f. of the velocity $U(t)$ at time t. The m.g.f. of $U(t)$

$$\phi(\theta;t) = \int\limits_{-\infty}^{\infty} e^{-u\theta} p(u;t)\,du,$$

then satisfies

$$\tfrac{1}{2}\sigma^2 \theta^2 \phi - \beta\theta \frac{\partial \phi}{\partial \theta} = \frac{\partial \phi}{\partial t}, \tag{97}$$

a first-order linear partial differential equation (a Lagrange equation) of the same type as equation (4.37). Equation (97) can also be obtained by using Bartlett's equation (62). The cumulant generating function $K(\theta;t) = \log \phi(\theta;t)$ also satisfies a Lagrange equation

$$\beta\theta \frac{\partial K}{\partial \theta} + \frac{\partial K}{\partial t} = \tfrac{1}{2}\sigma^2 \theta^2. \tag{98}$$

The solution to (98) for an initial velocity u_0 is

$$K(\theta;t) = -(u_0 e^{-\beta t})\theta + \frac{\sigma^2(1-e^{-2\beta t})}{2\beta}(\tfrac{1}{2}\theta^2), \tag{99}$$

which represents a normal distribution for $U(t)$ with

$$E\{U(t)\} = u_0 e^{-\beta t}, \qquad V\{U(t)\} = \frac{\sigma^2(1-e^{-2\beta t})}{2\beta}. \tag{100}$$

When $t \to \infty$ we get an equilibrium distribution for the velocity which is normal with zero mean and variance $\frac{1}{2}\sigma^2/\beta$. This is in accordance with Maxwell's law for the frequency distribution of the velocity of particles in a gaseous medium in equilibrium, namely that the cartesian components of velocity have independent normal distributions with zero mean and common variance. The process $U(t)$ is called the *Ornstein–Uhlenbeck (O.U.) process*. It is both Gaussian and Markovian. Unlike the Wiener process, it does not have independent increments.

To obtain the displacement distribution we observe that in a small time interval we have

$$dX(t) = U(t)\,dt, \qquad (101)$$

where $X(t)$ is the position of the particle at time t. Integrating formally, we obtain

$$X(t) - X(0) = \int_0^t U(s)\,ds. \qquad (102)$$

The expression on the right is a stochastic integral and may be interpreted in two ways. Firstly we can regard a realization $U(s)$ $(0 \leqslant s \leqslant t)$ as being drawn from an ensemble of possible realizations (the set of continuous functions on $(0,t)$). We call $U(s)$ $(0 \leqslant s \leqslant t)$ a random function, i.e. a function selected in some given random fashion from an ensemble of functions. Now each member of the ensemble is a continuous function and therefore the integral (102) exists and defines a random variable. A second interpretation of the stochastic integral (102) is by means of a familiar limiting process applied to random variables. For any finite set of points $\{s_j\}$ in an interval $[0,t]$, the joint distribution of $\{U(s_j)\}$ is known. Hence we can define the integral (102) as a limit in some sense (e.g. mean square limit or limit with probability one) of approximating Riemann sums; see, for example, Bartlett (1955, p. 141).

The displacement process $X(t)$ whose increments are defined by (101) is no longer a Markov process, but $X(t)$ and $U(t)$ jointly form a two-dimensional Markov process whose increments are defined by (95) and (101).

The second interpretation of the integral (102) enables us to regard it as the limit of a linear combination of dependent random variables. Hence the fact that $U(t)$ is Gaussian implies that $X(t) - X(0)$ is Gaussian. By extending the formulae for the mean and variance of a linear function $L = \sum a_k X_k$ of random variables X_k, namely

$$E(L) = \sum a_k E(X_k), \qquad V(L) = \sum_k \sum a_j a_k C(X_j, X_k),$$

we obtain for the stochastic integral (102)

$$E\{X(t) - X(0)\} = \int_0^t E\{U(s)\}\,ds, \tag{103}$$

$$V\{X(t) - X(0)\} = \int_0^t \int_0^t C\{U(s), U(s')\}\,ds\,ds'. \tag{104}$$

Suppose that the process $U(t)$ is in statistical equilibrium. Then $U(t)$ has zero mean and variance $\tfrac{1}{2}\sigma^2/\beta$ and

$$
\begin{aligned}
C\{U(s), U(s')\} &= E\{U(s)\,U(s')\} \\
&= E[U(s)\,E\{U(s')|U(s)\}] \quad (s' > s) \\
&= E[\{U(s)\}^2]\,e^{-\beta(s'-s)} = \frac{\sigma^2}{2\beta}e^{-\beta(s'-s)}, \tag{105}
\end{aligned}
$$

on using the result (100). In equilibrium, therefore the correlation coefficient between $U(s)$ and $U(s')$ $(s' > s)$ is $e^{-\beta(s'-s)}$. It follows from (103) that

$$E\{X(t) - X(0)\} = 0,$$

and from (104) and (105) that

$$V\{X(t) - X(0)\} = \frac{\sigma^2}{\beta^3}(\beta t - 1 + e^{-\beta t}). \tag{106}$$

Now $X(t) - X(0)$ has a normal distribution and for large t it is like a Wiener process in the sense that, as can be seen from (106), its variance is asymptotically proportional to t. On the other hand for small t

$$V\{X(t) - X(0)\} \sim \frac{\sigma^2}{2\beta}t^2,$$

in contrast to the Wiener process for which the increment has variance always proportional to the time interval. The process $X(t)$ therefore resolves the problem of the representation of Brownian motion indicated in Section 5.2.

5.9. Transformations of the Wiener process

Let $X(t)$ be the Wiener process with unit variance parameter, so that conditional on $X(0) = 0$, $X(t)$ is normally distributed with mean zero and variance t. Let $f(t) \geqslant 0$ be a continuous strictly increasing function and let $a(t)$ be a real continuous function. Then the process

$$Y(t) = a(t)\,X\{f(t)\}$$

is also a Gaussian diffusion process; we have simply transformed the scale of time measurement. We observe that

$$dY(t) = a'(t) X\{f(t)\} dt + a(t) dX\{f(t)\}$$

$$= \frac{a'(t)}{a(t)} Y(t) dt + a(t) Z\{f(t)\}\sqrt{\{f'(t) dt\}}, \tag{107}$$

where $Z(u)$ is a purely random Gaussian process with zero mean and unit variance. Comparison with (56) shows that conditional on $Y(t) = y$, $Y(t)$ has infinitesimal mean $\{a'(t)/a(t)\}y$ and infinitesimal variance $\{a(t)\}^2 f'(t)$. The process $Z(u)$ remains a purely random process under any continuous one-to-one transformation of the time scale.

Example 5.3. *The O.U. process as a transformation of the Wiener process.* We shall show that by appropriately scaling the Wiener process we can obtain the O.U. process. In fact, consider the process

$$Y(t) = e^{-\beta t} X(\tfrac{1}{2}\beta^{-1} \sigma^2 e^{2\beta t}) \quad (\beta > 0), \tag{108}$$

which is defined for $-\infty < t < \infty$. The result (107) in this case shows that the infinitesimal mean and variance of $Y(t)$ are respectively $-\beta y$ and σ^2 and so $Y(t)$ is the O.U. process. From (108) we see that $E\{Y(t)\} = 0$ and $V\{Y(t)\} = e^{-2\beta t}(\tfrac{1}{2}\beta^{-1}\sigma^2 e^{2\beta t}) = \tfrac{1}{2}\beta^{-1}\sigma^2$ and so $Y(t)$ is stationary. Hence the simple change of scale represented by (108) transforms the Wiener process into the O.U. process. Conversely if $Y(t)$ is the stationary O.U. process then for $t > 0$

$$X(t) = \left(\frac{\sigma^2}{2\beta t}\right)^{-\frac{1}{2}} Y\left\{\frac{1}{2\beta}\log\left(\frac{2\beta t}{\sigma^2}\right)\right\} \tag{109}$$

is the Wiener process.

Example 5.4. *Rapid oscillations of the Wiener and O.U. processes.* We shall show that both the Wiener process and the O.U. process have the following interesting property: when a passage of $X(t)$ through a value x occurs at time t_0 say, then $X(t)$ takes the value x infinitely often in the interval $(t_0 < t < t_0 + \epsilon)$ however small ϵ may be. Thus, for example, if $X(0) = 0$ then $X(t) = 0$ for an infinite number of values of t in an arbitrarily small time interval $0 < t < \epsilon$.

The above result can be deduced if we first show that for the Wiener process $X(t)$ in $0 \leqslant t < \infty$ every value x recurs infinitely often with probability one. That this is so follows from the fact, proved in (39) of Section 5.4, that a first passage from any value x_1 to any other value x_2 occurs with probability one. Now consider the two processes

$$W_1(t) = t^{-\frac{1}{2}} X(t) \quad \text{and} \quad W_2(t) = t^{\frac{1}{2}} X(1/t). \tag{110}$$

A simple calculation will show that the probability structures of these two processes are identical. Since $X(t)$ crosses the origin infinitely often

$(0 \leqslant t < \infty)$ so does $W_1(t)$ and so therefore does $W_2(t)$. Thus $X(1/t)$ crosses the origin infinitely often as $t \to \infty$ and so $X(t)$ does as $t \to 0$, which proves the result. That the O.U. process has the same property follows from the transformation (108).

This phenomenon affords an explanation, at any rate for the Wiener and O.U. processes, of the fact, mentioned in Section 5.7, that if we impose a boundary condition involving a visit to a barrier, the visit having finite duration with an exponential distribution, then we cannot have continuous movement (i.e. diffusion) away from the barrier. For diffusion away from the barrier implies an infinite number of returns to the barrier in a small time interval. But at each return we would have a visit of finite duration and so movement off the barrier would never occur. In effect the barrier becomes completely absorbing. Thus a jump away from the barrier is required after a visit of finite duration.

5.10. First passage times for homogeneous diffusion processes

The method used in Section 5.4 for finding the Laplace transforms of passage time distributions for a Wiener process may clearly be generalized to any time-homogeneous diffusion process. We proceed from the backward equation

$$\tfrac{1}{2}\alpha(x_0) \frac{\partial^2 p}{\partial x_0^2} + \beta(x_0) \frac{\partial p}{\partial x_0} = \frac{\partial p}{\partial t}, \qquad (111)$$

where $p = p(x_0, x; t)$ is the density function of $X(t)$ conditional on $X(0) = x_0$.

Let $-b$ and $a(-b < a)$ be two absorbing barriers and let $T = T(-b, a | x_0)$ be the time at which absorption at $-b$ or a occurs when $X(0) = x_0$ $(-b < x_0 < a)$. Thus T is the first passage time out of the interval $(-b, a)$ for an unrestricted process starting at the point x_0 inside the interval.

We now define $p(x_0, x; t)$ as the transition probability density of the process $X(t)$ restricted by the two absorbing barriers. Hence

$$1 - G(t) = \text{prob}(T > t) = \int_{-b}^{a} p(x_0, x; t) \, dx,$$

where $G'(t) = g(t) = g(t | x_0)$ is the p.d.f. of T. Methods similar to those used for the Wiener process in deriving equation (33) will show that the m.g.f.

$$\gamma(x_0) = g^*(s | x_0) = \int_{0}^{\infty} e^{-st} g(t | x_0) \, dt \qquad (112)$$

satisfies the differential equation

$$\tfrac{1}{2}\alpha(x_0) \frac{d^2\gamma}{dx_0^2} + \beta(x_0) \frac{d\gamma}{dx_0} = s\gamma \quad (-b < x_0 < a). \qquad (113)$$

The appropriate boundary conditions are

$$\gamma(-b) = \gamma(a) = 1, \tag{114}$$

since absorption occurs immediately if the process starts at $x_0 = -b$ or $x_0 = a$.

If absorption occurs, it must take place at either a or b. Thus we have that

$$g(t|x_0) = g_-(t|x_0) + g_+(t|x_0), \tag{115}$$

where $g_-(t|x_0)$ is the probability density of passing into $-b$ at time t and $g_+(t|x_0)$ that for a. Thus $g_-(t|x_0)$ is the probability density of passing into $-b$ at time t *before* having reached a, and similarly for $g_+(t|x_0)$. Thus if we solve (113) subject to the boundary conditions

$$\gamma(-b) = 1, \qquad \gamma(a) = 0 \tag{116}$$

then we obtain $g_-^*(s|x_0)$, the Laplace transform of a density which does not necessarily integrate to unity, while the boundary conditions

$$\gamma(-b) = 0, \qquad \gamma(a) = 1 \tag{117}$$

give $g_+^*(s|x_0)$. In particular the quantities $\pi_+(x_0) = g_+^*(0|x_0)$ and $\pi_-(x_0) = g_-^*(0|x_0)$ are the probabilities of being first absorbed at a and $-b$ respectively. These may be obtained by solving (113) with $s = 0$. Thus for example $\pi_+(x_0)$ satisfies the equation

$$\tfrac{1}{2}\alpha(x_0)\frac{d^2\pi_+}{dx_0^2} + \beta(x_0)\frac{d\pi_+}{dx_0} = 0 \quad (-b < x_0 < a) \tag{118}$$

subject to the conditions

$$\pi_+(a) = 1, \qquad \pi_+(-b) = 0. \tag{119}$$

For a reflecting barrier at $-b$, say, the equation (113) still determines the m.g.f. of the first passage time to a from x_0. The appropriate boundary conditions are then

$$\gamma(a) = 1, \tag{120}$$

$$\left(\frac{d\gamma}{dx_0}\right)_{x_0=-b} = 0. \tag{121}$$

The reason for condition (120) is the same as that for conditions (114). The following rough argument will indicate the reason for condition (121). Suppose the process is at the boundary $-b$ at time t. Then let the probability of still being at $-b$ at time $t + \Delta t$ be q and the probability of making an increment $\Delta x_0 = O\{(\Delta t)^{\frac{1}{2}}\}$ be p $(p+q=1)$. Let $T(x_0)$ be the first passage time from x_0 to a. Then $T(-b) = \Delta t + T_1(-b)$ with probability q, where $T_1(-b)$ is independent of $T(-b)$ with the same distribution, or

$T(-b) = \Delta t + T(-b + \Delta x_0)$ with probability p, where $T(-b + \Delta x_0)$ is independent of $T(-b)$. In terms of m.g.f.'s, we have that

$$\gamma(-b) = q\,e^{-s\Delta t}\gamma(-b) + p\,e^{-s\Delta t}\gamma(-b + \Delta x_0), \qquad (122)$$

or

$$\gamma(-b)(1 + s\Delta t + \ldots) = q\gamma(-b) + p\left\{\gamma(-b) + \Delta x_0\left(\frac{d\gamma}{dx_0}\right)_{x_0 = -b} + \ldots\right\}.$$

If $\Delta x_0 = O\{(\Delta t)^{\frac{1}{2}}\}$ then, letting $\Delta t \to 0$, we see that (121) must hold.

For a 'visiting' condition at the barrier $-b$ we may suppose that when the process reaches $-b$ it remains there for a time U having the p.d.f. $\lambda e^{-\lambda u}$ and then undergoes a jump transition to the point x $(-b < x < a)$, distributed with p.d.f. $f(x)$. Again we must solve (113) for the m.g.f. of the first passage time to a from x_0. The boundary condition at $x_0 = -b$ is clearly

$$\gamma(-b) = \left(\frac{\lambda}{\lambda + s}\right)\int_{-b}^{a} f(x)\,\gamma(x)\,dx \qquad (123)$$

since the passage time from $-b$ to a is the sum of the random variable U and the first passage time to a from the randomly distributed point x.

The equation (113) enables us to obtain equations for the moments of the first passage time distribution. For, since $\gamma(x_0) = g^*(s|x_0)$ is the m.g.f. of the first passage time T, we have

$$\gamma(x_0) = 1 + \sum_{n=0}^{\infty} (-s)^n m_n(x_0)/n!, \qquad (124)$$

where $m_n(x_0) = E(T^n)$. Substituting (124) in (113) and equating coefficients of powers of s, we obtain

$$\tfrac{1}{2}\alpha(x_0)\frac{d^2 m_n}{dx_0^2} + \beta(x_0)\frac{dm_n}{dx_0} = -nm_{n-1}. \qquad (125)$$

In particular, for the mean $m_1 = E(T)$,

$$\tfrac{1}{2}\alpha(x_0)\frac{d^2 m_1}{dx_0^2} + \beta(x_0)\frac{dm_1}{dx_0} = -1. \qquad (126)$$

Boundary conditions for (125) can be deduced from those of $\gamma(x_0)$.

In summary we may say that equation (113) is the fundamental equation for the m.g.f. of first passage times. One boundary is always an absorbing barrier while the other may be a barrier of any admissible type.

Example 5.5. The Wiener process with two absorbing barriers. For a Wiener process with drift μ and variance parameter σ^2 equation (113) becomes equation (33), namely

$$\tfrac{1}{2}\sigma^2\frac{d^2\gamma}{dx_0^2} + \mu\frac{d\gamma}{dx_0} = s\gamma. \qquad (127)$$

The general solution is (34) and if we take the lower boundary to be zero then the boundary conditions are

$$\gamma(0) = \gamma(a) = 1,$$

and we obtain

$$\gamma(x_0) = \frac{\{1 - e^{a\theta_2(s)}\} e^{x_0\,\theta_1(s)} - \{1 - e^{a\theta_1(s)}\} e^{x_0\,\theta_2(s)}}{e^{a\theta_1(s)} - e^{a\theta_2(s)}}. \tag{128}$$

To find the corresponding p.d.f. $g(t)$ we may use the results (78) or (81), replacing a, b and x by $a - x_0$, x_0 and $x - x_0$ respectively, and calculating

$$g(t) = -\frac{d}{dt} \int_0^a p(x;t)\,dx. \tag{129}$$

The probability of absorpion at a, given by (118) and (119) with $b = 0$, is

$$\pi_+(x_0) = \frac{\exp(-2\mu x_0/\sigma^2) - 1}{\exp(-2\mu a/\sigma^2) - 1}, \tag{130}$$

and the probability of absorption at zero is $\pi_-(x_0) = 1 - \pi_+(x_0)$.

Example 5.6. Wiener process with a reflecting and an absorbing barrier. If in the previous example we let a be a reflecting barrier then, to obtain the m.g.f. of the time T to absorption at zero, we must solve (127) subject to the conditions

$$\gamma(0) = 1, \qquad \gamma'(a) = 0.$$

Writing $\theta_1 = \theta_1(s)$, etc., the solution is

$$\gamma(x_0) = \frac{\theta_2\, e^{a\theta_2 + x_0\,\theta_1} - \theta_1\, e^{a\theta_1 + x_0\,\theta_2}}{\theta_2\, e^{a\theta_1} - \theta_1\, e^{a\theta_2}}. \tag{131}$$

In particular when $\mu = 0$ and $\sigma = 1$ then $\theta_2(s) = -\theta_1(s) = \sqrt{(2s)}$, and

$$\gamma(x_0) = \frac{\cosh\{(a - x_0)\,(2s)^{\frac{1}{2}}\}}{\cosh\{a(2s)^{\frac{1}{2}}\}}. \tag{132}$$

To find the p.d.f. corresponding to (131) or (132) we write $\gamma(x_0)$ in the following form, emphasizing now that we are dealing with a function of s,

$$\gamma(x_0) = g^*(s|x_0) = \sum_j \frac{\alpha_j}{s - s_j}, \tag{133}$$

where the s_j are the singularities of $g^*(s|x_0)$, all lying in the left half-plane. We then have for the density,

$$g(t|x_0) = \sum_j \alpha_j e^{s_j t}. \tag{134}$$

Note that the s_j do not depend on x_0, but the α_j do. Alternatively we may solve the forward equation to obtain an expression such as (81) and then use the formula (129).

Moments may be obtained by differentiation of (131) with respect to s or by using the result (126). We find, for example, that the mean time to absorption is

$$E(T|x_0) = \begin{cases} \dfrac{\sigma^2}{2\mu^2}\exp\left(\dfrac{2\mu a}{\sigma^2}\right)\left\{1 - \exp\left(-\dfrac{2\mu x_0}{\sigma^2}\right) - \dfrac{2\mu x_0}{\sigma^2}\exp\left(-\dfrac{2\mu a}{\sigma^2}\right)\right\} & (\mu \neq 0), \\[2ex] \dfrac{x_0(2a - x_0)}{\sigma^2} & (\mu = 0). \end{cases} \tag{135}$$

Example 5.7. The O.U. process with absorbing barriers. Suppose we take the parameters $\sigma^2 = 1$ and $\beta = 1$ in the O.U. process. Then equation (113) becomes

$$\frac{d^2\gamma}{dx_0^2} - 2x_0\frac{d\gamma}{dx_0} - 2s\gamma = 0. \tag{136}$$

For absorbing barriers at $-b$ and a the boundary conditions are $\gamma(-b) = \gamma(a) = 1$. The solution to (136) involves Weber's functions (see, for example, Whittaker and Watson, 1952, p. 347) and we shall not pursue it further. However, the probabilities of absorption are easily found to be

$$\pi_+(x_0) = \int_{-b}^{x_0} e^{x^2}\,dx \Big/ \int_{-b}^{a} e^{x^2}\,dx,$$
$$\pi_-(x_0) = 1 - \pi_+(x_0). \tag{137}$$

Letting $b \to \infty$ we obtain the result that for a single absorbing barrier at a the probability of absorption is

$$\pi_+(x_0) = 1 \quad (x_0 < a). \tag{138}$$

Hence the O.U. process is recurrent in the sense that any state is reached from any other state with probability 1.

5.11. Approximations to discrete processes by means of diffusion processes

At the beginning of this chapter we introduced a diffusion process by means of a limiting procedure applied to a Markov chain in which changes of state and the time intervals in which they occur are regarded as being very small. By this method we use our knowledge of discrete processes to throw light on diffusion processes. Another useful procedure, which can work well in particular instances, is the reverse one of using a diffusion process to study a discrete process. This procedure is useful because mathematical methods associated with the continuum (e.g. differential equations, integration) very often lend themselves more easily to analytical treatment than those associated with discrete coordinate axes (e.g. difference equations, summation).

In general a diffusion process is determined, apart from boundary conditions, by its infinitesimal mean and variance $\beta(x,t)$ and $\alpha(x,t)$ respectively. If we wish to approximate to a process with discrete states or in discrete time we must write down the equivalents of these quantities, namely the mean and variance of the increment in unit time. We may expect to get useful approximations for values of the time variable large compared with intervals between successive transitions.

Example 5.8 *The random walk.* Consider a general random walk X_t in discrete time defined by

$$X_t = X_{t-1} + Z_t \quad (t = 1, 2, \ldots),$$

where $\{Z_t\}$ is a sequence of independent, identically distributed random variables with mean μ and variance σ^2. Here the mean and variance of the increment per unit time are simply $E(Z_n) = \mu$ and $V(Z_n) = \sigma^2$, and this suggests taking a diffusion process with $\alpha(x) = \sigma^2$ and $\beta(x) = \mu$, i.e. the Wiener process, as an approximation to X_t. The fact that for large t the distribution function of the Wiener process approximates that of X is simply a statement of the central limit theorem.

Example 5.9. *The branching process.* Consider the branching process of Section 3.7. Let X_n denote the number of individuals in the nth generation and suppose that the number cf offspring per individual has mean μ and variance σ^2. The mean and variance of the increment $X_{n+1} - X_n$, conditional on $X_n = k$, are

$$
\begin{aligned}
E(X_{n+1} - X_n | X_n = k) &= k\mu - k = k(\mu - 1), \\
V(X_{n+1} - X_n | X_n = k) &= k\sigma^2.
\end{aligned}
\tag{139}
$$

Both these quantities are proportional to the population size k at time n. For an approximating diffusion process this suggests that we take a diffusion process $X(t)$ with infinitesimal mean $\beta(x,t) = \beta x$ and infinitesimal variance $\alpha(x,t) = \alpha x$, i.e. both proportional to x, the value of $X(t)$ at time t. For such a process the forward equation is, from (52),

$$\tfrac{1}{2}\alpha \frac{\partial^2}{\partial x^2} \{xp(x,t)\} - \beta \frac{\partial}{\partial x} \{xp(x,t)\} = \frac{\partial p}{\partial t} \quad (x > 0). \tag{140}$$

For the branching process the zero state is an absorbing state. For the process described by (140), the fact that $\alpha(x,t) = \alpha x$ vanishes at $x = 0$ makes the origin a natural absorbing barrier and (140) is an example of a singular diffusion equation in which we cannot impose a boundary condition at $x = 0$, i.e. the boundary condition is implicit in the equation (140).

A convenient method of solving (140) is to use Bartlett's equation (62) for the m.g.f. $\phi(\theta,t)$ of $X(t)$. In this case (62) becomes

$$\frac{\partial \phi}{\partial t} = \left\{ \theta \beta \frac{\partial}{\partial \theta} - \tfrac{1}{2} \theta^2 \alpha \frac{\partial}{\partial \theta} \right\} \phi \tag{141}$$

$$= (\beta \theta - \tfrac{1}{2} \alpha \theta^2) \frac{\partial \phi}{\partial \theta}. \tag{142}$$

This is another example of a Lagrange equation similar to (4.37). The solution for an initial value $X(0) = x_0$ is

$$\phi(\theta,t) = \exp \left\{ \frac{-x_0 e^{\beta t} \theta}{1 + (\tfrac{1}{2}\alpha/\beta)(e^{\beta t} - 1)\theta} \right\}, \tag{143}$$

from which we easily find that

$$E\{X(t)\} = x_0 e^{\beta t}, \qquad V\{X(t)\} = \frac{\alpha x_0}{\beta} e^{\beta t}(e^{\beta t} - 1). \tag{144}$$

For the branching process X_n, with k_0 individuals in the 0th generation we have from (3.76) and (3.77) that

$$E(X_n) = k_0 \mu^n, \qquad V(X_n) = \frac{k_0 \sigma^2}{\mu(\mu - 1)} \mu^n(\mu^n - 1). \tag{145}$$

The similarity between (144) and (145) is made apparent by the exponential functions of the time entering into both in a similar form.

To obtain the probability of absorption, i.e. $\text{prob}\{X(t) = 0\}$, we must examine (143) more closely. Let us write

$$A = A(x_0, t) = x_0 e^{\beta t}, \qquad B = B(t) = \frac{\alpha}{2\beta}(e^{\beta t} - 1).$$

Then

$$\phi(\theta,t) = \exp \left(\frac{-A\theta}{1 + B\theta} \right) \tag{146}$$

$$= \exp \left(-\frac{A}{B} \right) \sum_{n=0}^{\infty} \frac{1}{n!} \left(\frac{A}{B} \right)^n \frac{1}{(1 + B\theta)^n}. \tag{147}$$

Now $1/(1 + B\theta)^n$ is the m.g.f. of the gamma density $B^{-n} x^{n-1} e^{-x/B}/(n-1)!$ Thus $\phi(\theta,t)$ corresponds to a discrete probability $e^{-A/B}$ at the origin and an infinite mixture of gamma distributions. Hence the probability that absorption occurs before time t is

$$\text{prob}\{X(t) = 0\} = \exp\left(-\frac{A}{B}\right) = \exp\left\{-\frac{2\beta x_0 e^{\beta t}}{\alpha(e^{\beta t} - 1)}\right\}, \qquad (148)$$

and

$$\lim_{t \to \infty} \text{prob}\{X(t) = 0\} = \begin{array}{ll} \exp\left(-\dfrac{2\beta x_0}{\alpha}\right) & (\beta > 0), \\ 1 & (\beta \leqslant 0). \end{array} \qquad (149)$$

For the branching process with one initial individual the probability of ultimate extinction is ξ, the smallest positive root of the equation $z = G(z)$, where $G(z)$ is the p.g.f. of the number of offspring per individual; moreover $\xi < 1$ when $\mu > 1$. For k_0 initial individuals the probability of ultimate extinction is ξ^{k_0} since the descendants of any one initial individual are independent of those of any other. Thus in the discrete branching process $(\mu > 1)$ and in the diffusion process $(\beta > 0)$ the probability of ultimate absorption is an exponential function of the initial size, while for $\mu \leqslant 1$ and $\beta \leqslant 0$ in the respective processes, ultimate absorption is certain.

By changing the scale in the discrete process the resemblance between the discrete and continuous processes may be brought out even more strongly. Suppose each generation corresponds to a small time interval τ. In time t there will be $n = t/\tau$ generations. If we let

$$\mu - 1 = \beta\tau, \qquad \sigma^2 = \alpha\tau, \qquad (150)$$

then (139) becomes

$$E(X_{n+1} - X_n | X_n = k) = k\beta\tau, \qquad V(X_{n+1} - X_n | X_n = k) = k\alpha\tau, \qquad (151)$$

and (145) becomes

$$E(X_n) = k_0(1 + \beta\tau)^{t/\tau} \sim k_0 e^{\beta t} \quad (\tau \to 0), \qquad (152)$$

$$V(X_n) = \frac{k_0 \alpha\tau}{\beta\tau(1 + \beta\tau)}(1 + \beta\tau)^{t/\tau}\{(1 + \beta\tau)^{t/\tau} - 1\}$$

$$\sim \frac{\alpha k_0}{\beta} e^{\beta t}(e^{\beta t} - 1) \quad (\tau \to 0). \qquad (153)$$

Thus the formulae (145) for the branching process go over into the formulae (144) for the diffusion process.

5.12. Continuous and jump transitions

The processes considered so far in this chapter have been Markov processes with continuous transitions. Thus changes of state in a small time interval are always local and the sample functions are continuous;

hence in order to reach one given state from any other given state all intervening states have to be passed through. For the processes considered in Chapter 4, e.g. the birth–death process, the only transitions were jump transitions and each jump was a discrete random variable.

We now come to consider the most general type of Markov process in continuous time in which both continuous and jump transitions may occur.

Example 5.10. *Insurance risk.* We now consider a model in continuous time of the insurance risk problem, Example 2.1. Suppose that at time $t = 0$ the assets of an insurance company are denoted by x_0 money units. As time progresses the company receives income in various forms, e.g. from premiums and investments, and from time to time it pays out

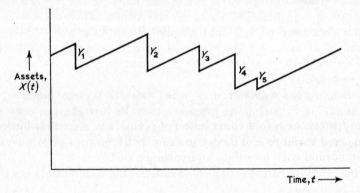

Fig. 5.1. Insurance risk problem. Assets $X(t)$ grow at a uniform rate between claims Y_1, Y_2, ..., which occur in a Poisson process.

claims. We suppose that the net income is deterministic and arrives at the rate of λ per unit time, while the claims occur in a Poisson process with rate μ, the magnitudes Y_1, Y_2, ... of the claims being independent random variables with common p.d.f. $a(y)$. Thus if $X(t)$ denotes the company's assets at time t, a realization of $X(t)$ will be of the form shown in Fig. 5.1, consisting of straight line segments of positive slope λ, corresponding to continuous transitions, and downward jumps of magnitude Y_1, Y_2, ... If at any time $X(t)$ becomes negative, the company is ruined, and one of the problems of interest is to compute the probability of ruin for given x_0 when λ, μ and $a(y)$ are known. The fact that claims occur in a Poisson process ensures that $X(t)$ is a Markov process.

Example 5.11. *Random walk in continuous time.* A particle starts at the origin and performs a random walk in continuous time in the following manner. The steps Y_1, Y_2, ... are independent random variables with p.d.f. $f(y)$ and they occur at the epochs of a Poisson process of rate α. Thus the first step occurs at time T_1, the second at time $T_1 + T_2$, and so on, where T_1, T_2, ... are independent random variables with the p.d.f. $\alpha e^{-\alpha t}$. If $X(t)$ denotes the position of the particle at time t, then $X(t)$ is a Markov process with only jump transitions and a typical realization of $X(t)$ will be a step function in which the jumps are independent random variables with p.d.f. $f(y)$. To find the distribution of $X(t)$ we note that, conditional on n steps occurring in time t, the characteristic function of $X(t)$ is

$$E(e^{i\theta X(t)} | n \text{ steps}) = \{\psi(\theta)\}^n,$$

where $\psi(\theta) = E(e^{i\theta Y})$ is the characteristic function of the step distribution $f(y)$. In this section, some extra generality is achieved by using characteristic functions rather than moment generating functions. The probability of n steps in time t is $e^{-\alpha t}(\alpha t)^n/n!$ and we therefore have that

$$\phi(\theta, t) = E(e^{i\theta X(t)}) = \sum_{n=0}^{\infty} \frac{1}{n!} e^{-\alpha t}(\alpha t)^n \{\psi(\theta)\}^n = \exp\left[\alpha t \{\psi(\theta) - 1\}\right]. \quad (154)$$

The p.d.f., $p(x,t)$, of $X(t)$ is therefore

$$p(x, t) = \sum_{n=0}^{\infty} \frac{1}{n!} e^{-\alpha t}(\alpha t)^n f_n(x), \quad (155)$$

where $f_n(x)$ is the n-fold convolution of $f(x)$. If the step distribution has mean μ and variance σ^2 it is easy to show that for large t, $X(t)$ is asymptotically normal with mean $\alpha t \mu$ and variance $\alpha t \sigma^2$.

From (154) we can see that $\phi(\theta, t)$ satisfies the differential equation

$$\frac{\partial \phi}{\partial t} = \alpha \{\psi(\theta) - 1\} \phi = -\alpha \phi(\theta, t) + \alpha \psi(\theta) \phi(\theta, t). \quad (156)$$

This may be written

$$\int_{-\infty}^{\infty} e^{i\theta x} \frac{\partial}{\partial t} p(x, t)\, dx = -\alpha \int_{-\infty}^{\infty} e^{i\theta x} p(x, t)\, dx$$

$$+ \alpha \int_{-\infty}^{\infty} e^{i\theta x} \left(\int_{-\infty}^{\infty} p(y, t) f(x - y)\, dy \right) dx,$$

or

$$\frac{\partial}{\partial t} p(x, t) = -\alpha p(x, t) + \alpha \int_{-\infty}^{\infty} p(y, t) f(x - y)\, dy. \quad (157)$$

This, for given α and $f(x)$, is an integro-differential equation for $p(x,t)$. In general, for Markov processes in continuous time with continuous and jump transitions, the transition probability functions will satisfy integro-differential equations.

We proceed to examine another example in more detail.

Example 5.12. *The Takács process.* Consider the single-server queueing system discussed in Example 3.10 in which customers arrive in a Poisson process with rate λ and their service-times are independently distributed with distribution function $B(x)$. We assume that $B(x)$ has a density $b(x) = B'(x)$. We found in Example 3.10 that we can construct an imbedded Markov chain for the number $N(t)$ of customers in the queue at time t by considering $N(t)$ at the successive instants at which customers leave the system. The process $N(t)$ itself, however, is not a Markov process in continuous time.

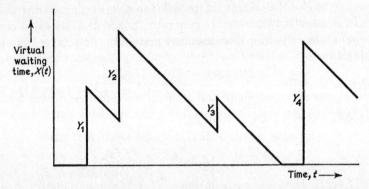

Fig. 5.2. A realization of the Takács process. The server is initially idle. Y_1, Y_2, ... are the service-times of the 1st, 2nd, ..., customers.

Following Takács (1955), we now show that by considering another variable, the so-called *virtual waiting time*, we can obtain a Markov process in continuous time with continuous and jump transitions. We define $X(t)$ to be the time which a customer who arrived at time t would have to wait until the commencement of his service. Alternatively $X(t)$ can be defined as the time, reckoned from the instant t onwards, which the customer at the end of the queue at time t will have to wait until the completion of his service. Thus, if the system is empty at time t, then $X(t) = 0$, and otherwise $X(t)$ decreases at unit rate except at the instants of arrival of customers when it jumps upwards by the amount of the arriving customer's service-time, this being a random variable drawn independently from the p.d.f. $b(x)$. A typical realization of $X(t)$ is shown in Fig. 5.2. We call $X(t)$ the Takács process or virtual waiting time process. The facts that $X(t)$ decreases in a deterministic manner in the intervals between jumps and that the jumps occur independently with a given distribution at the epochs of a Poisson process ensure that $X(t)$ is a Markov process.

At time t the distribution of $X(t)$ will consist of a discrete probability

$p_0(t)$ that $X(t) = 0$, i.e. that the system is empty, and a density $p(x,t)$ for $X(t) > 0$. Thus the distribution function of $X(t)$, given by

$$F(x,t) = p_0(t) + \int_0^x p(z,t)\, dz, \tag{158}$$

will have a jump discontinuity of magnitude $p_0(t)$ at $x = 0$ and will be continuous for $x > 0$.

In a small time interval $(t, t + \Delta t)$ the probability that no customers arrive is $1 - \lambda \Delta t + o(\Delta t)$. In this case

$$X(t + \Delta t) = \begin{cases} X(t) - \Delta t & (X(t) > \Delta t), \\ 0 & (0 \leqslant X(t) \leqslant \Delta t). \end{cases} \tag{159}$$

The probability that one new customer arrives in $(t, t + \Delta t)$ is $\lambda \Delta t + o(\Delta t)$. If this happens,

$$X(t + \Delta t) = X(t) + Y - \Delta t, \tag{160}$$

where Y, the service-time of the arriving customer, is drawn from the p.d.f. $b(y)$. We can therefore write down the following equations,

$$p(x, t + \Delta t) = p(x + \Delta t, t)\,(1 - \lambda \Delta t) + p_0(t)\, b(x)\, \lambda \Delta t$$

$$+ \left\{ \int_0^x p(x - y, t)\, b(y)\, dy \right\} \lambda \Delta t + o(\Delta t), \tag{161}$$

$$p_0(t + \Delta t) = p_0(t)\,(1 - \lambda \Delta t) + p(0, t)\, \Delta t (1 - \lambda \Delta t) + o(\Delta t), \tag{162}$$

which in turn lead to

$$\frac{\partial}{\partial t} p(x, t) = \frac{\partial}{\partial x} p(x, t) - \lambda p(x, t) + \lambda p_0(t)\, b(x) + \lambda \int_0^x p(x - y, t)\, b(y)\, dy, \tag{163}$$

$$p_0'(t) = -\lambda p_0(t) + p(0, t). \tag{164}$$

Equations (163) and (164) can be combined into a single equation for the distribution function $F(x, t)$,

$$\frac{\partial}{\partial t} F(x, t) = \frac{\partial}{\partial x} F(x, t) - \lambda F(x, t) + \lambda \int_0^x B(x - y)\, dF(y, t). \tag{165}$$

Equation (165) is called the integro-differential equation of Takács. We shall not attempt to solve (165) here.

However, if $X(t)$ is in statistical equilibrium then $p(x, t) = p(x)$, $p_0(t) = p_0$ and (163) and (164) become

$$0 = p'(x) - \lambda p(x) + \lambda p_0\, b(x) + \lambda \int_0^x p(x - y)\, b(y)\, dy, \tag{166}$$

$$0 = -\lambda p_0 + p(0). \tag{167}$$

Taking Laplace transforms in (166),

$$p^*(s) = \int_{0+}^{\infty} e^{-sx} p(x)\, dx, \qquad b^*(s) = \int_{0}^{\infty} e^{-sx} b(x)\, dx,$$

we have

$$0 = sp^*(s) - p(0) - \lambda p^*(s) + \lambda p_0 b^*(s) + \lambda p^*(s) b^*(s),$$

which gives

$$p^*(s) = \frac{p(0)\{1 - b^*(s)\}}{s - \lambda + \lambda b^*(s)}. \tag{168}$$

Let $\mu_b = -(d/ds)\{b^*(s)\}_{s=0}$ be the mean service-time. Then letting $s \to 0$ in (168), we obtain

$$p^*(0) = \frac{p(0)\mu_b}{1 - \lambda\mu_b}.$$

From (167) and the normalizing condition

$$p_0 + \int_{0}^{\infty} p(x)\, dx = p_0 + p^*(0) = 1,$$

we have that

$$p^*(s) = \frac{\lambda(1-\rho)\{1 - b^*(s)\}}{s - \lambda + \lambda b^*(s)} = \frac{\rho(1-\rho)\left[\{1 - b^*(s)\}/(s\mu_b)\right]}{1 - \rho[\{1 - b^*(s)\}/(s\mu_b)]} \tag{169}$$

where $\rho = \lambda\mu_b$ is the traffic intensity, i.e. the ratio of the mean service-time to the mean inter-arrival time. Also we have $p_0 = 1 - \rho$ and if we denote the m.g.f. of the equilibrium process by $w^*(s)$, we have that

$$w^*(s) = p_0 + p^*(s) = \frac{1-\rho}{1 - \rho g^*(s)}, \tag{170}$$

where

$$g^*(s) = \frac{1 - b^*(s)}{s\mu_b}.$$

Now $g^*(s)$ is itself a m.g.f. (cf. equations (9.2) and (9.13)) and for (170) to be a proper m.g.f. we must have $\rho < 1$, which is the condition for an equilibrium distribution to exist. The equilibrium distribution given by (170) has the form of a compound geometric distribution.

The mean waiting time in equilibrium is given by

$$E(X) = -w^{*\prime}(0) = \frac{\rho\mu_b(1 + c^2)}{2(1-\rho)}, \tag{171}$$

where c is the coefficient of variation of the service-time. Equation (171) is called the Pollaczek–Khintchine formula. If the service-time is exponentially distributed then $c = 1$ and $E(X) = \rho\mu_b/(1-\rho)$; if the service-time is constant and equal to μ_b, then $c = 0$ and $E(X) = \frac{1}{2}\rho\mu_b/(1-\rho)$. The ease with which (170) may be inverted to obtain an explicit form for the equilibrium waiting time distribution depends on the form of the service-time distribution. If, as in Examples 1.4 and 4.6, the latter is exponential with mean $1/\mu$, then $b^*(s) = \mu/(\mu+s)$, $\rho = \lambda/\mu$ and

$$w^*(s) = \frac{\mu-\lambda}{\mu}\left(1+\frac{\lambda}{\mu-\lambda+s}\right) \quad (\lambda < \mu), \tag{172}$$

which corresponds to a discrete probability $(\mu-\lambda)/\mu$ at the origin and an exponential density $\rho(\mu\text{-}\lambda)\ e^{-(\mu-\lambda)x}$ for $x > 0$.

5.13. Processes with independent increments

A process $X(t)$ is called a process with independent increments, or an additive process, if for $t' > t$, $X(t') - X(t)$ is independent of $X(t)$. Clearly, any process with independent increments is a Markov process. The process is homogeneous if the distribution of $X(t') - X(t)$ depends only on $t' - t$. The Wiener process, for example, is a diffusion process with independent increments; if $N(t)$ is the number of events in time t of a Poisson process, $N(t)$ is a discrete valued process with independent increments. The random walk in continuous time, described in Example 5.11, is a process with independent increments having only jump transitions. The superposition of a random walk of this type with a Wiener process will have the characteristic function

$$E\{e^{i\theta X(t)}\} = \exp\left(t[i\theta\mu + \tfrac{1}{2}(i\theta)^2\sigma^2 + \alpha\{\psi(\theta)-1\}]\right). \tag{173}$$

The function in square brackets may be regarded as the cumulant generating function per unit time of $X(t)$. The most general process with independent increments is a process in continuous time with continuous state space and it can be shown to consist of the superposition of a Wiener process and random walks in continuous time of the type described in Example 5.11 above. Thus if the motion of a particle is described by superposing on a Wiener process discrete jumps of different and independent types, where jumps of type j are independent random variables with characteristic function $\psi_j(\theta)$ and occur in a Poisson process of rate α_j $(j = 1,\ldots,n)$, then the characteristic function of the position $X(t)$ of the particle, conditional on $X(0) = 0$, is

$$E\{e^{i\theta X(t)}\} = \exp\left(t\left[i\theta\mu + \tfrac{1}{2}(i\theta)^2\sigma^2 + \sum_{j=1}^{n}\alpha_j\{\psi_j(\theta)-1\}\right]\right). \tag{174}$$

If $F_j(x)$ is the d.f. corresponding to $\psi_j(\theta)$, then

$$\sum_{j=1}^{n}\alpha_j\{\psi_j(\theta)-1\} = \sum_{j=1}^{n}\alpha_j\int_{-\infty}^{\infty}(e^{i\theta x}-1)\,dF_j(x)$$

$$= \int_{-\infty}^{\infty}(e^{i\theta x}-1)\,dG(x), \tag{175}$$

where $G(x) = \sum \alpha_j F_j(x)$ is a bounded non-decreasing function. The mean and variance of $X(t)$ are

$$E\{X(t)\} = t\left\{\mu + \int_{-\infty}^{\infty} x\, dG(x)\right\}, \tag{176}$$

$$V\{X(t)\} = t\left\{\sigma^2 + \int_{-\infty}^{\infty} x^2\, dG(x)\right\}. \tag{177}$$

The cumulant generating function per unit time has the form

$$i\theta\mu + \tfrac{1}{2}(i\theta)^2\sigma^2 + \int_{-\infty}^{\infty} (e^{i\theta x} - 1)\, dG(x), \tag{178}$$

the first two terms representing the continuous diffusion component and the integral term the discontinuous jump components. The most general form of this function for homogeneous processes is the Lévy–Khintchine form for the cumulant generating function of an infinitely divisible distribution given, for example, by Gnedenko and Kolmogorov (1954, p. 76). A distribution with characteristic function $\phi(\theta)$ is called infinitely divisible if $\{\phi(\theta)\}^t$ is a characteristic function for every $t > 0$. The characteristic function of a homogeneous additive process is always of the form $e^{tK(\theta)}$ and is therefore infinitely divisible. Essentially, an additive process is the continuous time analogue of the general random walk in discrete time.

We now consider a process with independent increments in the presence of absorbing barriers. The method of Wald's identity, which holds for the random walk in discrete time between absorbing barriers, can also be applied to the continuous time case. Thus suppose $X(t)$, with $X(0) = x_0$, is a homogeneous additive process with cumulant generating function $K(\theta)$ per unit time, and let $-b$ and a $(-b < x_0 < a)$ be two absorbing barriers. The continuous random variable T denotes the time at which $X(t)$ first reaches or jumps beyond one or other of the barriers. Then Wald's identity takes the form

$$E[e^{i\theta X(T)}\{e^{K(\theta)}\}^{-T}] = e^{i\theta x_0},$$

or

$$E[e^{i\theta X(T) - TK(\theta)}] = e^{i\theta x_0}. \tag{179}$$

To obtain the corresponding form in terms of m.g.f.'s rather than characteristic functions, we simply replace $i\theta$ by $-\theta$.

Example 5.13. *The busy period in the Takács process.* The Takács process of Example 5.12 can be regarded as an additive process in the presence of a reflecting barrier at the origin. If a customer arrives at time $t = 0$ and his service-time is x_0 then the time T until $X(t)$ is zero for the first time represents a busy period for the server. The succeeding idle period is the interval until the next customer arrives and will have the same exponential distribution as the inter-arrival times. Clearly T can be regarded as the time to absorption if we regard the origin as an absorbing barrier. The fact that $X(T) \equiv 0$, i.e. $X(t)$ can only change continuously in the negative direction and cannot therefore 'overshoot' the absorbing barrier, enables us to obtain exact results from Wald's identity. We have in this case

$$K(\theta) = -i\theta + \lambda \left\{ \int_0^\infty e^{i\theta x} b(x)\,dx - 1 \right\}. \tag{180}$$

The term $-i\theta$ corresponds to the diffusion component, in this case the deterministic decrease of $X(t)$ at unit rate, while the second term corresponds to the jump component, in this case the positive jumps with p.d.f. $b(x)$ occurring in a Poisson process with rate λ. Using the Laplace transform, i.e. replacing $i\theta$ by $-s$, we can write the right-hand side of (180) in the form

$$s + \lambda\{b^*(s) - 1\}. \tag{181}$$

If we assume that the queue is stable, i.e. $\lambda\mu_b < 1$, where

$$\mu_b = -(d/ds)\{b^*(s)\}_{s=0}$$

is the mean service-time, then Wald's identity becomes, on setting $X(T) = 0$,

$$E[e^{-T\{s + \lambda b^*(s) - \lambda\}}] = e^{-sx_0}. \tag{182}$$

If we now put

$$\omega = s + \lambda b^*(s) - \lambda, \tag{183}$$

then

$$E(e^{-\omega T}) = e^{-s(\omega)x_0}, \tag{184}$$

where $s(\omega)$ is the root of (183) which is positive when ω is positive and satisfies $s(0) = 0$. More precisely we may write (184) as

$$E(e^{-\omega T} | X(0) = x_0) = e^{-s(\omega)x_0}, \tag{185}$$

and since $X(0)$ has the p.d.f. $b(x_0)$ we have

$$f^*(\omega) = E(e^{-\omega T}) = \int_0^\infty e^{-s(\omega)x_0} b(x_0)\,dx_0$$

$$= b^*\{s(\omega)\}$$

$$= \frac{1}{\lambda}\{\omega - s(\omega) + \lambda\}, \tag{186}$$

which gives the moment generating function $f^*(\omega)$ of the random variable T.

To obtain moments of T we can differentiate (182) with respect to s and set $s = 0$. Thus for example

$$E[-T\{1 + \lambda b^{*\prime}(s)\}e^{-T\{s + \lambda b^{*}(s) - \lambda\}}|X(0) = x_0] = -x_0 e^{-sx_0},$$

and if we set $s = 0$ we obtain

$$E(T|x_0) = \frac{x_0}{1 - \lambda\mu_b} = \frac{x_0}{1 - \rho}, \tag{187}$$

and unconditionally,

$$E(T) = \frac{E\{X(0)\}}{1 - \rho} = \frac{\mu_b}{1 - \rho}. \tag{188}$$

Alternatively, we can use the Lagrange inversion formula (see, for example, Whittaker and Watson, 1952, p. 133) to express $s(\omega)$ as a power series in ω. The moments of T will then be given by (186). More directly, differentiating (183) with respect to ω we have

$$1 = s'(\omega) + \lambda b^{*\prime}(s)s'(\omega), \tag{189}$$

which gives

$$s'(0) = \frac{1}{1 + \lambda b^{*\prime}(0)} = \frac{1}{1 - \lambda\mu_b} = \frac{1}{1 - \rho}. \tag{190}$$

Similarly

$$s''(0) = \frac{-\lambda(\sigma_b^2 + \mu_b^2)}{(1 - \rho)^3}, \tag{191}$$

where σ_b^2 is the variance of the service-time distribution. From (186) we have

$$E(T) = \frac{1 - s'(0)}{\lambda} = \frac{\mu_b}{1 - \rho},$$

as before, and

$$V(T) = \frac{-s''(0)}{\lambda} - \{E(T)\}^2$$

$$= \frac{\sigma_b^2 + \rho\mu_b^2}{(1 - \rho)^3}. \tag{192}$$

5.14. Multidimensional processes

For a one-dimensional diffusion process $X(t)$ the transition p.d.f. $p(x_0, t_0; x, t)$, giving the p.d.f. of $X(t)$, conditional on $X(t_0) = x_0$, satisfies the forward and backward Kolmogorov equations (52) and (53). The first and second infinitesimal moments $\beta(x, t)$ and $\alpha(x, t)$, defined by the limits on the right-hand sides of (43) and (44) respectively, suffice to determine the process, apart from initial and boundary conditions. These results can be generalized to n-dimensional diffusion processes. We have already mentioned briefly one example of a two-dimensional process, namely the velocity and displacement of a particle whose velocity follows an Ornstein–Uhlenbeck process.

For an n-dimensional process $\mathbf{X}(t)$ with components $X_1(t), \ldots, X_n(t)$ we define the infinitesimal means for $i = 1, \ldots, n$ by

$$\beta_i(x_1, \ldots, x_n; t) = \lim_{\Delta t \to 0} \frac{1}{\Delta t} E[\{X_i(t + \Delta t) - X_i(t)\} | X_1(t) = x_i, \ldots, X_n(t) = x_n]$$

or using vector notation, $\mathbf{x} = (x_1, \ldots, x_n)$,

$$\beta_i(\mathbf{x}, t) = \lim_{\Delta t \to 0} \frac{1}{\Delta t} E[\{X_i(t + \Delta t) - X_i(t)\} | \mathbf{X}(t) = \mathbf{x}]. \tag{193}$$

The infinitesimal second moments are defined by

$$\alpha_{ij}(\mathbf{x}, t) = \lim_{\Delta t \to 0} \frac{1}{\Delta t} C[\{X_i(t + \Delta t) - X_i(t)\}, \{X_j(t + \Delta t) - X_j(t)\} | \mathbf{X}(t) = \mathbf{x}]$$
$$(i, j = 1, \ldots, n). \tag{194}$$

The forward Kolmogorov equation, or Fokker–Planck equation, then takes the form

$$\frac{1}{2} \sum_{j=1}^{n} \sum_{i=1}^{n} \frac{\partial^2}{\partial x_i \, \partial x_j} \{\alpha_{ij}(\mathbf{x}, t) \, p\} - \sum_{j=1}^{n} \frac{\partial}{\partial x_j} \{\beta_j(\mathbf{x}, t) \, p\} = \frac{\partial p}{\partial t}, \tag{195}$$

where $p(\mathbf{x}_0, t_0; \mathbf{x}, t)$ is the multivariate p.d.f. of $\mathbf{X}(t)$ conditional on $\mathbf{X}(t_0) = \mathbf{x}_0 = (x_{01}, \ldots, x_{0n})$. The backward Kolmogorov equation becomes

$$\frac{1}{2} \sum_{j=1}^{n} \sum_{i=1}^{n} \alpha_{ij}(\mathbf{x}_0, t) \frac{\partial^2 p}{\partial x_{0i} \, \partial x_{0j}} + \sum_{j=1}^{n} \beta_j(\mathbf{x}_0, t) \frac{\partial p}{\partial x_{0j}} = \frac{\partial p}{\partial t}. \tag{196}$$

Corresponding to the stochastic differential equation (56) we have, conditional on $\mathbf{X}(t) = \mathbf{x}$,

$$d\mathbf{X}(t) = \boldsymbol{\beta}(\mathbf{x}, t) \, dt + \mathbf{Z}(t) \, \mathbf{A}^{\frac{1}{2}}(\mathbf{x}, t) \sqrt{dt}, \tag{197}$$

where $\boldsymbol{\beta}(\mathbf{x}, t)$ is the row vector of $\beta_j(\mathbf{x}, t)$, $\mathbf{A}^{\frac{1}{2}}(\mathbf{x}, t)$ is the unique positive-definite matrix satisfying $\mathbf{A}^{\frac{1}{2}} \cdot \mathbf{A}^{\frac{1}{2}} = \mathbf{A}(\mathbf{x}, t) = [\alpha_{ij}(\mathbf{x}, t)]$ (Bellman, 1960, p. 92), and $\mathbf{Z}(t) = \{Z_1(t), \ldots, Z_n(t)\}$ is a vector of mutually independent purely random processes with zero mean and unit variance. For example, consider a three-dimensional Wiener process $\mathbf{X}(t) = \{X_1(t), X_2(t), X_3(t)\}$ with say $\mathbf{X}(0) = \mathbf{0}$. The increment $\mathbf{X}(t_0 + t) - \mathbf{X}(t_0)$ will be independent of $\mathbf{X}(t_0)$ and have a three-dimensional normal distribution with mean vector $t(\mu_1, \mu_2, \mu_3)$ and dispersion matrix $t\boldsymbol{\Sigma}$. The p.d.f. $p(\mathbf{x}, t)$ will satisfy the three-dimensional heat equation (forward equation)

$$\frac{1}{2} \sum_{i=1}^{3} \sum_{j=1}^{3} \sigma_{ij} \frac{\partial p}{\partial x_i \, \partial x_j} - \sum_{i=1}^{3} \mu_i \frac{\partial p}{\partial x_i} = \frac{\partial p}{\partial t}, \tag{198}$$

where $(\sigma_{ij}) = \boldsymbol{\Sigma}$.

Example 5.14. *The movement of larvae as a Wiener process.* After being hatched from the droppings of sheep or rabbits the larvae of the helminth

Trichostrongylus retortaeformis wander apparently at random. Broadbent and Kendall (1953) represented this as a two-dimensional Wiener process $\mathbf{X}(t) = (X_1(t), X_2(t))$ in which the components $X_1(t)$, $X_2(t)$ are independent with zero means and equal variance parameter σ^2. It follows from the preceding discussion that for $\mathbf{X}(0) = 0$, $\mathbf{X}(t)$ has the circular normal distribution in two dimensions. Thus if $R(t)$ is the radial distance of a larva from its starting point then $R^2(t) = X_1^2(t) + X_2^2(t)$ has an exponential distribution,

$$\text{prob}(R(t) < r) = 1 - \exp\left(\frac{-r^2}{2\sigma^2 t}\right). \tag{199}$$

The probability that a larva is in the annulus $(r, r+dr)$ at time t is

$$\frac{r}{\sigma^2 t} \exp\left(\frac{-r^2}{2\sigma^2 t}\right) dr. \tag{200}$$

If data are available on the observed radial distance of independent larvae at different times after starting, then equation (199) enables the model to be tested; for, according to (199), the logarithm of the proportion of larvae beyond r at time t, when plotted against r^2/t, should give a straight line. For further details and a statistical analysis, we refer to the above paper.

Bibliographic Notes

For the work of Einstein and Wiener mentioned in the introduction, see Einstein (1956) and Wiener (1923). The method of deriving the Fokker–Planck equation leading to equation (61) follows Lévy (1948, p. 65). A discussion of Feller's theory, mentioned at the end of Section 5.7, is given by Bharucha-Reid (1960). The transformations (108) and (109) are due to Doob (1942). The discussion of Example 5.4 follows Lévy (1948, p. 246). The discussion of Section 5.10 is based on the paper of Darling and Siegert (1953). Example 5.9 is due to Feller (1951). Studies of approximate methods for stochastic processes, including diffusion processes, are given by Daniels (1960) and Whittle (1957).

Exercises

1. By scaling the variables appropriately, show that the Ehrenfest model, Example 3.21, goes over into the O.U. process as the number of particles tends to infinity.

2. In a population of h individuals, each individual is one or other of two genotypes. At each unit of time one individual chosen at random dies and is replaced by a new individual whose genotype is determined at random from those existing before death. The number of individuals of, say, the

first genotype forms a Markov chain whose transition probabilities $p_{jk} = \text{prob}(X_n = k | X_{n-1} = j)$ satisfy

$$p_{j,j-1} = p_{j,j+1} = \frac{j(h-j)}{h^2}, \qquad p_{jj} = 1 - \frac{2j(h-j)}{h^2},$$

$$p_{jk} = 0 \quad (|j-k| > 1).$$

In a diffusion approximation for large h and small intervals between deaths, let $Y(t)$ denote the proportion at time t of individuals of the first genotype. Show that the approximation is governed by the diffusion equation

$$\frac{\partial f(y,t)}{\partial t} = \frac{\partial^2}{\partial y^2}\{y(1-y)f(y,t)\} \quad (0 < y < 1),$$

where $f(y,t)$ is the density function of $Y(t)$.

(Watterson, 1961)

3. Show that the equilibrium distribution for the Wiener process between two reflecting barriers is a truncated exponential distribution if the drift is non-zero and a uniform distribution if the drift is zero.

4. The displacement process $X(t)$ of a particle whose velocity $U(t)$ follows the O.U. process is defined by (102). Show that $X(t)$ is not a Markov process.

5. Show that the equilibrium distribution for the O.U. process between two reflecting barriers is a truncated normal distribution.

6. Consider the O.U. process defined by the stochastic differential equation

$$dU(t) = -\beta U(t) + \alpha^{\frac{1}{2}} dX(t),$$

where $X(t)$ is a Wiener process with unit variance parameter and zero drift. For a reflecting barrier at $-b < 0$ and an absorbing barrier at $a > 0$, show that the mean time to absorption, given $U(0) = u_0$ $(-b < u_0 < a)$, is

$$2\left(\frac{\pi}{\alpha\beta}\right)^{\frac{1}{2}} \int_{u_0}^{a} \left[\Phi\left\{\left(\frac{\alpha}{2\beta}\right)^{\frac{1}{2}}x\right\} - \Phi\left\{-\left(\frac{\alpha}{2\beta}\right)^{\frac{1}{2}}b\right\}\right] \exp\left(\frac{\beta}{\alpha}x^2\right) dx,$$

where $\Phi(x)$ is the normal probability integral.

7. An assembly line, moving with unit velocity, has items for service spaced along it. The single-server moves with the line while servicing an item and transfers instantaneously from completion of one item to service of the next. Service-times are independently distributed with exponential p.d.f. $\alpha e^{-\alpha x}$. The line moves into a fixed barrier and service becomes disabled if the server moves into the barrier. The server starts

on the first item when it is x_0 time units away from the barrier. Let $X(t)$ be the distance (in time units) of the server from the barrier at time t, so that $X(0) = x_0$. Show that if the spacings of the items are independent random variables then $X(t)$ follows the Takács process.

8. In the previous exercise, let N be the number of items completed before service becomes disabled. Using Wald's identity (179), or otherwise, obtain the following results.

(a) If the items have constant spacing ϵ, show that

$$\text{prob}(N = n) = \frac{x_0(x_0 + n\epsilon)^{n-1}}{n!}(\alpha e^{-\alpha\epsilon})^n e^{-\alpha x_0}.$$

Show further that N is finite with probability 1 if $\alpha\epsilon \leqslant 1$, and for $\alpha\epsilon < 1$ prove that

$$E(N) = \frac{\alpha x_0}{1 - \alpha\epsilon}, \qquad V(N) = \frac{\alpha x_0}{(1 - \alpha\epsilon)^3}.$$

(b) If the spacings are exponentially distributed with p.d.f. $\beta e^{-\beta x}$, show that

$$\sum_{n=0}^{\infty} z^n \text{prob}(N = n) = \exp\left(-\tfrac{1}{2}x_0[\alpha - \beta + \{(\alpha + \beta)^2 - 4\alpha\beta z\}^{\frac{1}{2}}]\right),$$

and that N is finite with probability 1 if $\alpha/\beta \leqslant 1$. For $\alpha/\beta < 1$, show that

$$E(N) = \frac{\alpha x_0}{1 - (\alpha/\beta)}, \qquad V(N) = \frac{\alpha x_0\{1 - (\alpha/\beta)^2\}}{\{1 - (\alpha/\beta)\}^3}.$$

(McMillan and Riordan, 1957)

9. In the diffusion approximation to the branching process, Example 5.9, show that the continuous part of the density corresponding to the m.g.f. (146) is

$$\exp\left(-\frac{x}{B}\right)\left(\frac{A}{xB^2}\right)^{\frac{1}{2}} I_1\left\{2\left(\frac{Ax}{B^2}\right)^{\frac{1}{2}}\right\},$$

where $I_1(z)$ is the modified Bessel function of the first kind defined by

$$I_1(z) = \sum_{n=0}^{\infty} \frac{(\tfrac{1}{2}z)^{2n+1}}{n!(n+1)!}.$$

10. By using the continuous-time version of Wald's identity (179), obtain the probability of absorption (130) for a Wiener process between two absorbing barriers.

11. In the insurance risk problem, Example 5.10, suppose that the

p.d.f. $a(y)$ of claims is the exponential distribution $\alpha e^{-\alpha y}$. Using Wald's identity (179), show that $\pi(x_0)$, the probability of ruin, is given by

$$\pi(x_0) = \begin{cases} 1 & (\mu \geqslant \alpha\lambda), \\ \exp\left\{\left(\dfrac{\mu}{\alpha}-\lambda\right)x_0\right\} & (\mu < \alpha\lambda). \end{cases}$$

12. Let $X(t)$ be a process with independent increments such that changes in $X(t)$ in the negative direction can only occur continuously. Changes in the positive direction may be arbitrary. (An example is the Tákacs process which changes continuously in the negative direction and by jumps in the positive direction.) Then $X(t)$ may be called 'skip-free' in the negative direction. Let $p(x;t)$ be the p.d.f. of $X(t)$ conditional on $X(0) = 0$. By using Wald's identity, or otherwise, show that the probability density $g(t)$ of the first passage time from $x_0 > 0$ to 0 is given by

$$g(t) = \frac{x_0}{t}p(-x_0;t).$$

(Keilson, 1963)

13. A particle in a plane makes steps of magnitude δ in a random direction. They occur randomly in time at rate λ. Find, in the limiting case for very small steps at increasing rate, the p.d.f. of the distance of the particle from the origin at time t.

(*Cambridge Diploma*, 1958)

14. Show that the diffusion approximation to the linear birth process, Example 4.1, is given by the formulae of Example 5.9 with $\sigma^2 = \mu = \lambda$, the birth rate per individual. Compare the exact and approximate solutions.

CHAPTER SIX

Non-Markovian Processes

6.1. Introduction

In the previous chapters we have discussed various types of Markov process. The detailed methods have differed depending on the nature of the state space and on whether the process is in discrete or in continuous time. However, the general principle is the same in all cases. In a Markov process, we can relate the probability distributions at different times, by using a simple transition probability distribution. Nearly always the transition probability distribution is specified fairly directly by the model under discussion. The procedure is discussed briefly in general terms in Chapter 1.

The Markov condition is crucial for the success of this method; the transition probabilities from a particular time can depend only on the state currently occupied and possibly also on time. One of the practical implications of the Markov condition for time-homogeneous processes is that, for processes with discrete states in continuous time, the underlying random variables that in effect generate the process are exponentially distributed. For example, in the linear birth–death process of Example 4.4 the times between successive births to an individual, and the time to death, are exponentially distributed. So too are the successive times spent in a particular state. The corresponding distribution in discrete time is the geometric.

Broadly speaking, a process with discrete states in which the underlying random variables are not exponentially, or geometrically, distributed cannot be dealt with directly by the methods of previous chapters. To analyse such processes we need somehow to reduce them to Markov processes and in the present chapter we outline a number of methods for attempting this. One of the methods to be discussed, the method of stages, has already been outlined in Section 4.6(i).

The essential difficulty with non-Markov processes can be illustrated by making a ponderous analysis of a mathematically trivial example. First, however, an important general definition is required. Let Y be a continuous positive random variable, which to be concrete, we shall call age to death. If at some instant an individual is alive and of age y, the limiting probability that death will occur in the next Δy is determined by

$$\lambda_f(y) = \lim_{\Delta y \to 0+} \frac{\text{prob}(y < Y < y + \Delta y \,|\, y < Y)}{\Delta y}. \tag{1}$$

This is called the *hazard function* or, for this specific application, the *age-specific death rate*. The suffix f is added to indicate the p.d.f. to which the hazard corresponds.

Let $\mathscr{F}(y)$ denote the survivor function of Y, i.e.

$$\mathscr{F}(y) = \text{prob}(Y > y) = \int_{y}^{\infty} f(u)\,du, \tag{2}$$

where $f(y)$ is the p.d.f. Now from the definition (1),

$$\lambda_f(y) = \frac{f(y)}{\mathscr{F}(y)}, \tag{3}$$

or equivalently

$$\mathscr{F}'(y) = -\lambda_f(y)\,\mathscr{F}(y), \tag{4}$$

whence

$$\mathscr{F}(y) = \exp\left\{-\int_{0}^{y} \lambda_f(u)\,du\right\}, \tag{5}$$

on using the initial condition $\mathscr{F}(0) = 1$. Therefore

$$f(y) = -\mathscr{F}'(y) = \lambda_f(y)\exp\left\{-\int_{0}^{y} \lambda_f(u)\,du\right\}. \tag{6}$$

These results are easily proved also by following the argument of Example 4.1, that is by relating $\mathscr{F}(y + \Delta y)$ to $\mathscr{F}(y)$, thereby obtaining the differential equation (4). Alternatively (4) follows directly from (3), on using the fact that $f(y) = -\mathscr{F}'(y)$.

It follows from these results that $\lambda_f(y)$ is constant if and only if the distribution of Y is exponential.

We can now discuss our example.

Example 6.1. *An inevitably fatal illness.* Suppose that an individual is born at $t = 0$ and let him remain in state 0 ('well') for a time X and then enter state 1 ('ill'). After a further time Y, spent entirely in state 1, the individual dies (enters state 2). Let X and Y be independent random variables with p.d.f.'s $\alpha e^{-\alpha x}$ and $f(y)$, respectively.

The properties of this process are very easily obtained directly, but it is instructive to try to follow the approach of Chapter 4, denoting by $p_i(t)$ the probability that state i is occupied at time t.

There is no difficulty about the equation for $p_0(t)$, namely

$$p_0(t + \Delta t) = p_0(t)(1 - \alpha\Delta t) + o(\Delta t),$$

leading to $p_0(t) = e^{-\alpha t}$.

However, to form a similar equation for $p_1(t+\Delta t)$, we need, as the coefficient of $p_1(t)$,

$$\text{prob(state 1 at } t+\Delta t | \text{state 1 at } t). \tag{7}$$

Now by the discussion leading to (3)–(6), the probability corresponding to (7) is known to be $1-\lambda_f(y)\,\Delta t+o(\Delta t)$ under the extra condition that the individual has been in state 1 for time y. Hence the required probability (7) is obtained by integrating $\lambda_f(y)$ over the distribution of the time that has been spent in state 1 up to time t. To obtain this latter distribution is, however, as complicated as solving the process as a whole. In other words, the transition probability (7) is approximately as difficult to find as the unknown properties of the process of interest, and this indeed is the general situation when random variables with non-exponential distributions are introduced into a process. Of course, if the distribution of Y is exponential, $\lambda_f(y)$ is constant and this difficulty does not arise. Note especially that there would be no difficulty in formulating an equation for $p_1(t+\Delta t)$ if the transition probabilities were dependent on time in an explicitly known way. In the present case the dependence is on the (random) duration in the current state, not on time as such.

One way of avoiding this difficulty which will be discussed in general terms in Section 6.3 is to augment the definition of the states of the system so that the value of the variables, in this case y, which control the transition rates are specified. Thus we can specify the system as being in state 0, in state $(1, y)$ or in state 2, where state $(1, y)$ is occupied when the individual is in state 1 and has been there a time y. The state space is now partly discrete and partly continuous.

Let $p_1(y;t)$ be defined by

$$p_1(y;t) = \lim_{\Delta y \to 0+} \frac{\text{prob\{state 1 occupied at } t \text{ having been entered in } (t-y-\Delta y,\ t-y)\}}{\Delta y}.$$

Then for $y > 0$

$$p_1(y+\Delta t; t+\Delta t) = p_1(y;t)\{1-\lambda_f(y)\,\Delta t\}+o(\Delta t), \tag{8}$$

and

$$p_1(0;t+\Delta t)\Delta t = p_0(t)\,\alpha\Delta t+o(\Delta t). \tag{9}$$

The essential point here is that in (8) the coefficient of $p_1(y;t)$ is now the conditional probability of not dying given y and is known in simple form. Note further that state $(1, y+\Delta t)$ with $y > 0$ can be entered only from the adjacent state $(1, y)$ and further that all individuals leaving state 0 enter state $(1, 0)$, thus accounting for (9).

A final equation follows because all transitions from state $(1,y)$, for any y, are into state 2. Thus

$$p_2(t+\varDelta t) = p_2(t) + \int_0^t p_1(y;t)\lambda_f(y)\,dy\varDelta t + o(\varDelta t). \tag{10}$$

It follows from (8)–(10) that in addition to $p_0(t) = e^{-\alpha t}$, we have that

$$\frac{\partial p_1(y;t)}{\partial y} + \frac{\partial p_1(y;t)}{\partial t} = -\lambda_f(y)\,p_1(y;t), \tag{11}$$

$$p_1(0,t) = \alpha e^{-\alpha t}, \tag{12}$$

$$\frac{dp_2(t)}{dt} = \int_0^t p_1(y;t)\lambda_f(y)\,dy. \tag{13}$$

Equation (11) is solved by defining by $P_1(y;t)$ by

$$p_1(y;t) = \exp\left\{-\int_0^y \lambda_f(u)\,du\right\}P_1(y;t)$$

and then showing from (11) that $P_1(y;t)$ must be a function of $t-y$. Hence, with the initial condition (12), we have that

$$p_1(y;t) = \alpha e^{-\alpha(t-y)}\exp\left\{-\int_0^y \lambda_f(u)\,du\right\} = \alpha e^{-\alpha(t-y)}\mathscr{F}(y) \tag{14}$$

and

$$p_2(t) = \int_0^t du \int_0^u e^{-\alpha(u-y)}\mathscr{F}(y)\lambda_f(y)\,dy = \int_0^t \alpha e^{-\alpha(t-u)}\{1-\mathscr{F}(u)\}\,du \tag{15}$$

where we have used the result (3) that $\mathscr{F}(y)\lambda_f(y) = f(y)$.

Now (14) and (15) can, of course, be obtained by a direct probabilistic argument; in particular (15) is simply $\text{prob}(X+Y < t)$. The instructive point is the necessity of an appropriately full definition of the state of the system if the methods based on transitions in a small time interval are to be applied.

An alternative simpler but much more specialized procedure is to use the 'stage' device illustrated in Example 4.10. In the simplest form of this method, we would take illness to consist of k fictitious stages, the durations of which are independently exponentially distributed with p.d.f.'s $\beta e^{-\beta y}$. The p.d.f. of Y is

$$\frac{\beta(\beta y)^{k-1}e^{-\beta y}}{(k-1)!}. \tag{16}$$

If the state of the system includes a specification of the stage of illness reached, the possible states are 0; $(1,1), \ldots, (1,k)$; 2. The methods for a discrete state Markov process can now be applied directly; the details are left as an exercise.

The essential idea is the same as in the previous approach; the state of the system must be so defined that the transition probabilities are known. With the distribution (16) for Y, a 'trick' definition of the state of the system enables this to be achieved more economically than in the general approach.

There are some general points following from this discussion. First, when investigating a new application it will nearly always be advisable to start with the simplest relevant model that forms a Markov process without the addition of supplementary variables, such as y in Example 6.1. Thus in a fairly complex queueing investigation with say several types of customer, we would take the arrivals to form independent Poisson processes with different rate parameters and the distributions of service-time exponential with different parameters. Equations for the process can then be formulated given, for example, certain priority rules between customers of different types. The general qualitative effect of departures from the Poisson process or from the exponential distributions of service-time would usually be clear. Having established the general properties of the system from the simple model, we can then proceed to examine either theoretically or by simulation the effect of non-Poisson arrival patterns or non-exponential distributions of service-time.

Secondly, we can review the methods for dealing with non-Markov processes. These are broadly as follows:

(a) The stage device, already illustrated in Example 4.10. In this, special types of distribution mathematically closely related to the exponential or geometric are used. Section 6.2 discusses the method in more general terms.

(b) The inclusion of supplementary variables, as illustrated in Example 6.1, to lead to a Markov process for arbitrary distributions of the underlying random variables. This will be discussed in Section 6.3.

(c) The use of an imbedded Markov process. This is not applicable to Example 6.1 but has been illustrated several times, notably in Examples 2.14, 3.10, 3.19. Instead of considering the process at the full set of time points, for example in continuous time, we consider the behaviour at a suitably selected sub-set of time points, so chosen that the resulting process is a Markov one. We shall give some more discussion and examples of this method in Section 6.4.

(d) Finally, it may be possible to consider a new quantity associated with a given physical process and chosen to give the properties of interest and to have the Markov property. Thus for the single-server queueing process with random arrivals we can consider the number of customers queueing or the virtual waiting time (Example 5.12). Both methods have their advantages, but the latter gives a Markov process in continuous time whereas the former does not. We shall not discuss this matter further, but it is clearly of great importance in formulating a model for a practical problem to specify the state of the process in terms that will give the properties of real significance and at the same time will allow as simple an analysis as possible.

We shall illustrate the various methods on variants of two problems, the single-server queue with random arrivals and the birth–death process.

6.2. The device of stages

We now consider in slightly more general terms the method of stages illustrated in Section 4.6(i). Suppose that we have a non-negative continuous random variable X representing, to be specific, the life-time of an individual. Let there be k stages, the time spent in the ith stage after each entry to that stage being exponentially distributed with parameter λ_i.

The two most important cases arise when the stages are taken (a) in series and (b) in parallel. In (a), the stages are traversed in order. The random variable, X, is then the sum of k independent random variables, and its distribution has Laplace transform

$$\prod \frac{\lambda_i}{\lambda_i + s}. \tag{17}$$

When the λ_i's are all different, (17) can be written in partial fraction form as

$$\sum \frac{\varpi_i \lambda_i}{\lambda_i + s},$$

so that the p.d.f. is

$$\sum \varpi_i \lambda_i e^{-\lambda_i x}. \tag{18}$$

Here $\sum \varpi_i = 1$, but the ϖ_i do not all lie in $(0,1)$, so that $\{\varpi_i\}$ is not a probability distribution. It is easily shown, for example by regarding the logarithm of (17) as a cumulant generating function, that the square of the coefficient of variation of the distribution is

$$\sum (1/\lambda_i)^2 / \{\sum 1/\lambda_i\}^2.$$

For non-negative λ_i, this is between 1 and $1/k$, the value 1 being attained when only one λ_i is non-zero and the value $1/k$ when all the λ_i are equal, as in the previous simple case of a gamma distribution.

When the stages are taken in parallel, we have a probability distribution $\{\pi_i\}$ $(i = 1, \ldots, k)$ and the random variable X is, with probability π_i, exponentially distributed with parameter λ_i. That is, only a single stage is used in any one realization of X. The p.d.f. of X is

$$\sum \pi_i \lambda_i e^{-\lambda_i x}. \tag{19}$$

This is formally equivalent to (18), but there is an essential difference in that in (19) the π_i are non-negative. It follows that the functions (19) all have a maximum at $x = 0$ and are decreasing convex functions of x. When there are just two stages, the p.d.f. is

$$\pi_1 \lambda_1 e^{-\lambda_1 x} + \pi_2 \lambda_2 e^{-\lambda_2 x}. \tag{20}$$

If $\rho = \lambda_1/\lambda_2$, the squared coefficient of variation is

$$\frac{2(\pi_1/\rho^2 + \pi_2)}{(\pi_1/\rho + \pi_2)^2} - 1, \tag{21}$$

where $\pi_1 + \pi_2 = 1$. Note in particular that if first we fix π_1 and take ρ very small, (21) approaches $2/\pi_1$. If now π_1 is taken small, a large coefficient of variation is obtained. In fact, if we wish to use a p.d.f. of the form (20) as an approximation, then with the adjustable parameters in (20) we can obtain any required mean, and any coefficient of variation greater than or equal to one; the remaining adjustable parameter can then be chosen in the light of some other feature of the distribution.

A further generalization is to have k stages, the ith stage being associated with an exponential distribution of parameter λ_i, say. A starting stage is selected in accordance with a given probability distribution. At the end of a term spent in the ith stage, the jth stage is entered with probability p_{ij} and the 'life' is ended with probability p'_i, with $p'_i + \sum p_{ij} = 1$. The process continues until 'life' is ended. The resulting distribution of life-time can be shown to have a rational Laplace transform, the degree of the denominator being at most k, the number of stages.

In practice, however, the two most important special cases are that treated previously, leading to the gamma distribution with index $k-1$, and (20). The first produces a range of distributions less dispersed than the exponential, the second a range with greater dispersion than the exponential and these two are enough for many practical purposes. Of course many extensions are possible; for example, by taking two simple series of stages in parallel, we can represent a mixture of two gamma distributions. In fact, any distribution can be approximated by a combination of stages in series and in parallel. To see this, consider first k stages in series, all with the same value of λ. If we let $k \to \infty$, $\lambda \to 0$ with $k/\lambda = \mu$ fixed, we obtain a point concentration of probability at μ. By taking a number of such arrangements in parallel, we can approximate any discrete distribution, and hence also any continuous distribution.

There is an exactly parallel discussion for distributions in discrete time, the basic distribution being the geometric instead of the exponential. The main practical advantage of the stage device is that it deals in a routine way with quite complex problems. Even if no simple explicit answer emerges, the numerical solution will be much simplified. If we take fairly simple problems, it will, however, often be possible to obtain a more general solution by other methods. This is the case with Example 6.1, which is included primarily to show how one physical process can be tackled from several different points of view.

In Example 4.10, the single-server queueing problem with random arrivals was solved when the distribution of service-time corresponds to k stages in series. The corresponding discussion for k stages in parallel is left as an exercise. As a further example of the stage device, we consider some problems connected with the birth–death process.

Example 6.2. *Birth–death process.* In Example 4.4 we showed that in a simple birth–death process in which the birth and death rates are λ and μ, the probability of ultimate extinction starting from n_0 individuals is

$$\begin{array}{ll} 1 & (\lambda \leqslant \mu), \\ (\mu/\lambda)^{n_0} & (\lambda > \mu). \end{array} \tag{22}$$

In this model the age of a particular individual at death is exponentially distributed with mean $1/\mu$. Suppose now that the distribution of age at death is not exponential. How is (22) affected? We shall concentrate on the probability of extinction, although our results would enable the distribution of time to extinction to be examined.

The model is now that for each individual the time from its birth to its death is a random variable with mean $1/\mu$ and p.d.f. $g(x)$, say, the random variables for different individuals being independent. Throughout the life of each individual, births occur at a constant probability rate λ. (At each birth, the 'age' of the offspring is zero and the 'age' of the parent is unchanged.) The qualitative effect on (22) can be guessed as follows. Ultimate extinction will occur with probability one unless the birth rate exceeds the death rate. If the distribution $g(x)$ is less dispersed than the exponential distribution, the process will be nearer to the corresponding deterministic process and hence the probability of ultimate extinction will be less than $(\mu/\lambda)^{n_0}$. Correspondingly if the distribution $g(x)$ is more dispersed than the exponential distribution, the probability of extinction will exceed $(\mu/\lambda)^{n_0}$.

Suppose first that

$$g(x) = \pi_1 \mu_1 e^{-\mu_1 x} + \pi_2 \mu_2 e^{-\mu_2 x}$$

with $\pi_1, \pi_2 > 0$, $\pi_1 + \pi_2 = 1$. That is, there are two stages in parallel, and we proceed as if there are two types of individual; with probability π_i, a new individual has an exponential distribution of life-time with parameter μ_i. Since we are interested in the probability of ultimate extinction, it is sensible to try using the backward equations. Let $p_{(ij)}(t)$ be the probability that the population is extinct by time t, starting from i individuals of type 1 and j of type 2. Because the offspring of different individuals are independent,

$$p_{(ij)}(t) = p_{(i0)}(t)\,p_{(0j)}(t) = \{p_{(10)}(t)\}^i\,\{p_{(01)}(t)\}^j. \tag{23}$$

It is thus enough to find $p_{(10)}(t)$ and $p_{(01)}(t)$, which we denote more simply by $p_{(1)}(t)$ and $p_{(2)}(t)$. To form the backwards equations, we consider the system at times $-\Delta t, 0, t$ and obtain the equation

$$p_{(1)}(t+\Delta t) = p_{(1)}(t)\,(1-\lambda\Delta t-\mu_1\,\Delta t)$$
$$+\lambda\Delta t\{\pi_1\,p_{(1)}^2(t)+\pi_2\,p_{(1)}(t)\,p_{(2)}(t)\}$$
$$+\mu_1\,\Delta t+o(\Delta t), \tag{24}$$

with a similar equation for $p_{(2)}(t+\Delta t)$. The non-linear term arises because, if there is a birth in $(-\Delta t, 0)$, we have, at time zero, either one individual of each type or two individuals of the first type. Thus

$$p'_{(1)}(t) = -(\lambda+\mu_1)\,p_{(1)}(t)+\lambda\{\pi_1\,p_{(1)}^2(t)+\pi_2\,p_{(1)}(t)\,p_{(2)}(t)\}+\mu_1, \tag{25}$$

with a similar equation for $p'_{(2)}(t)$.

Consider now the probability of ultimate extinction, obtained by letting $t\to\infty$. Denoting these limits by $p_{(1)}, p_{(2)}$, we have from (25) that

$$0 = -(\lambda+\mu_1)\,p_{(1)}+\lambda p_{(1)}(\pi_1\,p_{(1)}+\pi_2\,p_{(2)})+\mu_1,$$
$$0 = -(\lambda+\mu_2)\,p_{(2)}+\lambda p_{(2)}(\pi_1\,p_{(1)}+\pi_2\,p_{(2)})+\mu_2. \tag{26}$$

Since $p_{(1)}(t), p_{(2)}(t)$ are increasing functions of t, with $p_{(1)}(0) = p_{(2)}(0) = 0$, the probabilities of ultimate extinction are in fact the smallest non-negative roots of (26).

Now the probability of extinction starting from an individual chosen at random is $p = \pi_1 p_{(1)}+\pi_2 p_{(2)}$. We have from (26) that

$$p_{(1)} = \frac{\mu_1}{\lambda+\mu_1-\lambda p}, \qquad p_{(2)} = \frac{\mu_2}{\lambda+\mu_2-\lambda p},$$

from which it follows that

$$p = \frac{\pi_1\mu_1}{\lambda+\mu_1-\lambda p}+\frac{\pi_2\mu_2}{\lambda+\mu_2-\lambda p}.$$

This leads to the cubic equation in p

$$(p-1)\{\lambda^2 p^2-\lambda(\lambda+\mu_1+\mu_2)\,p+\lambda(\pi_1\mu_1+\pi_2\mu_2)+\mu_1\mu_2\} = 0. \tag{27}$$

The roots of the quadratic part are real and positive. The probability of ultimate extinction is thus the smaller of (a) unity and (b) the lesser root of the quadratic equation. A useful check is that if $\pi_1 = 1$ then $\pi_2 = 0$, or if $\mu_1 = \mu_2$, the result (22) is recovered.

It is straightforward to obtain numerical answers from (27). In particular, if $\pi_1 = \pi_2 = \frac{1}{2}$ and if the ratio of death rate to birth rate is θ, and if the squared coefficient of variation of time to death is $1 + \eta^2$, then

$$p = \theta + \tfrac{1}{2}\eta^2 \theta(1 - \theta) + O(\eta^4), \tag{28}$$

from which a quick assessment can be made of the effect of small departures from the exponential distribution.

This deals with overdispersion relative to the exponential distribution. To represent underdispersion, we take k stages in series, each with the same exponential parameter μ, leading to a gamma distribution of index k. If $p_{(i)}(t)$ is the probability of extinction at or before time t, starting from one individual in stage i, we have as in (24) and (25) the backward equations

$$p'_{(i)}(t) = -(\lambda + \mu)\, p_{(i)}(t) + \mu p_{(i+1)}(t) + \lambda p_{(i)}(t)\, p_{(1)}(t) \quad (i = 1, \ldots, k-1),$$
$$p'_{(k)}(t) = -(\lambda + \mu)\, p_{(k)}(t) + \mu + \lambda p_{(k)}(t)\, p_{(1)}(t). \tag{29}$$

As before, the probabilities, $p_{(i)}$, of ultimate extinction satisfy the limiting equations at $t \to \infty$. Put

$$G(z) = p_{(1)} + p_{(2)} z + \ldots + p_{(k)} z^{k-1}.$$

Then we have easily from the limiting form of (29) that

$$-(\lambda + \mu)\, G(z) + \mu \left\{ \frac{G(z)}{z} - \frac{p_{(1)}}{z} \right\} + \lambda p_{(1)}\, G(z) + \mu z^{k-1} = 0.$$

Thus

$$G(z) = \frac{\mu(p_{(1)} - z^k)}{(\lambda + \mu) z - \lambda p_{(1)} z - \mu}. \tag{30}$$

Now $G(z)$ is a polynomial and in particular is analytic for all z. Hence the numerator of (30) vanishes at the zero of the denominator, i.e.

$$p_{(1)}\{\lambda + \mu - \lambda p_{(1)}\}^k = \mu^k.$$

The ratio of death rate to birth rate is $\theta = (\mu/k)/\lambda$.

The probability of ultimate extinction starting from a single individual in stage 1 is thus the smallest non-negative root of the equation

$$p \left\{ 1 + \frac{1 - p}{k\theta} \right\}^k = 1. \tag{31}$$

In particular if $k \to \infty$, we obtain for a process with constant age at death that p satisfies

$$p \exp\{(1-p)/\theta\} = 1. \tag{32}$$

We have discussed this example at some length because it illustrates how departures from the simplest Markov model can be assessed approximately by entirely elementary methods. In a practical application it is unlikely that one would know the detailed form of the distribution precisely and the use of special forms chosen for mathematical convenience is likely to be adequate.

In fact the results for the probability of extinction, although not the differential equations for the distribution of time to extinction, can be obtained much more generally by a different argument. For, so far as ultimate extinction is concerned, the only quantity associated with an individual that we need consider is the total number, N, of offspring born to him. In fact, we can consider the process as going on in discrete time with generation number as the time variable, and then we have the multiplicative Markov chain of Section 3.7. The probability of ultimate extinction is obtained there in terms of the probability generating function of N. Results (27), (31) and (32) are special cases; for instance, when the age at death is constant and births occur in a Poisson process, N has a Poisson distribution and this leads directly to (32).

6.3. Supplementary variables

The most direct and straightforward way of dealing with processes that are not initially Markovian is to add a sufficient number of supplementary variables to the definition of the state of the process to make the process Markovian. We have already illustrated this method in a very simple case with Example 6.1. In this approach we get a Markov process in which the state space is multidimensional and often partly discrete and partly continuous.

Essentially the same technique is useful when we have a process which is Markovian in terms of one property, but for which some other property of the process is required.

Example 6.3. *A queueing problem.* Suppose that customers arrive at random at rate α in a single-server queue. Let the service-times of customers be independently distributed with p.d.f. $b(x)$. When $b(x)$ is an exponential distribution we can, as in Example 4.5, consider a Markov process in which the number of customers present defines the state. If $b(x)$ is not exponential, the probability of service being completed in $(t, t+\Delta t)$ depends on the length of time service has been in progress, in accordance with the arguments of (1)–(6). This system has been analysed also in Examples 3.10, 3.19 and 5.12.

Therefore, to apply the present approach, we define the state of the system to be

    0   if there are 0 customers present,

    (n, u) if there are n customers present $(n = 1, 2, \ldots)$,

and if the customer currently being served has been at the service point for a time u. Let $p_n(u;t)$ be the joint probability and probability density of the state (n, u) at time t. More explicitly, if U is the random variable corresponding to u,

$$p_n(u;t) = \lim_{\Delta u \to 0+} \frac{\text{prob}(u < U < u + \Delta u \text{ and } n \text{ customers present at } t)}{\Delta u}.$$

Then we can obtain the forward equations for the process by the usual argument; the discussion is an extension of that for Example 6.1.

The equations are for $n = 1, 2, \ldots$

$$p_0'(t) = -\alpha p_0(t) + \int_0^\infty p_1(u;t)\lambda_b(u)\, du,$$

$$\frac{\partial p_n(u;t)}{\partial u} + \frac{\partial p_n(u;t)}{\partial t} = -\{\lambda_b(u) + \alpha\}p_n(u;t) + \alpha p_{n-1}(u;t), \qquad (33)$$

$$p_n(0;t) = \int_0^\infty p_{n+1}(u;t)\lambda_b(u)\, du + \alpha p_0(t)\delta_{n1}.$$

In the second equation the term $\alpha p_{n-1}(u;t)$ is absent when $n = 1$. The function $\lambda_b(u)$ is the hazard function (1) associated with the distribution of service-time.

These equations have to be supplemented by an initial condition or, alternatively, if the equilibrium behaviour is under study, by a normalizing condition.

An interesting variant of the equations is required if we wish to study by this method the *busy period distribution*, i.e. the distribution of the time elapsing from the arrival of a customer, the queue being previously empty, to the first subsequent time at which the queue is empty. For this we consider the above process, modified only by making zero an absorbing state, i.e. by omitting the term $-\alpha p_0(t)$ in the first equation of (33). Then $p_0(t)$ is the probability that the busy period has ended at or before t, i.e. $p_0'(t)$ is the p.d.f. of the busy period. The busy period distribution has in fact been obtained by a different method in Example 5.13. We shall not consider it further here.

To solve the time-dependent equations (33) involves analytic considerations analogous to those sketched in the discussion of Section 4.6(iii); see Keilson and Kooharian (1960) for details. For simplicity we shall concentrate here on the equilibrium solution of (33). That is, we write $p_0(t) = p_0$, $p_n(u;t) = p_n(u)$. If then

$$G(z, u) = \sum_{n=1}^{\infty} z^n p_n(u),$$

we have, from the second equation of (33), that

$$\frac{\partial G(z, u)}{\partial u} = \{\alpha z - \alpha - \lambda_b(u)\} G(z, u). \tag{34}$$

The solution of (34) is

$$G(z, u) = G(z, 0) e^{\alpha u(z-1)} \exp\left\{-\int_0^u \lambda_b(v) \, dv\right\}$$

$$= G(z, 0) e^{\alpha u(z-1)} \{1 - B(u)\}. \tag{35}$$

The equilibrium forms of the other equations (33) are

$$\alpha p_0 = \int_0^{\infty} p_1(u) \lambda_b(u) \, du,$$

$$p_n(0) = \int_0^{\infty} p_{n+1}(u) \lambda_b(u) \, du + \alpha p_0 \delta_{n1},$$

and lead to the equation

$$\alpha z(1-z) p_0 + z G(z, 0) = \int_0^{\infty} \lambda_b(u) G(z, u) \, du. \tag{36}$$

Now if $b^*(s)$ is the Laplace transform of $b(x)$, we have from (35) and the definition $\lambda_b(u) = b(u)/\{1 - B(u)\}$, that the right-hand side of (36) is

$$G(z, 0) \, b^*(\alpha - \alpha z).$$

Thus (36) leads to

$$G(z, 0) = \frac{-\alpha z(1-z) p_0}{z - b^*(\alpha - \alpha z)}.$$

The marginal probability generating function of the number of customers present is

$$p_0 + \int_0^{\infty} G(z, u) \, du = p_0 \frac{b^*(\alpha - \alpha z)(1-z)}{b^*(\alpha - \alpha z) - z}, \tag{37}$$

agreeing with the result (3.119) obtained by solving the equilibrium equations of the imbedded Markov chain of Example 3.10. Finally the normalizing condition gives $p_0 = 1 - \alpha \mu_b$, where μ_b is the mean of the distribution $b(x)$.

Other properties of the system can be determined from the results given here. For example, given that a new customer arrives to find the system in state (n, u) his queueing-time will consist of the residual service-time of the customer already being served, which has p.d.f.

$$\frac{b(u+x)}{1-B(u)},$$

convoluted with the sum of $(n-1)$ service-times each having p.d.f. $b(x)$.

Equation (37) can be regarded as a generalization of (4.81) obtained by the method of stages when the distribution of service-time is such that $b^*(s) = (1+s/\beta)^{-k}$. In fact the stage device can be regarded in general as a convenient operational method for solving the integro-differential equations that arise with the method of supplementary variables.

Example 6.4. *Total number of individuals in birth–death process.* Consider the birth–death process of Example 4.4. Suppose that we are interested not in the number, N_t, of individuals alive at time t, but in the total number, M_t, of individuals born into the population at or before t, irrespective of whether they are still alive at t. This might be required, for example, in the study of a population of micro-organisms in which experimental counting procedures are used which do not distinguish between dead and alive organisms.

Since the offspring of different individuals are independent, there is no loss of generality in supposing that there is initially a single individual. A process defined in terms of M_t would be non-Markovian, since the transition rates at t are determined by N_t. Therefore we introduce N_t as a supplementary variable, i.e. we consider the two-dimensional process (M_t, N_t), with state space the set of integer pairs (m, n). If $p_{mn}(t)$ is the probability of the state at t, we have for the basis of the forward equations

$$p_{mn}(t+\Delta t) = p_{mn}(t)\{1-(n\lambda+n\mu)\,\Delta t\}$$
$$+ p_{m-1,\,n-1}(t)\,(n-1)\,\lambda\Delta t$$
$$+ p_{m,\,n+1}(t)\,(n+1)\,\mu\Delta t + o(\Delta t), \tag{38}$$

where λ and μ are the birth and death rates per individual.

If

$$G(w, z; t) = \sum p_{mn}(t)\,w^m\,z^n$$

is the joint probability generating function, we obtain (Kendall, 1949) from (38) the partial differential equation

$$\frac{\partial G(w, z; t)}{\partial t} = \{\lambda w z^2 - (\lambda+\mu)z + \mu\}\frac{\partial G(w, z; t)}{\partial z}. \tag{39}$$

The equation for N_t alone, studied in Example 4.4, is recovered by setting $w = 1$.

If initially $M_0 = N_0 = 1$, then $G(w, z; 0) = wz$. Equation (39) can be solved as a Lagrange differential equation. Let α, β be the roots of the quadratic equation, viz.,

$$\lambda w z^2 - (\lambda + \mu) z + \mu = 0.$$

It can then be shown that

$$G(w, z; t) = w \left\{ \frac{\alpha(\beta - z) + \beta(z - \alpha) e^{-\lambda w(\beta - \alpha)t}}{(\beta - z) + (z - \alpha) e^{-\lambda w(\beta - \alpha)t}} \right\}. \tag{40}$$

One special property that can be obtained from (40) is the distribution of the limiting value of M_t when $\lambda < \mu$, so that extinction is certain. Another general technique is illustrated in Exercise 7: to obtain the first few cumulants of M_t and N_t, we convert (39) into an equation for the cumulant generating function and then expand to obtain separate differential equations for the cumulants.

6.4. Imbedded Markov process

The final method to be considered for analysing a non-Markov process is the use of an imbedded Markov process. That is, we consider the process at a suitable discrete set of time points so chosen that the new discrete time process is Markovian. For this approach to be useful, not only must there be such a series of time points, but also it must be possible to determine the properties of interest from the behaviour of the imbedded process. Clearly it will not in general be possible to determine all properties of the process in continuous time from those of the imbedded process.

An important theoretical advantage of working in discrete time is that the very general and relatively elementary results of Chapter 3 can be used, for instance to examine the existence and uniqueness of equilibrium distributions. The corresponding rigorous treatment in continuous time is often much more difficult and the appropriate general theories are beyond the scope of this book.

The method has already been illustrated a number of times in previous chapters. The general idea is explained in terms of a queueing process in Example 1.4 and is mentioned in connexion with the birth–death process in Examples 4.4 and 6.2. Special imbedded processes connected with the single-server queueing process are analysed in Examples 2.14, 3.10 and 3.19.

Thus in the birth–death process we can consider the process at those instants at which a change in state occurs. With transition rates from state n of $\lambda_n = n\lambda$ for a birth and $\mu_n = n\mu$ for a death, the imbedded process is a simple random walk with zero as an absorbing barrier. For the generalized process with arbitrary λ_n, μ_n, the imbedded process is a random walk with state-dependent probabilities of $\lambda_n/(\lambda_n + \mu_n)$ for a jump upwards and $\mu_n/(\lambda_n + \mu_n)$ for a jump downwards. A second imbedded process associated rather more indirectly with the birth–death process is, as was mentioned in Example 6.2, the multiplicative process obtained by considering the total number of individuals of a particular generation as a stochastic process with generation number as the 'time' variable. Only very special properties of the original process can be recovered from those of this particular imbedded process, because of the overlap of generations in ordinary time.

Of the two imbedded processes associated with the single-server queueing process one, Example 2.14, is associated with the virtual waiting time process analysed in continuous time in Example 5.12. The other, Examples 3.10 and 3.19, is derived from the process defined by the number of customers in the system. We now consider this further.

Example 6.5. *The single-server queue.* We consider once more the equilibrium theory of the single-server queue in which customers arrive in a Poisson process of rate α and in which the service-times of different customers are independently distributed with distribution function $B(x)$. We need to consider the number of customers, N_b, arriving during the service of one customer. We have, as in (3.24), that

$$\text{prob}(N_b = r) = b_r = \int\limits_0^\infty \frac{e^{-\alpha x}(\alpha x)^r}{r!}\, dB(x),$$

since for given service-time x we have a Poisson distribution with mean αx. The probability generating function of N_b is thus

$$G_b(z) = b^*(\alpha - \alpha z), \tag{41}$$

assuming, without risk of loss of generality, that the distribution of service-time has a p.d.f. with Laplace transform $b^*(s)$.

If the distribution of service-time is not exponential, the process defined in terms of the number of customers currently in the system is not Markovian. As shown in Example 3.10 the process considered at the instants that the service of a customer ends forms a Markov chain, the transition matrix of which is given in (3.25). In Example 3.19, we used the process to illustrate methods for studying the nature of a Markov chain.

To find the equilibrium probability distribution, we have to find an invariant vector of this matrix. This is given in (3.119) in the form of a generating function. Alternatively we can argue directly from the defining equation of the process, namely that if X_n is the number of customers present immediately following the completion of service of the nth customer, then

$$X_{n+1} = \begin{cases} X_n - 1 + N_b & (X_n > 0) \\ N_b & (X_n = 0). \end{cases}$$

If $U(x)$ is the unit Heaviside function, defined by

$$U(x) = \begin{cases} 1 & (x > 0), \\ 0 & (x \leqslant 0), \end{cases}$$

then

$$X_{n+1} = X_n - U(X_n) + N_b.$$

For the equilibrium distribution, we have, on taking probability generating functions and using the independence of N_b and X_n, that

$$E(z^X) = E\{z^{X-U(X)}\} E(z^{N_b}). \tag{42}$$

Now if $\{\pi_r\}$ is the probability distribution of X,

$$E\{z^{X-U(X)}\} = \pi_0 + \sum_{r=1}^{\infty} \pi_r z^{r-1}$$

$$= \pi_0 + \frac{G_X(z) - \pi_0}{z}, \tag{43}$$

where $G_X(z)$ is the probability generating function of X. If we substitute (43) into (42), and use (41), we get that

$$G_X(z) = \frac{\pi_0(1-z)\,b^*(\alpha - \alpha z)}{b^*(\alpha - \alpha z) - z}, \tag{44}$$

agreeing with (3.119). Finally the normalizing condition, $G_X(1) = 1$, or a direct probability argument, gives that $\pi_0 = 1 - \alpha \mu_b$.

This is identical with the solution (37) obtained by the method of supplementary variables. At first sight this is remarkable, since the distributions refer to quite distinct random variables. Equation (37) refers to an arbitrary point in time, whereas (44) refers to the instant following the completion of a service operation. An explanation of why the imbedded equilibrium distribution is the same as the continuous time equilibrium distribution is given below.

The second point of interest about (44) is the occurrence of the term $b^*(\alpha - \alpha z)$. In our treatment of this problem by the method of stages, the special rational form of this entered after algebraic manipulation of the equilibrium equations. In the treatment by supplementary variables, $b^*(\alpha - \alpha z)$ entered again in a way having no direct probabilistic significance. Here, however, a probabilistic meaning is apparent from (41).

In a rather restricted set of problems, the relation between the equilibrium distribution in continuous time and that in the imbedded process can be shown as follows (Khintchine, 1932). Consider a process with states $\{0, 1, 2, \ldots\}$ such that transitions are one step up or down. Let two imbedded processes be formed by considering the state of the process

(i) immediately before a jump upwards;

(ii) immediately after a jump downwards.

Suppose that both these processes have equilibrium distributions, say $\{p_i'\}$ and $\{p_i''\}$. Then

$$p_i' \sim \frac{A_i'}{\sum A_j'}, \qquad p_i'' \sim \frac{A_i''}{\sum A_j''},$$

where A_i' is the number of transitions in a very long time period from state i to state $i+1$ and A_i'' is the number of transitions from $i+1$ to i. However $|A_i' - A_i''| \leqslant 1$ so the two equilibrium distributions are identical.

Now suppose further that the time instants in (i) form a Poisson process independently of the state of the process. Then the equilibrium distribution $\{p_i'\}$ is identical with the equilibrium distribution in continuous time. The same applies if the time instants (ii) form a Poisson process. That is, the imbedded equilibrium distributions (i) and (ii) are always the same, and agree with the distribution in continuous time if, and usually only if, one of the time sequences (i) and (ii) is a Poisson process.

In Example 6.5, the time instants (i) are the instants just before the arrival of a customer and do constitute a Poisson process.

Bibliographic Notes

The addition of supplementary variables to make the process have the Markov property is the most direct method of analysis and has an extensive history. Cox (1955a) examined a fairly general situation in which the supplementary variables are occupation times of discrete states and showed the relationship with the method of stages. Jensen (1948) has described Erlang's work on the method of stages and Jensen (1954) and Cox (1955b) have discussed extensions. Morse (1958) has given many examples of the application of the method of stages to operational research problems. References on imbedded chains are at the end of Chapter 3. All three methods discussed here have been widely used, for example in analysing special queueing systems; the book of Saaty (1961b) describes many of these investigations and has an extensive bibliography.

Exercises

1. A process alternates between two states 0 and 1, the successive times spent in state i being independently distributed with p.d.f. $f_i(x)$. If the

process starts with an interval in state i, find the probability that the system is in state i a time t later.

2. Set up the equilibrium equations for the single-server queueing process with an exponential distribution of service-time and for which the intervals between successive arrivals are independently distributed with a gamma distribution with integral index.

3. Construct an example based on Exercise 1 to provide a simple illustration of a process for which the equilibrium distribution in continuous time and the equilibrium distribution of an imbedded Markov chain are not the same.

4. Examine the nature of the hazard function (3) when the underlying distribution is

(a) of the gamma form;

(b) of the normal form;

(c) of the log normal form;

(d) of the Weibull form, with distribution function

$$1 - \exp\{-(\alpha x)^\beta\} \quad (x > 0; \; \alpha, \beta > 0).$$

5. Two similar machines are available for a certain task, although only one is used at a time. Initially both machines are in working order and machine 1 is used. So long as it is in use, a machine is subject to breakdowns in a Poisson process of rate α. After a breakdown the machine is not in working order until after a repair time, having the p.d.f. $f(x)$. When machine 1 fails, machine 2 is used. When this fails, machine 1, if repaired, is used. This continues until after a time T, say, neither machine is in working order and a system failure is said to occur. Study the distribution of T first by the methods of Section 6.3 and then, assuming $f(x)$ to have a convenient form, by the methods of Section 6.2.

(Gaver, 1963; Epstein)

6. For a random customer arriving in the system of Example 6.3, let U be the time for which the customer currently being served has been at the service point. If the system is empty, define $U = 0$. Use the formulae of Example 6.3 to examine the marginal distribution of U. Use a similar argument to examine the distribution of V, defined as the remaining time to be spent at the service point by the customer currently being served. Compare and comment on the relation between the two distributions.

7. Show that (39) leads to the following equation for the joint cumulant generating function

$$K(\theta_w, \theta_z; t) = \log G(e^{-\theta_w}, e^{-\theta_z}; t):$$

$$\frac{\partial K}{\partial t} = -\{\lambda e^{-\theta_w - \theta_z} - (\lambda + \mu) + \mu e^{\theta_z}\}\frac{\partial K}{\partial \theta_z}.$$

By expansion in powers of θ_z and θ_w, verify the result of Example 4.4 that $E(N_t) = n_0 e^{(\lambda - \mu)t}$, where n_0 is the number of individuals at $t = 0$ Show further that

$$\frac{dE(M_t)}{dt} = E(N_t).$$

Does this agree with deterministic theory?

CHAPTER SEVEN

Stationary Processes: Time Domain

7.1. Introduction

In the previous chapters we have repeatedly dealt with processes $\{X(t)\}$ having as $t \to \infty$ an equilibrium distribution, not depending on the initial conditions. Such an equilibrium distribution usually has the important property that if initially the process has this distribution, then the marginal distribution of $X(t)$ is the same for all $t > 0$ and is identical with the equilibrium distribution. In this particular sense the statistical fluctuations of the process are stationary in time and the equilibrium distribution is, in this context, called the stationary distribution of the process.

In this and the following chapter we consider in more detail processes whose statistical fluctuations are stationary in time. There is an extensive theory of such processes and here we only outline the main ideas. The text-book by Yaglom (1962) should be consulted for a very thorough introductory account of the theory.

One of the simplest stationary processes is the two-state Markov process in continuous time, Example 4.7. In this there are two states $(0, 1)$ and respective probabilities $\alpha \Delta t + o(\Delta t)$ and $\beta \Delta t + o(\Delta t)$ of transitions out of states 0 and 1, the Markov property holding. It is then easy to obtain the distribution of the process, (4.64), namely

$$
\begin{aligned}
p_0(t) &= \frac{\beta}{\alpha+\beta} + \left\{ p_0(0) - \frac{\beta}{\alpha+\beta} \right\} e^{-(\alpha+\beta)t}, \\
p_1(t) &= \frac{\alpha}{\alpha+\beta} + \left\{ p_1(0) - \frac{\alpha}{\alpha+\beta} \right\} e^{-(\alpha+\beta)t},
\end{aligned}
\tag{1}
$$

where $p_0(0)$, $p_1(0) = 1 - p_0(0)$ specify the distribution at $t = 0$. If now $p_0(0) = \beta/(\alpha+\beta)$, then for all $t \geqslant 0$

$$
p_0(t) = \frac{\beta}{\alpha+\beta}, \qquad p_1(t) = \frac{\alpha}{\alpha+\beta},
\tag{2}
$$

and the marginal distribution is stationary.

There is a much stronger sense, however, in which the process is stationary. Consider, for example, the joint distribution of the process at two time points $t, t+h$ $(h > 0)$. If $X(t)$ is a random variable indicating

272

the state occupied at t, we have, for instance, in virtue of the Markov property, that

$$\text{prob}\{X(t+h) = X(t) = 1\} = \text{prob}\{X(t) = 1\}$$
$$\times \text{prob}\{X(t+h) = 1 | X(t) = 1\}$$
$$= \frac{\alpha}{\alpha + \beta} p_1(h), \qquad (3)$$

where $p_1(h)$ is given by (1) with $p_1(0) = 1$. The significant point is that this is a function of time only through the separation h. A corresponding result applies to the other three probabilities,

$$\text{prob}\{X(t+h) = 1, X(t) = 0\}, \text{etc.}$$

More generally, the joint distribution of the process at several time points depends only on the lengths of the intervals between the time points, i.e. is unaffected by translating the time points as a set. This is the essence of the general definition of a stationary process, to be given in the next section.

The stationary process is generated in the above example by starting with the appropriate initial condition at $t = 0$. Alternatively we often find it convenient mathematically and sensible physically to regard the process as defined for all real t, i.e. as starting in the infinitely remote past.

All the processes previously discussed having an equilibrium distribution can be used to generate examples of stationary processes. The following further examples illustrate particularly applications in which stationary processes are important. We consider processes both in discrete time and in continuous time. Given a process in continuous time, we can always derive one in discrete time by sampling, for example at equally spaced time points.

Example 7.1. *Process control.* One important application of stationary processes is to the control of industrial processes. Suppose that X_n is the measured value at the nth time point of some property of the process which is required to control; of course, in some applications observations will be taken in conti̇ ıous time. Figure 7.1 shows observations at hourly intervals of the viscosity of the product in a chemical process; this type of variation is fairly typical of one common type of observed series.

A first, rather crude, formulation of the problem of control is as follows. It is required to control the process so that as far as possible X takes the constant value, a. To this end it is possible after the nth observation to adjust the process by any desired amount, i.e. in effect to add any desired constant to the resulting values of X_{n+1}, X_{n+2}, ... To achieve optimum control we can argue as follows. First we use the observations up to time n to obtain a predictor $\hat{X}_{n+1}$ of the value that would be taken at time $n+1$ if no adjustment were applied. We then adjust the process by $\hat{X}_{n+1} - a$. The resulting value at time $n+1$ will be off target by the error of prediction $X_{n+1} - \hat{X}_{n+1}$. After obtaining X_{n+1} we can adjust the process again depending on $\hat{X}_{n+2}$, and so on. Thus in this simple formulation the problems of prediction (or forecasting) and control are equivalent.

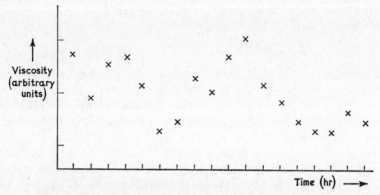

Fig. 7.1. Viscosity of chemical product at hourly intervals.

Often account has to be taken of the dynamics of the process. An adjustment made at time n may produce its effect gradually over an appreciable time rather than immediately. Another common complication is that control has to be based on many properties, not on one.

Example 7.2. *Economic time series.* Many types of economic time series are obtained at weekly, monthly or yearly intervals and can sometimes be regarded as realizations of a stochastic process in discrete time. Further, particularly if the series is examined only over a fairly limited time period, the structure of the series may appear reasonably constant in time and the stochastic process may then be assumed stationary. If the series is examined for a longer period, it may be possible to redefine the process to make it stationary; alternatively the process may be represented as the sum of a stationary process and a trend, for example a linear or quadratic function of time.

One trend-adjusted economic time series that has been much studied is Beveridge's trend-free wheat-price index, 1500–1869, a section of which is shown in Fig. 7.2. Most economic series are, of course, over time periods much shorter than this.

Time series at daily, weekly, monthly or annual intervals are of common occurrence also in meteorology, and certain other fields.

Example 7.3. Irregularity of textile yarns. If the thickness of a textile yarn is measured continuously along the length of the yarn, an irregular curve is obtained, of the general form already indicated in Fig. 1.5. Commonly such processes can be treated as stationary processes in continuous 'time'. Similar sets of data arise from many forms of automatic continuous recording.

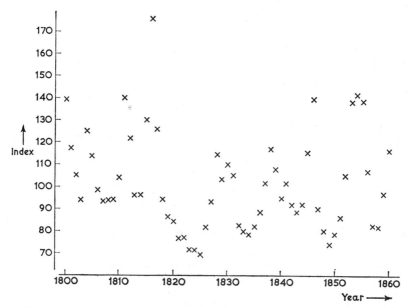

Fig. 7.2. Section of Beveridge's series of trend-free wheat-price indices.

In the textile example, we may wish to obtain convenient measures of the amount and type of irregularity, in order to be able to compare the effect on yarn irregularity of different methods of production. A further objective may be to relate the structure of the observed series to the physical mechanism producing the irregularity.

In studies of the irregularity of metal sheets, we would consider a process in two-dimensional 'time'.

7.2. Some definitions and special processes

We consider a process denoted by $\{X_n\}$ in discrete time or $\{X(t)\}$ in continuous time. For the reasons explained above, we often take the process as defined for $n = \ldots, -1, 0, 1, \ldots$ or $-\infty < t < \infty$. Further, in this chapter, we take the process to be one-dimensional and real-valued. In Chapter 8, we shall, however, consider complex valued processes; this is useful, for instance, when an A.C. signal is measured. The important extension to multivariate processes will be considered very briefly in Section 8.6.

The following definitions are for processes in discrete time; those in continuous time are exactly analogous. First, we call the process $\{X_n\}$ stationary in the strict sense if the sets of random variables

$$X_{n_1}, \ldots, X_{n_l} \tag{4}$$

and

$$X_{n_1+k}, \ldots, X_{n_l+k} \tag{5}$$

have the same joint distribution for every $n_1, \ldots, n_l, k$. Put another way, the joint distribution of (4) depends only on the intervals between the time points $n_1, \ldots, n_l$ and is unaffected by an arbitrary translation k of the time points. In particular, if $l = 1$ we have that X_n has the same distribution for all n, and if $l = 2$, we have that X_{n_1}, X_{n_2} have a joint distribution depending only on $n_1 - n_2$.

It is easily shown from (3) and its generalizations that the two-state Markov process discussed in Section 7.1 is strictly stationary in continuous time.

The definition just given is in some ways too general for practical use. For example, we would need a very large amount of data in order to examine at all searchingly whether an empirical series is stationary. To obtain a simpler definition, consider first an arbitrary Gaussian process, i.e. a process for which the joint distribution of $X_{n_1}, \ldots, X_{n_l}$ is multivariate normal. These multivariate normal distributions, of all orders, are completely characterized by

(a) the mean $E(X_n)$, as a function of n;
(b) the covariance matrix $\gamma(n_1, n_2) = C(X_{n_1}, X_{n_2})$ as a function of n_1, n_2.

Thus a Gaussian process is strictly stationary if and only if

(a) the mean $E(X_n)$ is constant;
(b) the covariance $\gamma(n_1, n_2)$ is a function of $n_1 - n_2$ only.

For any stationary process, we write

$$E(X_n) = \mu, \qquad C(X_{n+h}, X_n) = \gamma(h). \tag{6}$$

In particular $\gamma(0) = V(X_n)$ and is constant. The function $\gamma(h)$ is called the *autocovariance function*. The corresponding function $\rho(h) = \gamma(h)/\gamma(0)$ specifying the correlation coefficient between values of the process time points h apart is called the *autocorrelation function*. Clearly $\rho(0) = 1$ and for real valued processes $\rho(h) = \rho(-h)$. The argument h is often called the *lag*. When considering several processes, we add a suffix to $\gamma(h)$ or $\rho(h)$ to identify the series in question. writing, for example, $\gamma_x(h)$ and $\rho_x(h)$.

Quite generally, for processes that are not necessarily Gaussian, but have a finite covariance function, we call the process stationary to the second order, if (6) holds. Another equivalent term is stationary in the wide sense. That is, we require the stationary property only for $E(X_n)$ and for the covariance of pairs of values. A more general definition, which is sometimes required, is that a process is stationary of order p if all moments up to order p have the stationary property. Thus a process stationary of order three would satisfy, in addition to (6), the requirement that

$$E(X_{n_1} X_{n_2} X_{n_3}) = E(X_{n_1+k} X_{n_2+k} X_{n_3+k}) \tag{7}$$

for all n_1, n_2, n_3, k. In particular, the special case of (7) when $n_1 = n_2 = n_3$ requires that the third moment of the process is constant.

In what follows, we are mainly concerned with the autocovariance and autocorrelation functions of processes. Therefore second-order stationarity is enough for the validity of most of the following work. It will be clear from the context whenever a stronger assumption is required.

The autocorrelation function of the two-state Markov process of Section 7.1 is not of much direct interest, since the correlation coefficient is not a particularly useful method of describing the process. Nevertheless it is instructive to start with this simple process. We have that

$$E\{X(t)\} = \text{prob}\{X(t) = 1\} = \frac{\alpha}{\alpha+\beta},$$

$$V\{X(t)\} = \frac{\alpha\beta}{(\alpha+\beta)^2}.$$

In order to calculate $\rho(h)$, we need

$$\begin{aligned}
E\{X(t)\,X(t+h)\} &= \text{prob}\{X(t) = X(t+h) = 1\} \\
&= \text{prob}\{X(t) = 1\}\,\text{prob}\{X(t+h) = 1 | X(t) = 1\} \\
&= \frac{\alpha}{\alpha+\beta}\,p_1(h),
\end{aligned}$$

by (3). Hence, for $h \geqslant 0$,

$$\begin{aligned}
\gamma(h) &= C\{X(t+h), X(t)\} \\
&= \frac{\alpha\beta}{(\alpha+\beta)^2}\,e^{-(\alpha+\beta)h}.
\end{aligned}$$

Therefore for $h \geqslant 0$

$$\rho(h) = \gamma(h)/\gamma(0) = e^{-(\alpha+\beta)h}. \tag{8}$$

Note that because $\rho(h) = \rho(-h)$, it is enough to specify the function for $h \geqslant 0$. If, however, the autocorrelation function is required explicitly for all h, the form $e^{-(\alpha+\beta)|h|}$ must be used. We shall see later that the exponential autocorrelation function (8) is obtained widely for Markov processes. The analogous result in discrete time is that for a two-state Markov chain of Section 3.2, the autocorrelation function is

$$\rho(h) = \lambda^h \quad (h \geqslant 0), \tag{9}$$

where λ is a simple function of the transition matrix (Exercise 8).

Some important classes of processes will now be introduced by examples. Throughout $\{Z_n\}$ denotes a sequence of uncorrelated random variables of zero mean and constant variance, σ_z^2. We shall form processes by linear operations on the sequence $\{Z_n\}$. If, further, the $\{Z_n\}$ are independently normally distributed, our processes will be strictly stationary Gaussian processes. We shall, without loss of generality, arrange that $E(X_n) = 0$.

Example 7.4. Finite moving average. Let $a_0, \ldots, a_r$ be a set of constants and consider the process $\{X_n\}$ defined by

$$X_n = a_0 Z_n + \ldots + a_r Z_{n-r}. \tag{10}$$

If $\sum a_i = 1$, this is called a moving average with weights $\{a_0, \ldots, a_r\}$. In general,

$$E(X_n) = 0, \qquad V(X_n) = (a_0^2 + \ldots + a_r^2)\sigma_z^2$$

and

$$\gamma(h) = \begin{cases} \left(\sum_{s=0}^{r-h} a_s a_{s+h}\right)\sigma_z^2 & (h = 0, 1, \ldots, r), \\ 0 & (h = r+1, \ldots). \end{cases} \tag{11}$$

An important special case, the simple moving average, is obtained when $a_0 = \ldots = a_r = 1/(r+1)$, so that the process $\{X_n\}$ is a simple mean of $r+1$ of the uncorrelated random variables $\{Z_n\}$. Then it follows immediately from (10) and (11) that the autocorrelation function is, for $h \geqslant 0$,

$$\rho(h) = \begin{cases} 1 - \dfrac{h}{r+1} & (h = 0, \ldots, r), \\ 0 & (h = r+1, \ldots). \end{cases} \tag{12}$$

That is, the autocorrelation function is triangular, of extent determined by the length of the moving average.

Of course, a trivial special case of (12) is obtained when

$$a_1 = \ldots = a_r = 0.$$

The autocorrelation function of a series of uncorrelated observations is, by definition,

$$\rho(0) = 1, \qquad \rho(h) = 0 \quad (h = 1, 2, \ldots).$$

Functions like (10) will be considered later in connexion, for example, with the smoothing and prediction of series. We shall consider too the result of a linear averaging operation applied to a general stationary process rather than simply to an uncorrelated process.

Economic time series are quite often smoothed by taking a moving average. In physical processes in continuous time, the inertia of the recording apparatus introduces an averaging; for a recording apparatus with a 'linear response', the process observed will be a moving average of the underlying 'true' process.

Example 7.5. First-order autoregressive process. Suppose that the process $\{X_n\}$ is defined by the recurrence relation

$$X_n = \lambda X_{n-1} + Z_n, \tag{13}$$

the sequence $\{Z_n\}$ having the same properties as before. Equation (13) shows that, in statistical terminology, X_n has linear regression on X_{n-1} and $\{X_n\}$ is called a first-order autoregressive process. If (13) holds for all n, we can solve formally by successive substitution and write

$$X_n = \sum_{i=0}^{\infty} \lambda^i Z_{n-i}. \tag{14}$$

For a rigorous analysis, we would need to discuss carefully the sense in which the right-hand side of (14) converges to a well-defined random variable, and to justify our formal manipulations. The condition for (14) to define a random variable is in fact that the variance of the series on the right-hand side shall converge. Now, formally, $E(X_n) = 0$ and

$$V(X_n) = \sigma_z^2 \sum_{i=0}^{\infty} \lambda^{2i},$$

so that the required condition is $|\lambda| < 1$, when

$$V(X_n) = \frac{\sigma_z^2}{1-\lambda^2} = \sigma_x^2, \tag{15}$$

say. This does not depend on n.

Note that if $\lambda = 1$ and if the process starts at $n = 0$ with X_0 finite, (13) defines a random walk. This is not stationary since $V(X_n) \to \infty$ as $n \to \infty$.

One way of calculating the autocovariance function of the process is to work with the representation (14). In fact, proceeding formally, we have that for $h \geq 0$

$$E(X_{n+h} X_n) = E\left\{ \left(\sum_{j=0}^{\infty} \lambda^j Z_{n+h-j} \right) \left(\sum_{i=0}^{\infty} \lambda^i Z_{n-i} \right) \right\}$$

$$= \sigma_z^2 \sum_{i=0}^{\infty} \lambda^{h+i} \lambda^i$$

$$= \lambda^h \sigma_x^2, \tag{16}$$

by (15). The stationarity of the process follows because (16) does not depend on n, and the autocorrelation function is

$$\rho(h) = \lambda^h. \tag{17}$$

An alternative proof of (15) and (17) is instructive. Assume that the process is stationary. Now in the basic equation (13), X_{n-1} and Z_n are uncorrelated, since X_{n-1} is a function of $Z_{n-1}, Z_{n-2}, \ldots$ only. Hence if we take variances in (13), we have that

$$\sigma_x^2 = \lambda^2 \sigma_x^2 + \sigma_z^2,$$

proving (15).

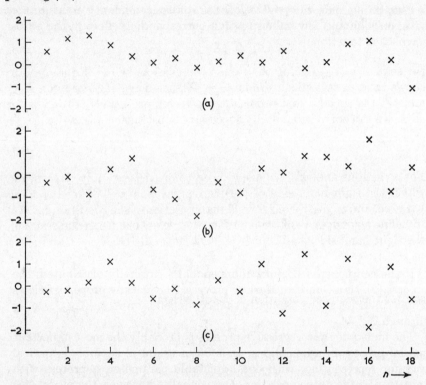

Fig. 7.3. Realizations of three first-order autoregressive processes. (a) $\lambda = 0.8$, (b) $\lambda = 0.3$, (c) $\lambda = -0.8$.

Now multiply (13) by X_{n-h} ($h \geqslant 1$) and take expectations, noting that X_{n-h} and Z_n are uncorrelated. We have that $\gamma(h) = \lambda\gamma(h-1)$, from which $\rho(h) = \lambda\rho(h-1)$, which taken with $\rho(0) = 1$, proves (17). An extension of the argument can be used to prove stationarity.

It is instructive to examine the connexion between the model (13), the qualitative appearance of the realizations, and the form of the autocorrelation function. Figure 7.3 gives realizations of (13) for $\lambda = 0.8$, 0.3,

-0.8 and Fig. 7.4 gives the corresponding autocorrelation functions for $\lambda = 0.8$, 0.3; that for $\lambda = -0.8$ is obtained by reversing the signs of alternate values in the first. The processes have been scaled so that $\sigma_x^2 = 1$.

By (13), each value X_n is a weighted combination of the previous value X_{n-1}, and a new random component Z_n, uncorrelated with the previous values in the process. If λ is near one, the first contribution is predominant, the realization changes rather smoothly, and the auto-correlation function dies slowly away to zero. If λ is near zero, the contribution Z_n is predominant in (13), the realization is nearly random and the autocorrelation function dies away rapidly to zero. Finally, if λ is near minus one, the realization has a strong tendency to alternate between high and low values, with a corresponding effect in the auto-correlation function.

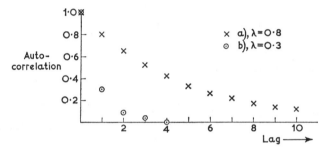

Fig. 7.4. Autocorrelation functions of two first-order autoregressive processes.
(a) $\lambda = 0.8$, (b) $\lambda = 0.3$.

If the components $\{Z_n\}$ are independently normally distributed, the process $\{X_n\}$ is a Gaussian Markov process. The Markov property follows from (13) which shows explicitly that X_n depends on $X_{n-1}, X_{n-2}, \ldots$ only through X_{n-1}.

The first-order autoregressive process is probably the most important special process in applications. It is often useful in a purely empirical way for representing, with one adjustable parameter, a process with smooth positive autocorrelation. Occasionally, however, the model (13) may be directly applicable.

Example 7.6. *Higher-order autoregressive processes.* A natural generalization of the previous example is to processes in which X_n has linear regression on $X_{n-1}, \ldots, X_{n-l}$, i.e. for which (13) is replaced by

$$X_n = \lambda_1 X_{n-1} + \ldots + \lambda_l X_{n-l} + Z_n. \tag{18}$$

Much the most important cases are $l = 1$, which we have already discussed, and $l = 2$. Since we shall give some generalities on autoregressive processes in Section 7, we concentrate here on the second-order process, $l = 2$.

Then the equations

$$X_n = \lambda_1 X_{n-1} + \lambda_2 X_{n-2} + Z_n \tag{19}$$

have the solution

$$X_n = w_0 Z_n + w_1 Z_{n-1} + w_2 Z_{n-2} + \ldots, \tag{20}$$

where, on substitution of (20) into (19) and comparing coefficients of $Z_n, Z_{n-1}, \ldots$, we have that

$$w_0 = 1, \qquad w_1 = \lambda_1 w_0, \qquad w_k = \lambda_1 w_{k-1} + \lambda_2 w_{k-2} \quad (k = 2, 3, \ldots). \tag{21}$$

The solution of the general difference equation has to be obtained subject to the initial conditions $w_0 = 1$, $w_1 = \lambda_1$. The required solution is

$$w_k = a_1 \xi_1^k + a_2 \xi_2^k, \tag{22}$$

where ξ_1, ξ_2 are the roots of the quadratic equation

$$\xi^2 - \lambda_1 \xi - \lambda_2 = 0 \tag{23}$$

and, in order to satisfy the initial conditions,

$$a_1 + a_2 = 1, \qquad a_1 \xi_1 + a_2 \xi_2 = \lambda_1. \tag{24}$$

Now, exactly as in the discussion of (15) in Example 7.5, we argue that (20) defines a random variable if the formal variance of (20), namely

$$\sigma_z^2 \sum_{k=0}^{\infty} w_k^2, \tag{25}$$

is finite. The condition for this is, from (22),

$$|\xi_i| < 1 \quad (i = 1, 2). \tag{26}$$

This can be expressed in terms of λ_1, λ_2; see Exercise 12. Further, we have from (22) that, (26) being satisfied,

$$\sum_{k=0}^{\infty} w_k^2 = \frac{a_1^2}{1 - \xi_1^2} + \frac{2 a_1 a_2}{1 - \xi_1 \xi_2} + \frac{a_2^2}{1 - \xi_2^2}$$

which, after some appreciable calculation, gives that

$$\sigma_x^2 = \frac{(1 - \lambda_2)}{(1 + \lambda_2)\{(1 - \lambda_2)^2 - \lambda_1^2\}} \sigma_z^2. \tag{27}$$

To calculate the autocorrelation coefficients, we can argue directly from the recurrence equation (19). Assuming stationarity, we multiply (19) by X_{n-1} and take expectations, noting that $E(X_{n-1} Z_n) = 0$. We get

$$\gamma(1) = \lambda_1 \gamma(0) + \lambda_2 \gamma(1),$$

i.e.
$$\rho(1) = \frac{\lambda_1}{1 - \lambda_2}. \tag{28}$$

Similarly, if we multiply (19) by X_{n-h} $(h = 2, 3, \ldots)$, take expectations, and divide by $\gamma(0)$, we have that

$$\rho(h) = \lambda_1 \rho(h-1) + \lambda_2 \rho(h-2).$$

Comparing with (21)–(23), we have that

$$\rho(h) = b_1 \xi_1^h + b_2 \xi_2^h,$$

where $b_1 + b_2 = 1$, $b_1 \xi_1 + b_2 \xi_2 = \lambda_1/(1 - \lambda_2)$.

The most interesting case is when the roots ξ_1, ξ_2 are complex, i.e. when $\lambda_1^2 + 4\lambda_2 < 0$. Then

$$(\xi_1, \xi_2) = \sqrt{(-\lambda_2)}\,(e^{i\beta}, e^{-i\beta}),$$

where

$$\tan \beta = \frac{\lambda_1}{\sqrt{\{-(\lambda_1^2 + 4\lambda_2)\}}}.$$

Then

$$\rho(h) = \{\sqrt{(-\lambda_2)}\}^h \frac{\cos(h\beta + \psi)}{\cos \psi}, \tag{29}$$

where ψ is a constant chosen to agree with (28).

Thus the autocorrelation function is a damped harmonic function.

The second-order autoregressive process was first introduced as a model in discrete time for the motion of a damped pendulum subject to random impulses.

A common modification to these processes is to have a superimposed random term, which may, for example, be an error of observation. Thus associated with the first-order process $\{X_n\}$ of Example 7.5, we can define a new process

$$Y_n = X_n + U_n,$$

where $\{U_n\}$ is a sequence of uncorrelated random variables of zero mean and variance σ_u^2, and uncorrelated with the process $\{Z_n\}$. Since $V(Y_n) = \sigma_x^2 + \sigma_u^2$ and $C(Y_n, Y_{n+h}) = C(X_n, X_{n+h})$ $(h = 1, 2, \ldots)$, the autocorrelation function for the $\{Y_n\}$ process is

$$\rho_y(h) = \begin{cases} 1 & (h = 0) \\ \alpha \lambda^h & (h = 1, 2, \ldots), \end{cases} \tag{30}$$

where $\alpha = \sigma_x^2/(\sigma_x^2 + \sigma_u^2)$. When $\alpha = 1$, we recover the autocorrelation function for the first-order autoregressive process.

Note carefully the distinction between the effects of the two sequences of uncorrelated random variables $\{Z_n\}$, $\{U_n\}$. A particular Z_n affects X_n and Y_n and all subsequent observations, i.e. is 'incorporated into' the process. A particular U_n, on the other hand, affects only the corresponding Y_n.

Equation (30) illustrates the general effect of a superimposed random error; the autocorrelation function for non-zero lags is reduced by a constant factor.

Example 7.7. An overlap problem. The following is a representation of an effect sometimes found, for example, when material is divided into batches. Suppose that the amount of product in the nth batch is originally Y_n, but that when the material is divided into sections, an amount U_n is transferred from the nth batch to the $(n+1)$th. Then X_n, the final amount in the nth batch, is

$$X_n = Y_n - U_n + U_{n-1}. \tag{31}$$

It would often be reasonable to assume that the process $\{U_n\}$ is uncorrelated with constant mean and constant variance σ_u^2, and that the processes $\{Y_n\}$, $\{U_n\}$ are uncorrelated. Then for the autocovariances, we have that

$$\gamma_x(h) - \gamma_y(h) = \begin{cases} 2\sigma_u^2 & (h = 0), \\ -\sigma_u^2 & (h = 1), \\ 0 & (h = 2, \dots). \end{cases} \tag{32}$$

If, further, the process $\{Y_n\}$ is uncorrelated, the autocorrelation function of $\{X_n\}$ is $\rho_x(1) = -\rho'$, $\rho_x(h) = 0$ $(h = 2, 3, \dots)$, where $\rho' = \sigma_u^2/(2\sigma_u^2 + \sigma_y^2)$.

It may be realistic to suppose that the process $\{Y_n\}$ is itself autocorrelated, often with positive correlation between adjacent values. This would tend to mask the negative correlation just noted.

The processes discussed so far are such that the random disturbance introduced into the process at one time may possibly influence all subsequent values of the process, as in (14). Nevertheless the influence decreases rapidly as the time point gets more and more remote. Put another way, a knowledge of X_{n-k} for a large value k is of no use for predicting the value of X_n. We call such processes *purely indeterministic*; a formal definition will be given later.

By contrast, we consider some stationary processes for which knowledge of X_{n-k} for large k does help in predicting X_n.

Example 7.8. A simple cyclic model. It might be thought that the simplest example of a stationary process showing long range regularity is a deterministic sinusoidal oscillation

$$X_n = a \cos(n\omega_0). \tag{33}$$

However, (33) is not a stationary process in the sense of our definition, because the (degenerate) probability distribution of X_n changes with n. We easily convert (33) into a stationary process by introducing a single random variable V, uniformly distributed over $(0, 2\pi)$ and writing

$$X_n = a \cos (n\omega_0 + V). \tag{34}$$

It is now easy to show that (34) defines a stationary process and, the mean being zero, that the autocovariance function is

$$\frac{a^2}{2\pi} \int_0^{2\pi} \cos (n\omega_0 + v) \cos (n\omega_0 + h\omega_0 + v) \, dv,$$

leading to the autocorrelation function

$$\cos (h\omega_0). \tag{35}$$

Now although (34) is technically a stochastic process, it is unlike the other processes considered so far in this book in that any particular realization is a deterministic function; once enough values of $\{X_n\}$ have been observed to fix the phase V, all subsequent values of the process are determined. The process is called *completely deterministic*. Note in particular that for ω_0 a rational multiple of π, the autocorrelation function takes the values ± 1 for an infinite set of values of the lag h.

The process (34) becomes more interesting and no longer purely deterministic if there is a random error added to (34). Thus we consider a new process

$$X_n = a \cos (n\omega_0 + V) + U_n, \tag{36}$$

or more generally

$$X_n = \sum_{i=1}^{p} a_i \cos (n\omega_{0i} + V_i) + U_n. \tag{37}$$

That is, the process is a linear combination of sinusoidal oscillations with random phases, plus an uncorrelated 'error' term. In (37) the process $\{U_n\}$ consists of uncorrelated random variables of mean zero and constant variance σ_u^2, the random variables $V_1, \ldots, V_p$ are independently rectangularly distributed over $(0, 2\pi)$ and $a_1, \ldots, a_p$ and $\omega_{01}, \ldots, \omega_{0p}$ are constants. The autocorrelation function of (37) is

$$b \sum a_i^2 \cos (h\omega_{0i}), \tag{38}$$

where

$$b^{-1} = \sum a_i^2 + \sigma_u^2.$$

The processes (36) and (37) are sometimes called systems with hidden periodicities. In the earliest statistical work on stationary time series, such models were widely used. The simplest deterministic 'stationary' function being a cosine wave, the hope was that simple stationary time series could be represented by a combination of a small number of cosine terms plus random error. Such models are intermediate between purely indeterministic ones and purely deterministic ones; values of the process in one time period are of some use in predicting the process a long time ahead, but do not allow exact prediction unless the $\{U_n\}$ are zero. In recent work on time series, there is more emphasis on purely indeterministic processes, but the model (37) remains important and will be considered repeatedly in this and the next chapter.

7.3. Some general results about stationary processes

(i) THE WOLD DECOMPOSITION

We now deal in outline with some general results about stationary processes in discrete time. Most of the essential ideas have been introduced in the previous Section 7.2; in particular, examples have been given there of moving average processes, of autoregressive processes and of periodic processes. The object here is to show how an arbitrary process can be represented in terms of the above special processes. This representation is obtained in (43) and (44) below.

Consider a stationary process $\{X_n\}$ with zero mean and variance σ^2. Take a large integer p and examine the joint distribution of the random variables $X_n, X_{n-1}, \ldots, X_{n-p}$. We are really concerned with the limit as $p \to \infty$ and this raises problems of rigour concerned, essentially, with setting up the theory of linear regression for infinitely many random variables; we shall not attempt a rigorous development.

From the theory of the linear least squares regression of X_n on X_{n-1}, $\ldots, X_{n-p}$, we have (Cramér, 1946, p. 302) that

$$X_n = (\text{linear combination of } X_{n-1}, \ldots, X_{n-p}) + Z_{n,p} \qquad (39)$$

where $Z_{n,p}$ is the residual about the regression.‡ It is known that the residual $Z_{n,p}$ has zero mean and is uncorrelated with the 'independent' variables $X_{n-1}, \ldots$ If we pass formally to the limit $p \to \infty$, we can replace (39) by

$$X_n = R_n + Z_n, \qquad (40)$$

where the residual Z_n is such that $E(Z_n) = 0$ and

$$C(Z_n, X_{n-k}) = 0 \quad (k = 1, 2, \ldots) \qquad (41)$$

and R_n, the regression component, is a linear combination of X_{n-1}, $X_{n-2}, \ldots$ We call Z_n the *innovation* at time n. It is the component of X_n

‡ We appeal here to the theory of linear regression for arbitrary random variables of finite variance; the regression equation is the linear combination of $X_{n-1}, \ldots,$ that has smallest expected squared error for predicting X_n.

that is uncorrelated with all previous values of the process. It can be shown that R_n and Z_n are uniquely defined, although the expression of R_n as a linear combination of X_{n-1}, X_{n-2}, ... may, in certain cases, have alternative forms. Since Z_{n-1} is a linear combination of $(X_{n-1}, ...)$, we have that $C(Z_n, Z_{n-1}) = 0$ and, in general, the process $\{Z_n\}$ of innovations is a series of uncorrelated random variables of zero mean and, because of stationarity, having constant variance.

One important mathematical interpretation of (40) is to think of each random variable X as a point in an infinite dimensional vector space (Hilbert space), the scalar product of two vectors X and Y being defined as $E(XY) = C(X, Y)$, the squared length of a vector, X, being $E(X^2)$. The interpretation of (40) is that X_n is resolved into two components, one, R_n, in the sub-space spanned by X_{n-1}, X_{n-2}, ... and the other, Z_n, orthogonal to that sub-space. This orthogonality is expressed by (41).

We now consider a series of linear regressions, namely of

$$X_n \quad \text{on} \quad X_{n-1}, X_{n-2}, ...,$$

then of

$$X_n \quad \text{on} \quad X_{n-2}, X_{n-3}, ...,$$

and of

$$X_n \quad \text{on} \quad X_{n-q}, X_{n-q-1}, ..., \tag{42}$$

and so on. Let τ_q^2 be the residual variance for (42). Clearly τ_q^2 is a non-decreasing bounded sequence and therefore tends to a limit as $q \to \infty$.

If this limit is $V(X)$, we call the process *purely indeterministic*; linear regression based on the 'remote past' of the process is useless for predicting the current value by linear regression methods. The other extreme case is when $\tau_q^2 = 0$ ($q = 1, 2, ...$). Then X_n can be predicted perfectly by a linear combination of values in the 'remote past' of the process. We call such a process *purely deterministic*. In intermediate cases, when $0 < \lim \tau_q^2 < V(X)$, we argue as follows. Decompose X_n into two components V_n and W_n, being the limiting regression and residual terms in (42) as $q \to \infty$. Then $C(V_n, W_n) = 0$ by the orthogonality of regression and residual terms. It can be shown further that the whole process $\{V_n\}$ is uncorrelated with the process $\{W_n\}$. The process $\{W_n\}$ by itself is purely indeterministic, whereas $\{V_n\}$, being a function of the 'remote past' is purely deterministic. Finally, we decompose the purely indeterministic component, W_n. The innovation Z_n is uncorrelated with all previous values in the process, and hence, in particular, with V_n which depends only on the 'remote past'. So too are all innovations Z_{n-1}, Z_{n-2}, ... Hence any linear combination of innovations is uncorrelated with V_n. Conversely it can be shown that the space spanned by Z_n, Z_{n-1}, ... is the sub-space of all random variables orthogonal to the space of the $\{V_n\}$, i.e. that any random variable orthogonal to the process $\{V_n\}$ can be expressed as a linear combination of the Z's.

Thus, finally, we can write

$$X_n = V_n + W_n, \tag{43}$$

where

$$W_n = b_0 Z_n + b_1 Z_{n-1} + b_2 Z_{n-2} + \dots, \tag{44}$$

with $b_0 = 1$. We repeat that $\{Z_n\}$ is a process of uncorrelated random variables. The constancy of the coefficients $b_0, b_1, \dots$ in (44) follows from the stationarity of the process.

This result is known as Wold's decomposition theorem. An arbitrary stationary process can be expressed as the sum of two processes uncorrelated with one another, one purely deterministic and one purely indeterministic. Further, the purely indeterministic component, which is the essentially stochastic part, can be written as a linear combination of the innovation process $\{Z_n\}$, which is a sequence of uncorrelated random variables.

In particular, if the original process is purely indeterministic, only the component (44) arises. That is, any purely indeterministic process can be represented as a moving average of an uncorrelated series. We have had a number of examples of such representations in the previous section. In most practical cases, the purely deterministic component will be represented as the sum of a number of sinusoidal terms, as in Example 7.8.

These representations involve only first- and second-moment properties. For example, different innovations Z_n and Z_m are always uncorrelated, but are independent only in special circumstances. If, however, the process is Gaussian, so that the joint distribution of any set of X's is multivariate normal, the linear regression relationships hold with independent normally distributed residuals and equations such as (43) and (44) hold with V_n, W_n independently normally distributed.

For processes generated in a very non-linear way, the Wold decomposition may be very artificial and of little direct interest. As an extreme example, results of Exercise 8.1 show that it is possible to have processes that are purely indeterministic in the above sense, but which can be predicted exactly by a suitable non-linear predictor.

The linear representations, and in particular the equation

$$X_n = b_0 Z_n + b_1 Z_{n-1} + \dots$$

for the purely indeterministic process, are of much greater interest if different Z_n's are independent and not merely uncorrelated. We then call the process a *linear process*.

(ii) AUTOCORRELATION FUNCTION OF GAUSSIAN MARKOV PROCESSES

A further application of the connexion with the theory of multiple linear correlation and regression is to establish the form of the autocorrelation function for a Gaussian stationary Markov process. For in a Gaussian process of zero means, the distribution is determined by the covariance matrix. Now the Markov condition is that the distribution of X_n given X_{n-1} does not depend on X_{n-h} ($h = 2, 3, \ldots$). This is equivalent to requiring that the partial correlation coefficient of X_n and X_{n-h} given X_{n-1} is zero for $h = 2, 3, \ldots$ Now if $\rho(X, Z)$ denotes the correlation coefficient of X and Z, the required partial correlation coefficient is (Cramér, 1946, p. 306)

$$\frac{\rho(X_n, X_{n-h}) - \rho(X_n, X_{n-1})\,\rho(X_{n-h}, X_{n-1})}{\sqrt{[\{1 - \rho^2(X_n, X_{n-1})\}\{1 - \rho^2(X_n, X_{n-h})\}]}}. \tag{45}$$

The numerator of this is, for a stationary process,

$$\rho(h) - \rho(1)\,\rho(h-1)$$

and vanishes for all $h = 2, 3, \ldots$ if and only if

$$\rho(h) = \{\rho(1)\}^h. \tag{46}$$

The corresponding result in continuous time, proved by an identical argument, is that $\rho(h) = e^{-\lambda h}$. We have already had a number of special illustrations of this result.

Now for non-Gaussian processes the argument breaks down, because the form of the conditional distributions is not determined by the partial correlation coefficients. There exist stationary non-Gaussian processes having an autocorrelation coefficient different from (46). For example, if X_n is a stationary Gaussian Markov process, $Y_n = X_n^3$ is a stationary non-Gaussian process with autocorrelation function not of the form (46); see Exercise 14. Conversely, it is clear that if (46) is satisfied, so that the partial correlation coefficients (45) all vanish, this, being a second-moment property only, cannot in general imply that the whole distribution of X_n given X_{n-1} is independent of X_{n-h}. Nevertheless, if we restrict ourselves to prediction by linear relationships, (45) and (46) is the nearest we can get to a Markov property and processes with autocorrelation functions (46) are sometimes called Markov processes in the wide sense.

(*iii*) RELATION BETWEEN AUTOREGRESSIVE AND MOVING AVERAGE REPRESENTATIONS

Now consider further the representation (44) of a purely indeterministic process. This is a moving average representation, so called because (44) is a moving average, in general of infinite extent, of the process $\{Z_n\}$ of innovations. The expression (40), however, suggests an autoregressive representation in which X_n is expressed as a linear combination of Z_n and the previous values $\ldots, X_{n-2}, X_{n-1}$. We now investigate formally the relation between these two representations.

Suppose that

$$a_0 X_n + a_1 X_{n-1} + \ldots = Z_n, \tag{47}$$

and

$$X_n = b_0 Z_n + b_1 Z_{n-1} + \ldots, \tag{48}$$

where we may take $a_0 = b_0 = 1$. These are respectively the autoregressive and moving average representations of the process.

Equation (47) is to be considered as a recurrence equation defining X_n in terms of X_{n-1}, X_{n-2}, ... and the innovation Z_n. Equation (48) is the corresponding expression for X_n in terms of the sequence of innovations.

Quite formally, we can define generating functions

$$X(\zeta) = \sum_{n=-\infty}^{\infty} X_n \zeta^n, \qquad Z(\zeta) = \sum_{n=-\infty}^{\infty} Z_n \zeta^n,$$

$$A(\zeta) = \sum_{n=0}^{\infty} a_n \zeta^n, \qquad B(\zeta) = \sum_{n=0}^{\infty} b_n \zeta^n.$$

Then from (47) and (48)

$$A(\zeta) X(\zeta) = Z(\zeta), \qquad X(\zeta) = B(\zeta) Z(\zeta).$$

Thus

$$A(\zeta) B(\zeta) = 1; \tag{49}$$

this is a key equation connecting the autoregressive and moving average representations.

A sufficient condition for the validity of these formal manipulations is that (48) should define a random variable with finite variance, i.e. that $\sum b_n^2 < \infty$. This implies that $B(\zeta)$ is analytic inside the circle $|\zeta| = 1$. For nearly all the cases we consider, $B(\zeta)$ has no poles on the unit circle and correspondingly, the zeros of $A(\zeta)$ lie outside the circle $|\zeta| = 1$. This last condition has a simple interpretation when $A(\zeta)$ is a polynomial, i.e. when the autoregression is finite. For then (47) is a finite difference equation with constant coefficients, whose solution has a form determined by the zeros of $A(\zeta)$. The condition that this solution is finite is easily seen to be that all roots have modulus greater than one.

Now these representations, being based on linear regression theory, are uniquely determined by the autocovariance function, $\gamma_x(h)$, of the original process. To examine this relation, we have first from (48) that

$$\gamma_x(h) = \sigma_z^2 \sum_{i=0}^{\infty} b_i b_{i+h}. \tag{50}$$

We introduce an autocovariance generating function

$$\Gamma_x(\zeta) = \sum_{h=-\infty}^{\infty} \gamma_x(h) \zeta^h. \tag{51}$$

It turns out rather more convenient to use the Laurent series in the form (51) with a doubly infinite sum instead of a Taylor series; since $\gamma_x(h) = \gamma_x(-h)$, we have that

$$\Gamma_x(\zeta) = \Gamma_x(1/\zeta).$$

In general, for the processes we consider, (51) will converge in an annulus containing the unit circle.

We have from (50) that

$$\Gamma_x(\zeta) = \sigma_z^2 B(\zeta) B(1/\zeta) \tag{52}$$

$$= \sigma_z^2 \{A(\zeta) A(1/\zeta)\}^{-1}, \tag{53}$$

where the series $A(\zeta)$, $B(\zeta)$ involve non-negative powers of ζ only.

The variance σ_x^2 of $\{X_n\}$, being the coefficient of ζ^0 in (52), can be written

$$\sigma_x^2 = \sigma_z^2 \cdot \frac{1}{2\pi i} \int_C \frac{B(\zeta) B(1/\zeta)}{\zeta} \, d\zeta, \tag{54}$$

where C can be taken as the unit circle.

Thus it is straightforward in principle to determine the autocovariance function given either of the representations (47) or (48). The converse problem of determining the representations from the autocovariance function can be regarded as that of splitting the function $\Gamma_x(\zeta)$ into two appropriate factors, one analytic inside the unit circle, the other analytic outside the unit circle. Such a split is called a Wiener–Hopf factorization and is not unique unless conditions are placed on the zeroes of the factors. However, if $\Gamma_x(\zeta)$ is a rational function of ζ, then we can recognize the appropriate factorization by separating off a suitable portion having poles outside the unit circle and identifying this with $B(\zeta)$. Then the uniqueness follows from the symmetry of the Laurent expansion of $\Gamma_x(\zeta)$ and from the condition that $B(0) = 1$; for further discussion, see Whittle (1963, Section 2.8).

We shall return to this in the next chapter and for the present consider only the simplest example of the formulae, namely the first-order auto-regressive process of Example 7.5, $\gamma_x(h) = \sigma_x^2 \lambda^{|h|}$. The representations (47) and (48) take the forms

$$X_n - \lambda X_{n-1} = Z_n, \qquad X_n = Z_n + \lambda Z_{n-1} + \lambda^2 Z_{n-2} + \dots,$$

so that

$$A(\zeta) = 1 - \lambda\zeta, \qquad B(\zeta) = 1 + \lambda\zeta + \lambda^2 \zeta^2 + \dots,$$

and the identity (49) is immediately verified. Note that the condition for stationarity is $|\lambda| < 1$, so that the zero of $A(\zeta)$ must be outside the unit circle. The autocovariance generating function (51) is

$$\sigma_x^2 \left\{ \sum_{h=0}^{\infty} \lambda^h \zeta^h + \sum_{h=1}^{\infty} \lambda^h \zeta^{-h} \right\} = \frac{\sigma_x^2 (1 - \lambda^2) \zeta}{(1 - \lambda \zeta)(\zeta - \lambda)}. \tag{55}$$

This has annulus of convergence $\lambda < |\zeta| < 1/\lambda$.

If, given the function on the right of (55), we wished to find the moving average and autoregressive representations, we would write the function

$$\frac{\sigma_x^2 (1 - \lambda^2)}{(1 - \lambda \zeta)(1 - \lambda/\zeta)}.$$

The factor $1/(1 - \lambda \zeta)$ corresponds to a pole outside the unit circle and therefore contributes to $B(\zeta)$; the factor $1/(1 - \lambda/\zeta)$ likewise contributes to the second factor. Hence

$$B(\zeta) = \sigma_x (1 - \lambda^2)^{\frac{1}{2}} / \{ \sigma_z (1 - \lambda \zeta) \},$$
$$A(\zeta) = \sigma_z (1 - \lambda \zeta) / \{ \sigma_x (1 - \lambda^2)^{\frac{1}{2}} \}.$$

Now we have normalized $B(\zeta)$ by requiring that $b_0 = 1$, and so we must have

$$\sigma_z^2 = \sigma_x^2 (1 - \lambda^2)$$

and $B(\zeta) = 1/(1 - \lambda \zeta)$, $A(\zeta) = 1 - \lambda \zeta$, thus checking the general formulae.

(iv) ERGODIC THEOREMS

To complete this section we consider a different general topic. In several previous chapters, we have looked at ergodic properties of processes, in particular at the identity of the equilibrium probability of a state and the proportion of a long time period that a single realization spends in that state. This equivalence is often of crucial importance in applications. There are very general ergodic theorems for stationary processes asserting the equivalence of probability averages with long-run averages taken over a single realization. Examples for a general stationary process $\{X_n\}$, are the identity in some sense as $n \to \infty$ of

(a) $E(\overline{X}_n) = \mu$ and $\dfrac{1}{n} \displaystyle\sum_{i=1}^{n} X_i = \overline{X}_n$,

(b) $V(X_n) = \sigma^2$ and $\dfrac{1}{n} \displaystyle\sum_{i=1}^{n} (X_i - \overline{X}_n)^2$,

(c) $C(X_n, X_{n+h}) = \gamma(h)$ and $\dfrac{1}{n} \displaystyle\sum_{i=1}^{n-h} (X_i - \overline{X}_n)(X_{i+h} - \overline{X}_n)$.

The results for (a) are a direct generalization of the classical laws of large numbers in the theory of probability which would apply when the

X's are independent. Following the method used in previous chapters, we can prove simple versions of the required convergence, involving convergence in probability, by showing that, if $\sum \gamma_x(h) < \infty$, then

$$E(\overline{X}_n) = \mu, \qquad \lim_{n \to \infty} V(\overline{X}_n) = 0.$$

We can then apply Chebychev's inequality. The corresponding argument for the two cases, (b) and (c), requires some restrictions on the fourth moment properties of the process. We shall not go into further details here.

7.4. Processes in continuous time

In the previous sections, we have mostly dealt with stationary processes in discrete time. When we pass to continuous time, many of the previous definitions generalize in a straightforward way, although some new problems arise in connexion with limiting operations.

Let $\{X(t)\}$ be a stationary process with

$$E\{X(t)\} = 0, \qquad V\{X(t)\} = \sigma^2, \qquad C\{X(t+h), X(t)\} = \gamma(h),$$

and with autocorrelation function $\rho(h) = \gamma(h)/\sigma^2$.

Each realization is a function of the real variable t and it is therefore natural to consider introducing formally such things as

$$\frac{dX(t)}{dt} = \dot{X}(t), \int_a^b X(t)\,dt, \text{ etc.}$$

These are random variables varying from realization to realization. Broadly speaking, we can manipulate derivatives and integrals freely, exactly as in differentiating and integrating ordinary functions, provided that the mechanism generating the process is 'smooth' enough. The only case where appreciable regularity conditions are required is in connexion with differentiation and this we now consider briefly.

First consider the simpler idea of the continuity of $\{X(t)\}$ at a particular time. A natural definition is to require

$$\lim_{\Delta t \to 0} \{X(t+\Delta t) - X(t)\} = 0,$$

in some sense. We could require convergence in probability (weak convergence), or almost certain convergence (strong convergence), but the most convenient requirement is in fact to require convergence in mean square, i.e.

$$\lim_{\Delta t \to 0} E\{X(t+\Delta t) - X(t)\}^2 = 0. \tag{56}$$

This is equivalent to requiring that $2\sigma^2\{1 - \rho(\Delta t)\} \to 0$, i.e. that $\rho(h)$ is continuous at $h = 0$.

To deal similarly with differentiation, it would be natural to require the existence of a random variable $\dot{X}(t)$ such that

$$\lim_{\Delta t \to 0} E\left\{\dot{X}(t) - \frac{X(t+\Delta t) - X(t)}{\Delta t}\right\}^2 = 0. \tag{57}$$

The conditions for this can be deduced from a version of Cauchy's general principle of convergence, for convergence in mean square. Instead we shall proceed informally. Let

$$\dot{X}(t, \Delta t) = \frac{X(t+\Delta t) - X(t)}{\Delta t},$$

a well-defined random variable for all $\Delta t \neq 0$. Then we have by elementary calculation that

$$E\{\dot{X}(t, \Delta t)\} = 0, \qquad C\{X(t+h), \dot{X}(t, \Delta t)\} = \sigma^2 \frac{\rho(h - \Delta t) - \rho(h)}{\Delta t},$$

$$V\{X(t, \Delta t)\} = \sigma^2 \frac{2 - 2\rho(\Delta t)}{(\Delta t)^2} = \sigma^2 \frac{2\rho(0) - \rho(\Delta t) - \rho(-\Delta t)}{(\Delta t)^2},$$

$$C\{\dot{X}(t+h, \Delta u), \dot{X}(t, \Delta t)\} =$$
$$\sigma^2 \frac{\rho(h + \Delta u - \Delta t) - \rho(h + \Delta u) - \rho(h - \Delta t) + \rho(h)}{\Delta t \Delta u}.$$

Now it is reasonable to expect all these expressions to tend to finite limits, if $\dot{X}(t, \Delta t)$ tends to a well-defined random variable $\dot{X}(t)$ as $\Delta t \to 0$. For $V\{\dot{X}(t, \Delta t)\}$ to converge, $\rho(h)$ must be differentiable twice at $h = 0$; it can be shown that it is then differentiable twice for all h. The limiting forms of the above results are then that

$$E\{\dot{X}(t)\} = 0, \qquad\qquad C\{X(t), \dot{X}(t)\} = 0,$$

$$V\{\dot{X}(t)\} = -\sigma^2 \rho''(0), \qquad C\{X(t+h), \dot{X}(t)\} = \sigma^2 \rho'(h), \tag{58}$$

$$C\{\dot{X}(t+h), \dot{X}(t)\} = -\sigma^2 \rho''(h).$$

In particular the autocorrelation function of the process $\{\dot{X}(t)\}$ is $\rho''(h)/\rho''(0)$. These results can be extended to higher derivatives.

As an example, suppose that $\rho(h) = e^{-\lambda|h|}$. The Ornstein–Uhlenbeck process of Chapter 5 is a particular process with this $\rho(h)$. The autocorrelation function is not differentiable twice (or even once) at $h = 0$ and hence the process $\{X(t)\}$ is not differentiable. The reason is essentially that

$$E\{X(t + \Delta t) - X(t)\}^2 = 2\sigma^2\{1 - \rho(\Delta t)\} \sim 2\sigma^2 \lambda \Delta t,$$

so that the value of $X(t + \Delta t) - X(t)$ will be roughly of the order of the standard deviation, i.e. of order $\sqrt{\Delta t}$ in probability. Hence the value of

$\{X(t+\Delta t) - X(t)\}/\Delta t$ will be of order $1/\sqrt{\Delta t}$ and hence will not tend to a finite limit as $\Delta t \to 0$.

A general point about these definitions is that they refer to continuity and differentiability at a *single* value of t, not to the continuity and differentiability of a realization over an interval. Results concerning behaviour of a realization over an interval are available but are much more difficult and will not be considered here.

As an example of the application of (58), we consider some descriptive statistical properties of stationary Gaussian processes.

Example 7.9. *Frequency of upcrosses in a stationary Gaussian process.* Consider a stationary Gaussian process of mean zero with autocorrelation function $\rho(h)$, assumed twice differentiable, so that $\{\dot{X}(t)\}$ exists in mean square. We shall assume further that the sample realizations are differentiable. We say that an upcross over the mean occurs whenever the realization passes through zero with positive slope. Let $q_+\Delta t$ be the probability of such an upcross in $(t, t+\Delta t)$.

Let $p(x,y)$ be the joint probability density function of $X(t)$ and $\dot{X}(t)$; this does not involve t because the process is stationary. Now $X(t)$ and $\dot{X}(t, \Delta t)$ have a bivariate normal distribution for every $t, \Delta t$. Hence $p(x,y)$ is bivariate normal with mean $(0,0)$ and, by (58), covariance matrix

$$\sigma^2 \begin{bmatrix} 1 & 0 \\ 0 & -\rho''(0) \end{bmatrix}.$$

Take Δt so small that the realization is effectively linear in the interval $(t, t+\Delta t)$. If $X(t) = x$, $\dot{X}(t) = y$, the local equation of the realization is $X(t+\tau) = x + y\tau$ and this intersects the zero level at $\tau = -x/y$. The condition for an upcross in $(t, t+\Delta t)$ is thus

$$y > 0, \qquad -y\Delta t < x < 0. \qquad (59)$$

Therefore

$$q_+ \Delta t = \int\limits_0^\infty dy \int\limits_{-y\Delta t}^0 dx\, p(x,y),$$

i.e.

$$q_+ = \int\limits_0^\infty y p(0, y)\, dy.$$

Since

$$p(x,y) = \frac{1}{2\pi\sigma^2 \sqrt{\{-\rho''(0)\}}} \exp\left[-\frac{x^2}{2\sigma^2} - \frac{y^2}{2\sigma^2\{-\rho''(0)\}} \right],$$

it follows that

$$q_+ = \frac{\sqrt{\{-\rho''(0)\}}}{2\pi}.$$

There is an equal probability $q_- \Delta t$ for downcrosses and hence a total probability $q \Delta t$ for crossing the zero line given by

$$q = \frac{\sqrt{\{-\rho''(0)\}}}{\pi}. \tag{60}$$

The total number of crosses of the mean in a long time t_0 will be approximately $q t_0$.

The result (60) is qualitatively reasonable. The larger $-\rho''(0)$ is, the more rapidly does the autocorrelation function $\rho(h)$ die away from its unit value at $h = 0$ and hence the larger the local random component of variation. Since the proportion of a long time period spent near zero is fixed by the marginal normal distribution of $X(t)$, it follows that the frequency of crosses should increase with $-\rho''(0)$.

Note that dimensional analysis establishes that if

$$q \propto \{-\rho''(0)\}^k,$$

then $k = \frac{1}{2}$. If, further, the result is to hold for a pure cosine wave, the constant of proportionality must be $1/\pi$. That is, we could have guessed at (60) from very elementary considerations; this sort of simple argument is not to be despised, especially for getting rough answers to difficult problems.

If we apply (60) to a process with autocorrelation function $e^{-\lambda|h|}$, we get formally $q = \infty$. The qualitative meaning of this is that, because of the high rate at which $\rho(h)$ decreases, corresponding to a continuous non-differentiable function, the realizations have a rapid short-term 'jitter'. When the process is near to the zero level, it therefore crosses and recrosses that level infinitely often. In Example 5.4 this fact was proved in detail for the Ornstein–Uhlenbeck process. Of course, even if one had a physical process with this autocorrelation function, any recording device used would smooth the process somewhat, so that this mildly pathological behaviour is not really observable in practice.

Consider now the analogues in continuous time of the canonical representations of Section 7.3. We have first to establish the analogue of the sequence $\{Z_n\}$ of uncorrelated random variables of zero mean and constant variance that are basic to the representations in discrete time.

For this, we can consider formally a process $\{Z(t)\}$, with

$$E\{Z(t)\} = 0, \qquad C\{Z(t), Z(u)\} = 0, \quad (t \neq u). \tag{61}$$

Such a process is called *pure noise* and has already been used in Chapter 5; it is not physically realizable, since the construction of a realization would involve the choice of a separate uncorrelated random variable for each of a continuum of time points, leading to an exceedingly discontinuous function. A process approximating to (61) can, however, be obtained by considering either

(a) a process in continuous time with an autocorrelation function, say $e^{-\lambda|h|}$, and letting $\lambda \to \infty$; or

(b) an uncorrelated process in discrete time at time intervals Δt and letting $\Delta t \to 0$.

We shall, however, always deal with the pure noise process in forms analogous to the linear combinations

$$\sum b_i Z_{n-i} \tag{62}$$

of (48), in discrete time. We could try to define integrals of the form

$$\int b(v) Z(t-v) \, dv, \tag{63}$$

but instead it is more convenient to work with the Stieltjes integral form

$$\int b(v) \, dU(t-v), \tag{64}$$

defined as the limit (in mean square) of the usual sequence of sums.

The increments of the process $\{U(t)\}$ in (64) over small time intervals are to correspond to the innovation process $\{Z_n\}$ in (62). This correspondence is achieved by requiring that for any $t_1 < t_2 < t_3 < t_4$

$$E\left\{\int_{t_1}^{t_2} dU(v)\right\} = 0, \qquad V\left\{\int_{t_1}^{t_2} dU(v)\right\} = (t_2 - t_1)\sigma_z^2,$$

$$C\left\{\int_{t_1}^{t_2} dU(v), \int_{t_3}^{t_4} dU(v)\right\} = 0. \tag{65}$$

It is often convenient to rewrite (65) in the form

$$E\{\Delta U(t)\} = 0, \qquad V\{\Delta U(t)\} = \sigma_z^2 \Delta t, \qquad E\{\Delta U(t_1)\Delta U(t_2)\} = 0$$

$$(t_1 \neq t_2), \quad (66)$$

$\Delta U(t)$ denoting formally the increment in $(t, t + \Delta t)$, and the time intervals involved in the covariance being non-overlapping.

Thus, so far as first- and second-moment properties are concerned, $\{U(t)\}$ has the form of a Wiener process without drift; a process satisfying the conditions (65) is sometimes called an orthogonal process, or a process with orthogonal increments.

The extension of Wold's decomposition theorem to continuous time is that an arbitrary stationary process $\{X(t)\}$ can be split into two uncorrelated components, one purely deterministic (i.e. linearly predictable from the 'remote past' of the process) and the other purely indeterministic. The latter component can be written in the form

$$\int_0^\infty b(v)\,dU(t-v),\tag{67}$$

where $\{U(t)\}$ is a process with orthogonal increments, $U(t)$ playing the part of the cumulative sum of all innovations up to t.

The autocovariance function of the component (67) is found by considering the integral as a limiting sum and applying (66). We have that for lag $h \geqslant 0$

$$C\left\{\int_0^\infty b(w)\,dU(t+h-w),\ \int_0^\infty b(v)\,dU(t-v)\right\}$$

$$= E\left\{\int_0^\infty \int_0^\infty b(w)\,b(v)\,dU(t+h-w)\,dU(t-v)\right\}$$

$$= \sigma_z^2 \int_0^\infty b(v+h)\,b(v)\,dv,\tag{68}$$

only the line where $t-v = t+h-w$ contributing a non-zero amount to the expectation. Equation (68) is directly analogous to (50) in discrete time.

Finally, to complete the parallel with the theory in discrete time, we consider briefly the analogue of autoregressive processes, dealing for simplicity only with a first-order process, Example 7.5. We can regard this process as defined by a first-order difference equation with constant coefficients in which the 'forcing function' on the right-hand side is an uncorrelated process. The analogue in continuous time is thus the first-order differential equation with constant coefficients forced by pure noise, i.e. a process defined by

$$\dot{X}(t) + aX(t) = \dot{U}(t),\tag{69}$$

where $\dot{U}(t)$ is the formal derivative of $U(t)$. In physical contexts, (69) is called the Langevin equation. In Chapter 5 we have considered equations similar to (69) but expressed in slightly different notation. Indeed it is physically more meaningful to write (69) in the form

$$dX(t) + aX(t)\,dt = dU(t).$$

We are interested in a solution of (69) for all real t representing a stationary process. The solution of (69) is

$$X(t) = \int_0^\infty e^{-av}\,dU(t-v).\tag{70}$$

and this represents a stationary process if $a > 0$.

The physical interpretation of (70) is that the response of (69) to a unit impulse at $t = 0$ is e^{-at}; since the system is linear the full solution for $X(t)$ is the sum of the contributions from all impulses prior to t.

It is easily shown from (68) that the autocovariance function of the process $\{X(t)\}$ is exponential, a reflection of the Markovian form of the process (see (45) and (46)).

More generally, if we have a linear operator $\mathscr{L}$ and consider the equation

$$\mathscr{L}X(t) = \dot{U}(t), \tag{71}$$

there will be a solution, under suitable conditions, of the form (67) for a suitable function $b(v)$ and the preceding theory applies.

7.5. Prediction theory

(i) FORMULATION OF PROBLEMS OF PREDICTION, FILTERING AND REGULATION

One of the main applications of the theory of stationary processes is in connexion with the control of systems such as industrial processes. Three closely related problems arising in this connexion are those of prediction, of filtering and of regulation.

We consider a stationary process $\{X_n\}$; we take discrete time for simplicity, but this choice makes no essential difference. The broad distinction between the above three types of problem is as follows. In a prediction problem we observe ..., X_{n-2}, X_{n-1}. It is required, for some $k \geqslant 0$, to compute a function $\hat{X}_{n+k}$ of the observations that will in some sense be close to the as yet unobserved value X_{n+k}. Very often, we predict one step ahead, $k = 0$.

In the filtering problem, we assume that the process $\{X_n\}$ is the sum of two component stationary processes,

$$X_n = W_n + Z_n,$$

where W_n is called the *signal* and Z_n the *noise*. Note that noise in this context is not necessarily pure noise! We assume that the main properties of the processes $\{W_n\}$ and $\{Z_n\}$ are known. For instance, we might assume the two processes to be mutually uncorrelated and to have known autocovariance functions, $\gamma_w(h)$ and $\gamma_z(h)$. In particular, the process $\{Z_n\}$ might be assumed uncorrelated. It is required to construct a function $\hat{W}_n$ of the observations ..., X_{n-1}, X_n, X_{n+1}, ..., that will in some sense be close to the value W_n. That is, it is required to estimate the signal, given observations on signal plus noise. In different versions of the problem, we may observe only ..., X_{n-1}, X_n, or, alternatively, only some finite set of values.

Finally, in the problem of regulation, it is possible to adjust the process just before each observation is obtained. For example, suppose that at time n the process would, in the absence of adjustment, take the value x. If, just before time n, we apply an adjustment c_n, the achieved value will be $x - c_n$. The object is to specify c_n as a function of ..., X_{n-2}, X_{n-1}, so that the achieved values are, in some sense, as near as possible to target values, say zero. If adjustments can be made without cost, this is equivalent to the prediction problem, in that the optimum adjustment is clearly the predicted value of X_n. More generally, however, there will be a cost attached to each adjustment and this must be balanced against the cost of not achieving the target. Then the problems of prediction and regulation are not the same.

The monograph by Whittle (1963) gives a concise discussion of these problems. Middleton (1960) has given a very thorough account of statistical communication theory, one of the most fruitful fields of application of the ideas of this and the next chapter. It may, however, help readers more familiar with the statistical topics of acceptance sampling and of charts for process control to consider these as illustrations of respectively filtering and prediction.

Example 7.10. *Acceptance sampling and quality control.* Suppose that we have a series of batches for sentencing and that W_n measures the true quality of the nth batch. We observe not W_n but $X_n = W_n + Z_n$, where the 'noise' Z_n is a sampling error. Our object is to use the observations to reach an appropriate decision about each batch in turn, i.e. to filter the signal $\{W_n\}$ from the noise. Usually in acceptance sampling we consider a two-decision problem in which a choice has to be made between acceptance and rejection. The above formulation of the filtering problem is, however, in terms of point estimation. For Gaussian processes, the difference is minor.

Simple acceptance sampling rules base the decision for the nth batch solely on X_n, but when the properties of successive batches are correlated, it will be more efficient to use also the information contained in observations on neighbouring batches. If each batch is to be sentenced as it is obtained, we require a function of ..., X_{n-1}, X_n that will be close to W_n. If deferred sentencing is allowed, we can use observations after, as well as before, X_n.

In acceptance sampling, the object is to examine the 'current' batch. In process control, on the other hand, we use the observations to control material to be produced in the immediate future. This is a problem of regulation, or, if the cost of making adjustments is negligible, a problem of prediction. We are no longer interested as such in the material recently produced. In the older work on statistical quality control, attention was concentrated very largely on the last observation, but in general all recent observations will contribute some relevant information and subsequent developments have used this fact (Barnard, 1959; Roberts, 1959; Box and Jenkins, 1962, 1963; Bather, 1963).

Here we concentrate on the prediction problem, in particular illustrating the application of the representations of Section 7.3. Suppose, therefore, that in the stationary process $\{X_n\}$, with mean zero, variance σ_x^2, autocovariance function $\gamma_x(h)$ and autocorrelation function $\rho_x(h)$, we observe ..., X_{n-2}, X_{n-1}. We restrict attention to problems where

(a) only linear predictors are considered, i.e. $\hat{X}_{n+k}$ is constrained to be a linear combination of ..., X_{n-2}, X_{n-1};

(b) it is required to minimize the mean square error of prediction, $E(\hat{X}_{n+k} - X_{n+k})^2$;

(c) the mean, variance and autocorrelation function of the process are known.

These restrictions may all be serious in applications.

(ii) DIRECT APPROACH TO PREDICTION PROBLEMS

First we outline a simple approach which is useful when, for example, $\rho_x(h)$ is given empirically. Consider first finite predictors

$$\hat{X}_{n+k} = l_1 X_{n-1} + l_2 X_{n-2} + \ldots + l_r X_{n-r}; \tag{72}$$

it is easily shown that the mean square error is increased by adding a non-zero constant term to (72). Now

$$E\{\hat{X}_{n+k} - X_{n+k}\}^2 = \sigma_x^2 \left\{ 1 - 2 \sum_{i=1}^{r} l_i \rho_x(k+i) + \sum_{i,j=1}^{r} l_i l_j \rho_x(i-j) \right\}.$$

Hence to make the mean square error stationary, and in fact to minimize it, we differentiate with respect to the l_i in turn, obtaining the equation

$$
\begin{bmatrix}
1 & \rho_x(1) & \ldots & \rho_x(r-1) \\
\rho_x(1) & 1 & \ldots & \rho_x(r-2) \\
\vdots & \vdots & & \vdots \\
\rho_x(r-1) & \rho_x(r-2) & \ldots & 1
\end{bmatrix}
\begin{bmatrix}
l_1 \\
\vdots \\
l_r
\end{bmatrix}
=
\begin{bmatrix}
\rho_x(k+1) \\
\vdots \\
\rho_x(k+r)
\end{bmatrix}. \tag{73}
$$

This has the general form of a set of least-squares equations. If we are interested in a closed expression for the predictor we may need to let $r \to \infty$.

As a special case of (73), consider again the first-order autoregressive process, Example 7.5, for which $\rho_x(h) = \lambda^{|h|}$. It is easily verified that, for all r, the solution of (73) is $l_1 = \lambda^{k+1}$, $l_i = 0$ ($i = 2, \ldots$), so that the optimum linear predictor of X_{n+k} is λ^{k+1}, X_{n-1}. The conclusion that only the last available observation is used is a direct consequence of the way in which in the autoregressive representation the process can be regarded as built up by combining the last observation with the current innovation. In particular, for prediction one step ahead, the optimum linear predictor of X_n is λX_{n-1}.

(*iii*) USE OF AUTOREGRESSIVE AND MOVING AVERAGE REPRESENTATIONS

To obtain the optimum linear predictor in a form more meaningful than (73), we use the autoregressive and moving average representations in terms of the process $\{Z_n\}$ of innovations. We consider purely indeterministic processes, since the prediction of the deterministic component, once its form is established, is easy. We have then that

$$
\begin{aligned}
X_n &= Z_n - a_1 X_{n-1} - a_2 X_{n-2} - \dots, \\
X_n &= Z_n + b_1 Z_{n-1} + b_2 Z_{n-2} + \dots
\end{aligned}
\tag{74}
$$

From the way these representations were obtained, the part omitting Z_n is the linear combination of $\dots, X_{n-2}, X_{n-1}$ that best predicts X_n in the mean square sense. This gives the required predictor as a combination either of $\dots, X_{n-2}, X_{n-1}$ or of $\dots, Z_{n-2}, Z_{n-1}$. Normally we want the first. Hence if the autoregressive representation can be found, or is given from the way the process is set up, the optimum linear one-step ahead predictor can be written down directly as a combination of $\dots, X_{n-2}, X_{n-1}$. If the moving average representation is known, the simplest procedure will usually be to determine the autoregressive representation by solving (49). It is not difficult to show by using (73) and the relations (52) and (53) involving the autocovariance generating function that this procedure is equivalent to the use of (73), for $k = 0$, $r \to \infty$.

Suppose now that we want the linear predictor of X_{n+k} for $k > 0$. Now

$$
X_{n+k} = Z_{n+k} + \dots + b_{k+1} Z_{n-1} + b_{k+2} Z_{n-2} + \dots,
$$

and, by the same argument as before, the required predictor is

$$
b_{k+1} Z_{n-1} + b_{k+2} Z_{n-2} + \dots
\tag{75}
$$

We normally need to express this as a combination of $\dots, X_{n-2}, X_{n-1}$, and this is achieved by substitution from (74). Then (75) becomes

$$
c_1^{(k)} X_{n-1} + c_2^{(k)} X_{n-2} + \dots,
\tag{76}
$$

where

$$
c_j^{(k)} = b_{k+1} a_{j-1} + b_{k+2} a_{j-2} + \dots + b_{k+j}.
\tag{77}
$$

This can be expressed most neatly by introducing the generating functions

$$
B_k(\zeta) = \sum_{i=1}^{\infty} b_{k+i}\, \zeta^i, \qquad C_k(\zeta) = \sum_{i=1}^{\infty} c_i^{(k)}\, \zeta^i.
\tag{78}
$$

Equation (77) is equivalent to

$$C_k(\zeta) = A(\zeta) B_k(\zeta). \tag{79}$$

Returning to the first-order autoregressive process, we have that

$$A(\zeta) = 1 - \lambda\zeta, \qquad B_k(\zeta) = \zeta\lambda^{k+1}/(1 - \lambda\zeta),$$

so that $C_k(\zeta) = \zeta\lambda^{k+1}$, showing again that the optimum linear predictor is $\lambda^{k+1} X_{n-1}$.

(iv) PREDICTION OF A NON-STATIONARY SERIES

Finally we discuss briefly simple examples illustrating two new points. First we may occasionally have a given predictor and wish to find the process, if any, for which it is optimum. This can be done by reversing the argument of the last sub-section. When we do this, we may obtain a formal autoregressive representation having a generating function $A(\zeta)$ with a zero on the unit circle. By the discussion of Section 7.3(iii), this cannot represent a stationary process, but we shall see that it may represent a type of non-stationary process for which simple prediction is possible.

Consider first the predictor $\hat{X}_n = X_{n-1}$, i.e. the predictor is simply the last observation. Then this is the optimum predictor if and only if

$$X_n = X_{n-1} + Z_n, \tag{80}$$

where Z_n is the innovation at time n. The process $\{X_n\}$ is thus a random walk, and is non-stationary. If we regard (80) as a formal autoregressive equation, we have $a_1 = -1$, $a_2 = a_3 = \ldots = 0$, so that

$$A(\zeta) = 1 - \zeta$$

and has a zero on the unit circle. Note that $B(\zeta) = (1 - \zeta)^{-1}$, which on expansion leads to $X_n = Z_n + Z_{n-1} + \ldots$, the solution of (80).

Next consider the predictor

$$\hat{X}_n = (1 - \alpha)(X_{n-1} + \alpha X_{n-2} + \alpha^2 X_{n-3} + \ldots), \tag{81}$$

which is called an *exponentially weighted moving average* and has been quite widely applied in connexion with sales forecasting, etc.

Note that the sum of the coefficients is one, so that if the mean of the process drifts to a new level consistent prediction is obtained. This is an important feature in practice. The value of the predictor is most easily computed from the recurrence equation

$$\hat{X}_n = (1 - \alpha) X_{n-1} + \alpha\hat{X}_{n-1}.$$

The predictor (81) is optimum for the process

$$X_n = (1 - \alpha)(X_{n-1} + \alpha X_{n-2} + \ldots) + Z_n. \tag{82}$$

For this

$$A(\zeta) = 1 - (1-\alpha)\,\zeta - \alpha(1-\alpha)\,\zeta^2 - \dots$$

$$= \frac{1-\zeta}{1-\alpha\zeta}.$$

Thus $B(\zeta) = (1-\alpha\zeta)/(1-\zeta)$, whence $b_n = 1-\alpha$ $(n = 1, 2, \dots)$, $b_0 = 1$. Therefore, formally,

$$X_n = Z_n + (1-\alpha)\,(Z_{n-1} + Z_{n-2} + \dots),$$

as is easily proved directly from (82).

This, while not stationary, is converted into a stationary process by differencing

$$\nabla X_n = X_n - X_{n-1} = Z_n - \alpha Z_{n-1}. \tag{83}$$

The process $\{\nabla X_n\}$ is thus stationary with variance $\sigma_z^2(1+\alpha^2)$ and with the first autocorrelation coefficient $-\alpha$, and the remainder all zero.

Another representation of the process $\{X_n\}$ is as follows. Let $\{U_n\}$ and $\{Y_n\}$ be two mutually uncorrelated processes, each of uncorrelated random variables of zero means and variances σ_u^2, σ_y^2. Let

$$X_n = (Y_n + Y_{n-1} + \dots) + U_n,$$

so that X_n can be regarded as a random walk formed from the steps $\{Y_n\}$ plus an 'error of observation' U_n. Then

$$\nabla X_n = Y_n + U_n - U_{n-1} \tag{84}$$

and it is easily verified that if $\sigma_y^2 = (1-\alpha)^2\sigma_z^2$, $\sigma_u^2 = \alpha\sigma_z^2$, the first- and second-moment properties of (83) and (84) are identical. If the processes are both Gaussian the two representations are completely equivalent.

To sum up, we have shown that the exponentially weighted moving average (81) is optimum for predicting a process consisting of a random walk plus an error of observation. For further discussion, see Box and Jenkins (1962).

Bibliographic Notes

Autoregressive processes were introduced by Yule (1927) in connexion with the analysis of sun-spot data. Before that, the main theoretical model was the scheme of hidden periodicities. The decomposition theorem was given in a book by Wold, the second edition of which (Wold, 1954) is a valuable introduction to the theory in discrete time. The general theory of prediction is associated with the names of Kolmogorov and Wiener; their work is best approached through the book of Yaglom (1962). Doob (1953) has given a more mathematical account. Wiener's own work on prediction theory is set out in a book (Wiener, 1949).

Parzen (1961) has emphasized the connexion with reproducing kernel Hilbert spaces. Cramér (1961) has discussed representation theorems and prediction theory for non-stationary processes and Wiener and Masani (1957, 1958) have developed the theory of prediction for multivariate processes.

The books of Bendat (1958), Laning and Battin (1956), Middleton (1960) and Solodovnikov (1960) are particularly concerned with engineering applications. Descriptive properties of Gaussian processes were considered in two major papers by Rice (1944, 1945).

Exercises

1. Prove that for a stationary process $\{X_n\}$ with autocovariance function $\gamma_x(h), v_m = V\{(X_1 + \ldots X_m)/m\}$ is given by

$$v_m = \frac{\gamma_x(0)}{m} + \frac{2}{m^2} \sum_{h=1}^{m-1} (m-h)\gamma_x(h).$$

Show that this tends to zero as $m \to \infty$ if

$$\frac{1}{m} \sum_{h=1}^{m} \gamma_x(h) \to 0,$$

and hence complete the proof of the convergence in probability of $(X_1 + \ldots + X_m)/m$, sketched in Section 7.3(iv).

Prove that the autocovariance function is proportional to the second differences of the sequence $m^2 v_m$ $(m = 1, 2, \ldots)$.

2. For a stationary process $\{X(t)\}$, define

$$v(l) = V\left\{\frac{1}{l} \int_0^l X(t)\,dt\right\}.$$

Prove that

$$v(l) = \frac{2}{l^2} \int_0^l (l-h)\gamma_x(h)\,dh,$$

$$\gamma_x(h) = \frac{1}{2}\frac{d^2}{dh^2}\{h^2 v(h)\}.$$

3. Use the results of the previous exercise to show that if $\gamma_x(h) \sim ah^{-b}$ $(0 < b < 1)$ as $h \to \infty$, then $v(l) \sim al^{-b}/(1-b)^{-1}(1-\tfrac{1}{2}b)^{-1}$. By comparing $v(2l)$ and $v(l)$, prove that for large l the correlation coefficient between the means of adjacent lengths l is $2^{1-b} - 1$. Investigate the conditions on $\gamma_x(h)$ under which the means of adjacent long lengths are uncorrelated and show that these conditions hold for all the special processes considered in the chapter.

4. A new stationary process $\{Y_n\}$ is formed by taking first differences of a given stationary process $\{X_n\}$, i.e. $Y_n = X_n - X_{n-1}$. Obtain the autocovariance function $\{Y_n\}$ in terms of that of $\{X_n\}$ and examine in particular the case $\gamma_x(h) = \lambda^{|h|}$. Discuss the results qualitatively.

5. Obtain the autocovariance function and the autoregressive and moving average representations of a process defined by a first-order difference equation forced by a first-order autoregressive process, i.e. a process $\{X_n\}$ defined by

$$X_n - \lambda X_{n-1} = U_n, \qquad U_n - \kappa U_{n-1} = V_n,$$

where $\{V_n\}$ is a stationary process of uncorrelated random variables of zero mean.

6. Consider the process $X_n = X$, where X is a random variable with a known probability distribution. (That is each realization is a constant.) Show that the process is stationary. What is the autocorrelation function? Does an ergodic theorem hold for the mean of a large number of values? What are the components in the Wold decomposition?

7. Develop for the second-order autoregressive process the relations between the autocovariance generating function and the generating functions for the autoregressive and moving average representations. Show how to determine the latter from the autocovariance generating function.

8. Find the autocovariance function for a two-state ergodic Markov chain in which the two states are labelled 0 and 1. Develop from first principles the autoregressive representation for such a process, and comment on its usefulness.

9. Prove that for an arbitrary process $\{X_n\}$ the predictor of X_n, given $\ldots, X_{n-2}, X_{n-1}$, that minimizes the mean square error is the mean of the conditional distribution of X_n given $\ldots, X_{n-2}, X_{n-1}$.

10. Prove that the Ornstein–Uhlenbeck process of Chapter 5 taken with an appropriate initial condition, is a stationary process with exponential autocorrelation function.

11. If

$$Y_n = \sum_{r=-\infty}^{\infty} w_r X_{n-r},$$

obtain the autocovariance generating function of the process $\{Y_n\}$ in terms of that of $\{X_n\}$. Check the result by investigating special cases.

12. Construct for the second-order autoregressive process of Example 7.6 a diagram showing (a) the region in the (λ_1, λ_2) plane for which the process is stationary and (b) the region for which the roots (ξ_1, ξ_2) are complex.

13. Discuss why the two-state Markov process in continuous time has an exponential autocorrelation function even though the argument of Section 7.3(ii) applies only to Gaussian processes.

14. Let $\{X(t)\}$ be a stationary Gaussian process of zero mean. Find the autocorrelation functions of the process $\{X^2(t)\}$ and $\{X^3(t)\}$ and examine the special case when $\{X(t)\}$ is a Markov process. [The relevant moments can be obtained from the joint characteristic function of $X(t)$ and $X(t+h)$.]

15. Suppose that to predict $X(t+h)$ based on observations up to time t of a stationary process of zero mean in continuous time the linear function

$$\int_0^\infty X(t-u)\,c(u)\,du$$

is used. Evaluate the mean square error in terms of the autocovariance function of the process. By considering a new predictor with $c(u)$ replaced by $c(u) + \epsilon c_1(u)$, for small ϵ, show that the mean square error is stationary if for all $v > 0$

$$\gamma(v+k) = \int_0^\infty c(u)\,\gamma(v-u)\,du.$$

(Wiener, 1949)

CHAPTER EIGHT

Stationary Processes: Frequency Domain

8.1. Introduction

In the previous chapter, we have discussed stationary processes $\{X_n\}$ placing emphasis on the autocovariance function $\gamma_x(h)$ and on representations of the process in terms of a stationary sequence $\{Z_n\}$ of uncorrelated random variables. In the present chapter, we develop a complementary approach to the theory of stationary processes, based on harmonic analysis.

There is one fairly minor generalization of the work in the previous chapter that we need first. We considered there real-valued processes. Now in the context of harmonic analysis it is convenient, and gives some extra generality, to allow the process to be complex-valued. This enables us, for instance, to deal directly with A.C. signals. We have then to define expectations, variances and covariances for complex-valued random variables and this is done as follows. Let $X = X_{(1)} + iX_{(2)}$, $Y = Y_{(1)} + iY_{(2)}$, where $X_{(1)}$, $X_{(2)}$, $Y_{(1)}$, $Y_{(2)}$ are real and the suffixes are bracketed to emphasize that they do not indicate a time variable. Then we define, as one would expect,

$$E(X) = E(X_{(1)}) + iE(X_{(2)}).$$

Now suppose $E(X) = E(Y) = 0$. We define the covariance by

$$C(X, Y) = E(X\overline{Y}), \tag{1}$$

where $\overline{Y}$ denotes the conjugate complex of Y, i.e. $\overline{Y} = Y_{(1)} - iY_{(2)}$. Thus

$$C(Y, X) = \overline{C(X, Y)}, \tag{2}$$

i.e. the covariance matrix of a set of multivariate complex random variables is Hermitian in general, rather than symmetric. In particular

$$V(X) = E(X\overline{X}) = E(X_{(1)}^2 + X_{(2)}^2) \tag{3}$$

is real and positive.

The physical justification for defining variances and covariances so that (3) holds is that expected squared modulus is the natural scalar measure of variability for a complex-valued random variable of zero mean. More specifically, if $X_{(1)} + iX_{(2)}$ represents an A.C. signal, (3) is proportional to the mean power generated.

308

If $X = X_{(1)} + iX_{(2)}$, $Y = Y_{(1)} + iY_{(2)}$, we have that

$$C(X, Y) = E(X_{(1)} Y_{(1)} + X_{(2)} Y_{(2)}) + iE(X_{(2)} Y_{(1)} - X_{(1)} Y_{(2)}).$$

Now $C(X, Y) = 0$ does not imply that the components $X_{(1)}$, $X_{(2)}$, $Y_{(1)}$, $Y_{(2)}$ are separately uncorrelated. Hence to avoid possible confusion, we shall say that X and Y are *orthogonal* if $C(X, Y) = 0$.

In general

$$V(X + Y) = V(X) + C(X, Y) + C(Y, X) + V(Y),$$

so that for orthogonal random variables

$$V(X + Y) = V(X) + V(Y).$$

Suppose now that $\{X_n\}$ is a stationary complex-valued process of mean zero and autocovariance function

$$\gamma_x(h) = E(X_{n+h} \bar{X}_n) = \overline{\gamma_x(-h)}. \tag{4}$$

We can carry over the arguments of Section 7.4 to obtain autoregressive and moving average representations in terms of a stationary process of innovations $\{Z_n\}$, which are orthogonal random variables of zero mean, and for which therefore

$$E(Z_{n+h} \bar{Z}_n) = 0 \quad (h \neq 0). \tag{5}$$

If we interpret the regression calculations geometrically, we take (1) as defining the scalar product of two vectors. The squared length of a vector X is thus $E(X\bar{X})$; the regression equations are selected to minimize the squared length of the deviation from regression.

In the preceding chapter, then, a central role was played by the process of innovations and its use to represent a general class of processes. In the present chapter we consider another type of canonical representation, this time in terms of a set of harmonic processes. That is, the basic process is, for fixed ω,

$$R e^{i\omega t} \quad (-\infty < t < \infty) \tag{6}$$

in continuous time and

$$R e^{i\omega n} \quad (n = \ldots, -1, 0, 1, \ldots) \tag{7}$$

in discrete time, where R is a complex-valued random variable. In (6), ω can be any real number, but in (7) there is no loss of generality in taking $-\pi \leqslant \omega < \pi$, since for all n and integral k, $e^{i(\omega + 2k\pi)n} = e^{i\omega n}$.

We first examine the conditions under which (7) represents a second-order stationary process. The arguments apply equally in discrete and in continuous time. The expectation of (7) is independent of n if and only if $E(R) = 0$, and this we assume from now on. We have that

$$E(R e^{i\omega(n+h)} \bar{R} e^{-i\omega n}) = e^{i\omega h} E(R\bar{R}), \tag{8}$$

and this is automatically independent of n, so that a stationary autocovariance function exists provided that $E(R\bar{R}) < \infty$.

Now consider a process formed by combining two processes (6), for different values of ω, i.e. consider a process

$$R_1 e^{i\omega_1 n} + R_2 e^{i\omega_2 n}, \tag{9}$$

where R_1 and R_2 are random variables. To ensure the stationarity of the mean of (9), we take $E(R_1) = E(R_2) = 0$. The covariance of values h apart is

$$E(R_1 e^{i\omega_1(n+h)} \bar{R}_1 e^{-i\omega_1 n}) + E(R_1 e^{i\omega_1(n+h)} \bar{R}_2 e^{-i\omega_2 n})$$

$$+ E(R_2 e^{i\omega_2(n+h)} \bar{R}_1 e^{-i\omega_1 n}) + E(R_2 e^{i\omega_2(n+h)} \bar{R}_2 e^{-i\omega_2 n}).$$

To represent a stationary autocovariance function this must be independent of n. For $\omega_1 \neq \omega_2$, this happens if and only if

$$E(R_1 \bar{R}_2) = E(R_2 \bar{R}_1) = 0,$$

i.e. if and only if R_1 and R_2 are orthogonal.

More generally, if the ω_j are unequal, and the R_j are random variables, the finite sum

$$R_1 e^{i\omega_1 n} + \ldots + R_k e^{i\omega_k n} \tag{10}$$

represents a stationary process if and only if $E(R_j) = 0$, $C(R_j, R_l) = 0$ ($j \neq l$). Further, the autocovariance function of (10) is

$$\sum_{j=1}^{k} e^{i\omega_j h} E(R_j \bar{R}_j) \tag{11}$$

and in particular the variance of (10) is

$$\sum_{j=1}^{k} E(R_j \bar{R}_j). \tag{12}$$

Thus the variance of R_j can be regarded as determining the contribution of the component at frequency ω_j to the total variance of the process.

Now if these representations were restricted to finite sums (10) they would be of rather limited interest, since for one thing the process (10) is clearly purely deterministic; in fact, k suitably spaced observations on the process would be enough to determine the realized values of R_1, ..., R_k.

Suppose, however, that formally we let $k \to \infty$ in (10). In discrete time, we take a set of ω_j's covering densely the interval $[-\pi, \pi)$. In continuous time we cover in the limit the whole real axis. Then the limiting form of (10) is obtained by considering processes

$$X_n = \int_{-\pi-0}^{\pi} e^{i\omega n} dS(\omega) \quad (n = \ldots, -1, 0, 1, \ldots) \tag{13}$$

in discrete time, where $S(\omega)$ is a stochastic process defined over $[-\pi, \pi)$, and

$$X(t) = \int_{-\infty}^{\infty} e^{i\omega t} dS(\omega) \quad (-\infty < t < \infty) \tag{14}$$

in continuous time, where $S(\omega)$ is now a stochastic process defined over $(-\infty, \infty)$. At discontinuity points we shall, by convention, take $S(\omega)$ to be continuous on the right.

For example, if we take the ω_j's to be equally spaced over $[-\pi, \pi)$, with $\omega_j = j\Delta\omega$, we consider (13) to be the limit as $\Delta\omega \to 0$ of

$$\sum_{j=-\pi/\Delta\omega}^{\pi/\Delta\omega} e^{i\omega_j n} R_j, \tag{15}$$

where

$$R_j = \int_{\omega_j - \frac{1}{2}\Delta\omega}^{\omega_j + \frac{1}{2}\Delta\omega} dS(\omega). \tag{16}$$

There is a corresponding result for (14). Now the requirement on the R_j in order that the sum (15) defines a stationary process is that the R_j have mean zero and are orthogonal. That is, from (16), the increments of the process $S(\omega)$ must have zero expectation and must be orthogonal over non-overlapping intervals:

$$E\left\{\int_{\omega_1}^{\omega_2} dS(\omega)\right\} = 0,$$

$$E\left\{\int_{\omega_1}^{\omega_2} dS(\omega) \int_{\omega_3}^{\omega_4} d\bar{S}(\omega)\right\} = 0 \quad (\omega_1 < \omega_2 < \omega_3 < \omega_4). \tag{17}$$

In virtue of (17), we call $S(\omega)$ a process with *orthogonal increments*.

It follows directly from (17) that, taking processes in discrete time for definiteness, we have for $\omega_1 < \omega_2$

$$V\{S(\omega_2)\} = V\left\{\int_{-\pi}^{\omega_1} dS(\omega) + \int_{\omega_1}^{\omega_2} dS(\omega)\right\} = V\{S(\omega_1)\} + V\left\{\int_{\omega_1}^{\omega_2} dS(\omega)\right\}. \tag{18}$$

Thus $V\{S(\omega)\}$ is a non-decreasing real-valued function of ω, say $G(\omega)$. Because of our convention about the definition of $S(\omega)$ at discontinuity points, $G(\omega)$ will be continuous on the right. This function completely specifies the second-order properties of the process $S(\omega)$. We can regard $dG(\omega)$ as the variance of an infinitesimal increment $dS(\omega)$. Further, by (12) and (17)

$$V(X_n) = V\left\{\int_{-\pi}^{\pi} dS(\omega)\right\} = G(\pi) \tag{19}$$

in discrete time, and $V\{X(t)\} = G(\infty)$ in continuous time, where we have, without loss of generality, taken $G(-\pi-0) = 0$, in discrete time and $G(-\infty) = 0$ in continuous time. Thus if we define $F(\omega) = G(\omega)/\sigma_x^2$, it has the mathematical properties of a cumulative distribution function defined over $[-\pi, \pi)$ or $(-\infty, \infty)$. We call $F(\omega)$ the *spectral distribution function*. If we wish to stress that $F(\omega)$ or $G(\omega)$ refer to the process $\{X_n\}$, we write $F_x(\omega)$ or $G_x(\omega)$.

By (13) and (18), $F(\omega_0)$ is the proportion of the total variance of the process (13) or (14) contributed by harmonic components with $\omega \leqslant \omega_0$. That is, the spectral distribution function specifies how the total variance of the process $\{X_n\}$ or $\{X(t)\}$ is sub-divided among the orthogonal components making up the representations (13) and (14).

We have reached the position that by analogy with the finite sums

$$R_1 e^{i\omega_1 n} + \ldots + R_k e^{i\omega_k n},$$

with random coefficients we may consider processes having the structure

$$X_n = \int_{-\pi}^{\pi} e^{i\omega n} dS(\omega) \quad \text{or} \quad X(t) = \int_{-\infty}^{\infty} e^{i\omega t} dS(\omega), \qquad (20)$$

where to ensure stationarity, the stochastic process $S(\omega)$ specifying the coefficients of the harmonic terms has orthogonal increments, i.e. obeys (17). Further the second-order properties of $S(\omega)$ are specified by a function $G(\omega)$, or, equivalently, by $G(\pi)$ (or $G(\infty)$), which is the total variance σ_x^2 of the process, and by the spectral distribution function $F(\omega) = G(\omega)/\sigma_x^2$. We call the right-hand sides of (20) the *spectral representations* of the processes $\{X_n\}$ and $\{X(t)\}$.

Now in the next section, we shall study the properties of processes defined by (20). Later, in Section 8.4, a proof will be outlined that all stationary second-order processes can be represented in the form (20), so that the class of processes represented by (20) is, except for the generalization to complex-valued processes, the same as that of Chapter 7.

We therefore have two complementary general representations of stationary processes, both in terms of a set of orthogonal random variables. That of Chapter 7 is the natural one for considering the evolution of the process in time; the associated second-order function is the autocovariance function. We call an analysis in terms of the concepts of Chapter 7 an analysis in the *time domain*. The representation (20) is the natural one when either the separate harmonic components have a physical significance, or, broadly speaking, when the practical effect of the process is most conveniently measured separately for the different harmonic components. The associated second-order function is the function $G(\omega)$, or the spectral distribution function. We call an analysis in terms of these concepts an analysis in the *frequency domain*.

8.2. The spectral representation

We now consider properties of processes defined by (20). We discuss processes in discrete time; the theory in continuous time involves only the replacement of the finite interval $[-\pi, \pi)$ for ω by $(-\infty, \infty)$. The essential properties of $S(\omega)$ are summarized formally in the equations

$$E\{dS(\omega)\} = 0, \qquad C\{dS(\omega_1), dS(\omega_2)\} = E\{dS(\omega_1)\,d\bar{S}(\omega_2)\} = 0,$$
$$V\{dS(\omega)\} = E\{|dS(\omega)|^2\} = dG(\omega), \tag{21}$$

where $\omega_1 \neq \omega_2$. We have already seen in (19) that in discrete time, $V(X_n) = G(\pi)$, and this is obtained directly from (21) by noting that

$$V(X_n) = E(X_n \bar{X}_n) = E\left\{\int_{-\pi}^{\pi} e^{i\omega_1 n} dS(\omega_1) \int_{-\pi}^{\pi} e^{-i\omega_2 n} d\bar{S}(\omega_2)\right\} = \int_{-\pi}^{\pi} dG(\omega),$$

since pairs $\omega_1 \neq \omega_2$ contribute zero to the expectation.

An extension of this argument gives the autocovariance function. For

$$\gamma(h) = E(X_{n+h} \bar{X}_n) = E\left\{\int_{-\pi}^{\pi} e^{i\omega_1(n+h)} dS(\omega_1) \int_{-\pi}^{\pi} e^{-i\omega_2 n} d\bar{S}(\omega_2)\right\}$$

$$= \int_{-\pi}^{\pi} e^{i\omega h} dG(\omega), \tag{22}$$

since, again, only contributions where $\omega_1 = \omega_2$ need be considered. In terms of the autocorrelation function $\rho(h) = \gamma(h)/\sigma_x^2$ and spectral distribution function $F(\omega) = G(\omega)/\sigma_x^2$, (22) becomes

$$\rho(h) = \int_{-\pi}^{\pi} e^{i\omega h} dF(\omega). \tag{23}$$

For a process in continuous time, the range of integration is, of course, $(-\infty, \infty)$. Thus $\rho(h)$ has the mathematical form of the characteristic function of a distribution.

It follows that the spectral distribution function $F(\omega)$ has two interpretations. First it gives the proportion of the variance of X_n contributed by components in various ranges of ω, i.e. gives the probabilistic properties of the components in a Fourier analysis of the process itself. Secondly, by (23), it gives directly the components in a Fourier analysis of the autocorrelation function.

The above results are for complex-valued processes. It is of interest to see the special form taken when the process is real-valued. The process $\{X_n\}$ is real-valued if and only if the terms corresponding to ω and $-\omega$ are complex conjugates, i.e. if for $\omega \neq \pi$

$$dS(\omega) = d\bar{S}(-\omega).$$

Then the process can be written

$$X_n = \int_0^\pi \cos{(n\omega)}\,dS_1(\omega) + \int_0^\pi \sin{(n\omega)}\,dS_2(\omega), \qquad (24)$$

where $\{S_1(\omega)\}$ and $\{S_2(\omega)\}$ are real-valued uncorrelated processes each with uncorrelated increments and characterized by the same variance and spectral distribution function $F_{(1)}(\omega)$, defined over $[0,\pi)$. In terms of the complex representation (20), the condition $dS(\omega) = d\bar{S}(-\omega)$, shows that the spectral distribution function $F(\omega)$ over $[-\pi,\pi)$ must be symmetric except for any discrete component at $\omega = -\pi$, which we have by convention placed there rather than at $\omega = \pi$. Clearly $F_{(1)}(\omega) = 2F(\omega) - 1$. It is usually simplest for general work to retain the complex-valued representation (20). Note that the symmetry of the distribution $F(\omega)$ implies, by (23), that $\rho(h) = \rho(-h)$. This, combined with the result $\rho(h) = \bar{\rho}(-h)$ which holds for all processes, implies that the autocorrelation function is real.

Now, in general, $F(\omega)$, being a distribution function, can be decomposed into a discrete and an (absolutely) continuous component; it is mathematically possible to have also a singular continuous component, but this possibility is of no practical importance.

Suppose first that $F(\omega)$ is a discrete distribution. That is, there is a finite or enumerable set of values ω_1, ω_2, ... at which $F(\omega)$ has positive increments $f_1, f_2, \ldots$ with $\sum f_j = 1$, and otherwise $F(\omega)$ is constant. Then the process (20) has the form

$$X_n = \sum e^{i\omega_j n} R_j, \qquad (25)$$

where $E(R_j) = 0$, $V(R_j) = f_j \sigma_x^2$, $C(R_j, R_l) = 0$ $(j \neq l)$. Such a process is said to have a *discrete* (or line) spectrum. The variance of the term $R_j e^{i\omega_j n}$ is $f_j \sigma_x^2$ and f_j is the proportion of the total variance of the process contributed by this term.

The autocovariance function, or the autocorrelation function, can be calculated from (23) or from (25). We have that

$$\rho(h) = \sum e^{i\omega_j h} f_j. \qquad (26)$$

Thus

$$f_j = \lim_{a \to \infty} \frac{1}{2a} \sum_{h=-a}^{a} e^{-i\omega_j h} \rho(h), \qquad (27)$$

the limit being zero when ω_j is replaced by a value of ω not in the set ω_1, ω_2, ...; see, for example, Parzen (1960, p. 402).

The processes (25) are completely deterministic in the sense of Chapter 7. Further, the autocorrelation function (26) does not tend to zero as $h \to \infty$. In fact, it can be shown that the autocorrelation function takes for large h values arbitrarily close to plus and minus one. Note that the f_j's cannot be found from a single realization, no matter how long. From the analysis of the whole of a single realization, only the realized values $R_1, R_2, \ldots$ can be determined.

We now consider processes for which the spectral distribution function $F(\omega)$ is (absolutely) continuous with a spectral density function $f(\omega)$. Then (23) becomes

$$\rho(h) = \int_{-\pi}^{\pi} e^{i\omega h} f(\omega) \, d\omega. \tag{28}$$

Thus the $\rho(h)$ for $h = \ldots, -1, 0, 1, \ldots$ are proportional to the Fourier coefficients of $f(\omega)$. Therefore

$$f(\omega) = \frac{1}{2\pi} \sum_{h=-\infty}^{\infty} e^{-i\omega h} \rho(h). \tag{29}$$

The corresponding results for processes in continuous time are

$$\rho(h) = \int_{-\infty}^{\infty} e^{i\omega h} f(\omega) \, d\omega \tag{30}$$

and

$$f(\omega) = \frac{1}{2\pi} \int_{-\infty}^{\infty} e^{-i\omega h} \rho(h) \, dh, \tag{31}$$

the latter following from the inversion formula for the Fourier integral (Cramér, 1946, p. 94).

It follows on applying the Riemann–Lebesgue lemma (Titchmarsh, 1948, p. 11) to (28) and (30) that $\rho(h) \to 0$ as $h \to \infty$; this is in contrast to the behaviour for processes with a discrete spectrum. It can be shown by integration by parts that if $f(\omega)$ is well-behaved near $\omega = 0$, $\rho(h)$ will tend to zero rapidly as $h \to \infty$. This connexion between the behaviours as $\omega \to 0$ and $h \to \infty$ arises because the contributions to the spectral representation from the region near $\omega = 0$ represent slow 'long-term' fluctuations, and the smaller the proportion of variance contributed by such fluctuations, the more rapidly will $\rho(h)$ tend to zero as $h \to \infty$.

In general, we can decompose a process into two components, one with a discrete spectrum and the other with a continuous spectrum. These components are uncorrelated, since they are generated by non-overlapping values of ω.

The most important examples of processes with a discrete spectrum are the combinations of a small number of sinusoidal terms. We now give examples of processes with continuous spectra. In fact, we can take any function having the properties of a probability density function for $f(\omega)$

and thereby obtain a possible spectral density function. If $f(\omega) = f(-\omega)$, the corresponding process defined by (20) can be real-valued.

The formulae given above relating $F(\omega)$ and $\rho(h)$ show how, in principle, to determine spectral representations for the processes considered in Chapter 7, where we obtained the autocorrelation function for a number of special processes. We invert the key equation (23) to obtain the function $F(\omega)$ corresponding to a given function $\rho(h)$. If the resulting $F(\omega)$ is a distribution function, we have the required representation. Actually, we shall see in Section 8.4 that all functions $\rho(h)$ corresponding to a second-order stationary process will yield a proper spectral representation, so that (23) is a characterization of those functions which can be autocorrelation functions. That is, if the $F(\omega)$ corresponding to a given function $\rho(h)$ were not a distribution function, the $\rho(h)$ could not be an autocorrelation function.

The general mathematical theorem here, called the Wiener–Khintchine theorem, is that any stationary process has an autocorrelation function of the form (23) and that, conversely, given a function $\rho(h)$ representable as (23), there exists a stationary process with $\rho(h)$ as autocorrelation function.

Suppose first that $f(\omega) = (2\pi)^{-1}$ over $[-\pi, \pi)$; then by (28), $\rho(h) = 0$ ($h \neq 0$). Thus the uniform spectral density corresponds to a stationary orthogonal process. The corresponding pure noise process in continuous time is improper, as we saw in Section 7.4. It can be approximated however by a process with a uniform spectral density over a long interval $(-\omega_0, \omega_0)$, where ω_0 is large. For this, the autocorrelation function is, by (30),

$$
\begin{aligned}
\rho(h) &= \frac{1}{2\omega_0} \int_{-\omega_0}^{\omega_0} e^{i\omega h}\, d\omega \\
&= \frac{\sin(\omega_0 h)}{\omega_0 h}.
\end{aligned}
\tag{32}
$$

As $\omega_0 \to \infty$, this tends to zero for $h \neq 0$.

Consider next a process in discrete time with autocorrelation function $\lambda^{|h|}$ with $|\lambda| < 1$. Then by (29) the corresponding spectral density function is

$$
\begin{aligned}
f(\omega) &= \frac{1}{2\pi} \left\{ 1 + \sum_{h=1}^{\infty} \lambda^h e^{i\omega h} + \sum_{h=1}^{\infty} \lambda^h e^{-i\omega h} \right\} \\
&= \frac{1 - \lambda^2}{2\pi(1 - \lambda e^{i\omega})(1 - \lambda e^{-i\omega})} \\
&= \frac{1 - \lambda^2}{2\pi(1 - 2\lambda \cos\omega + \lambda^2)} \quad (-\pi \leqslant \omega < \pi).
\end{aligned}
\tag{33}
$$

Figure 8.1 shows this function for $\lambda = 0 \cdot 2$ and $0 \cdot 8$.

The exponential autocorrelation function for a process in continuous time, $e^{-\lambda|h|}$, leads, by (31), to the spectral density function, defined for all real ω,

$$f(\omega) = \frac{1}{2\pi} \int\limits_{-\infty}^{\infty} e^{-\lambda|h|} e^{-i\omega h} \, dh$$

$$= \frac{\lambda}{\pi(\lambda^2 + \omega^2)} . \tag{34}$$

Note especially that in (33), ω varies over $[-\pi, \pi)$, whereas in (34) the corresponding range is $(-\infty, \infty)$.

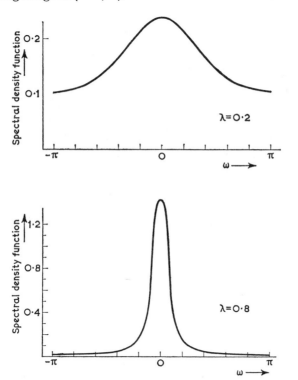

Fig. 8.1. Spectral density function for first-order autoregressive process.

There is a close connexion between (33) and the corresponding autocorrelation generating function obtained from the autocovariance generating function by putting $\sigma_x^2 = 1$ in (7.55); in fact the autocorrelation generating function is

$$\frac{(1 - \lambda^2)}{(1 - \lambda\zeta)(1 - \lambda/\zeta)},$$

where ζ is the argument of the generating function.

In general, we relate the autocovariance generating function $\Gamma(\zeta)$ of (7.51) and the density $g(\omega)$ corresponding to the function $G(\omega) = \sigma_x^2 F(\omega)$ occurring in the spectral representation. The relation follows directly from the definition (7.51)

$$\Gamma(\zeta) = \sum_{h=-\infty}^{\infty} \gamma(h)\, \zeta^h$$

and the inversion formula obtained by multiplying (29) by σ_x^2, namely

$$g(\omega) = \frac{1}{2\pi} \sum_{h=-\infty}^{\infty} e^{-i\omega h}\, \gamma(h).$$

Thus

$$g(\omega) = \frac{\Gamma(e^{-i\omega})}{2\pi}. \tag{35}$$

Now for real-valued processes the results (7.52) and (7.53), namely

$$\Gamma(\zeta) = \sigma_z^2 B(\zeta) B(1/\zeta) = \sigma_z^2 \{A(\zeta) A(1/\zeta)\}^{-1},$$

where σ_z^2 is the variance of the innovation process, express $\Gamma(\zeta)$ in terms of the functions $A(\zeta)$ and $B(\zeta)$ specifying the autoregressive and moving average representations of the process. Thus

$$g(\omega) = \frac{\sigma_z^2}{2\pi A(e^{i\omega}) A(e^{-i\omega})} \tag{36}$$

$$= \frac{\sigma_z^2 B(e^{i\omega}) B(e^{-i\omega})}{2\pi}. \tag{37}$$

If the spectral density function $f(\omega)$ is required, (36) and (37) must be divided by σ_x^2.

It follows from (36) and (37) that both a finite autoregressive process and a finite moving average process have spectral densities that are rational functions of $e^{i\omega}$. As a special case, consider a process formed from moving totals of extent k, i.e. a process defined by

$$X_n = Z_n + \ldots + Z_{n-k+1}.$$

Then $B(\zeta) = 1 + \zeta + \ldots + \zeta^{k-1} = (1 - \zeta^k)/(1 - \zeta)$. Therefore

$$g(\omega) = \frac{\sigma_z^2 (1 - e^{ik\omega})(1 - e^{-ik\omega})}{2\pi (1 - e^{i\omega})(1 - e^{-i\omega})}.$$

Since $V(X_n) = k\sigma_z^2$, it follows that the spectral density function for the process $\{X_n\}$ is

$$\frac{(1 - e^{ik\omega})(1 - e^{-ik\omega})}{2\pi k(1 - e^{i\omega})(1 - e^{-i\omega})} = \frac{\sin^2 (\tfrac{1}{2}k\omega)}{2\pi k \sin^2 (\tfrac{1}{2}\omega)}. \tag{38}$$

The autocorrelation function for the simple moving average process in continuous time is $1 - |h|/k$ $(|h| \leqslant k)$ and therefore, by (31), the corresponding spectral density function is

$$
\begin{aligned}
f(\omega) &= \frac{1}{2\pi} \int_{-k}^{k} e^{-i\omega h} \left(1 - \frac{|h|}{k}\right) dh \\
&= \frac{1}{2\pi} \int_{0}^{k} (e^{i\omega h} + e^{-i\omega h}) \left(1 - \frac{h}{k}\right) dh \\
&= \frac{1 - \cos(\omega k)}{\pi \omega^2 k}.
\end{aligned}
\tag{39}
$$

8.3. Linear operations on stationary processes

In many applications of the theory of stationary processes, we are concerned with the effect of forming a new stationary process, say $\{Y_n\}$ or $\{Y(t)\}$, by linear operation on a given process $\{X_n\}$ or $\{X(t)\}$. For example, in an industrial context we might be concerned with the relation between the variation of some material going into a process, considered as defining a time series, and the variation of the output, considered as defining a second time series. Further, we have had already several important examples of linear operations. These include the moving average and autoregressive processes, which are obtained by linear operations on an uncorrelated process and the derivative process $\{\dot{X}(t)\}$ of Section 7.4.

The theory in continuous time is rather more interesting and so we consider that first. Let $\mathscr{L}$ denote any stationary linear operation on the process $\{X(t)\}$ producing a new process $\{Y(t)\}$,

$$Y(t) = \mathscr{L}\{X(t)\}.$$

Examples are

(a) $Y(t) = \dot{X}(t)$;

(b) $Y(t) = c_0 X(t) + c_1 X(t - r_1) + \ldots$;

(c) $Y(t) = \int_{0}^{\infty} X(t - u) \, dC(u)$;

(d) $\dot{Y}(t) + a_1 Y(t) = X(t)$.

In (d), $\mathscr{L}$ is the operator inverse to $(d/dt + a_1)$; in the other examples, $\mathscr{L}$ is defined directly by the right-hand side of the relevant equation. The essential properties of the operator $\mathscr{L}$ are its stationarity, i.e. that if $T_h X(t) = X(t + h)$, then $\mathscr{L}\{T_h X(t)\} = T_h \mathscr{L}\{X(t)\}$, and its linearity,

$$\mathscr{L}\{X_1(t) + X_2(t)\} = \mathscr{L}\{X_1(t)\} + \mathscr{L}\{X_2(t)\}.$$

Now consider the spectral representation

$$X(t) = \int\limits_{-\infty}^{\infty} e^{i\omega t} dS_x(\omega),$$

where we have attached a suffix x to the process $\{S(\omega)\}$ to emphasize its connexion with $\{X(t)\}$. Now because of the linearity of the operator $\mathscr{L}$

$$Y(t) = \mathscr{L}\left\{\int\limits_{-\infty}^{\infty} e^{i\omega t} dS_x(\omega)\right\}$$

$$= \int\limits_{-\infty}^{\infty} [\mathscr{L}\{e^{i\omega t}\}] dS_x(\omega). \tag{40}$$

Further, for all linear operators $\mathscr{L}$,

$$\mathscr{L}\{e^{i\omega t}\} = l(\omega) e^{i\omega t}, \tag{41}$$

where $l(\omega)$ is complex. That is, the effect of the operator $\mathscr{L}$ on the harmonic oscillation $e^{i\omega t}$ is to produce a harmonic oscillation of the same frequency but in general with different amplitude and phase. Thus

$$Y(t) = \int\limits_{-\infty}^{\infty} e^{i\omega t} l(\omega) dS_x(\omega)$$

$$= \int\limits_{-\infty}^{\infty} e^{i\omega t} dS_y(\omega) \tag{42}$$

say. The process $\{S_y(\omega)\}$ is clearly a process with orthogonal increments, since $\{S_x(\omega)\}$ is, and hence (42) gives the spectral representation of the process $\{Y(t)\}$.

The formal equation

$$dS_y(\omega) = l(\omega) dS_x(\omega) \tag{43}$$

is of fundamental importance.

By definition the function $G_y(\omega)$ associated with the process $\{Y(t)\}$ is given by

$$dG_y(\omega) = E\{dS_y(\omega) d\bar{S}_y(\omega)\}$$

$$= E\{l(\omega) \bar{l}(\omega) dS_x(\omega) d\bar{S}_x(\omega)\}$$

$$= |l(\omega)|^2 dG_x(\omega).$$

For processes with a continuous spectrum, we have that

$$g_y(\omega) = |l(\omega)|^2 g_x(\omega)$$

$$= L(\omega) g_x(\omega), \tag{44}$$

say. Here $l(\omega)$ is called the *transfer function* associated with the linear operation $\mathscr{L}$. Note that $L(\omega)$ depends only on the amplitude change in (41) and takes no account of the phase change at ω. The importance of (44) lies in its showing in very simple form the dependence of the properties of the process $\{Y(t)\}$ on the properties of the initial process $\{X(t)\}$ and of the operator $\mathscr{L}$. Since the spectral densities $f_y(\omega)$ and $f_x(\omega)$ are equal to $g_y(\omega)/\sigma_y^2$ and $g_x(\omega)/\sigma_x^2$, it follows from (44) that

$$
\sigma_y^2 = \int_{-\infty}^{\infty} L(\omega)\, g_x(\omega)\, d\omega
$$

$$
= \sigma_x^2 \int_{-\infty}^{\infty} L(\omega)\, f_x(\omega)\, d\omega \tag{45}
$$

and that

$$
f_y(\omega) = \frac{\sigma_x^2}{\sigma_y^2} L(\omega)\, f_x(\omega). \tag{46}
$$

As an example suppose $\mathscr{L} = d/dt$, so that the process $\{Y(t)\}$ is that previously denoted $\{\dot{X}(t)\}$. Then

$$
\mathscr{L}\{e^{i\omega t}\} = i\omega\, e^{i\omega t},
$$

so that $l(\omega) = i\omega$, $L(\omega) = \omega^2$. Hence

$$
g_y(\omega) = \omega^2 g_x(\omega). \tag{47}
$$

Thus, in particular,

$$
\sigma_y^2 = \int_{-\infty}^{\infty} \omega^2 g_x(\omega)\, d\omega
$$

$$
= \sigma_x^2 \int_{-\infty}^{\infty} \omega^2 f_x(\omega)\, d\omega. \tag{48}
$$

Now the autocorrelation function of $\{X(t)\}$ is given by (30) as

$$
\rho(h) = \int_{-\infty}^{\infty} e^{i\omega h} f(\omega)\, d\omega.
$$

If we differentiate twice with respect to h, we have that

$$
\rho''(h) = -\int_{-\infty}^{\infty} \omega^2 e^{i\omega h} f(\omega)\, d\omega. \tag{49}
$$

Then, by (48)

$$
\sigma_y^2 = \sigma_x^2 \{-\rho''(0)\}; \tag{50}
$$

this was proved by a more direct method in (7.58). Also, it follows easily from (47), (49) and the relation connecting the autocorrelation function and the spectral density function that the autocorrelation function of the process $\{Y(t)\}$ is $\rho''(h)/\rho''(0)$.

The qualitative interpretation of (47) is that a component of $\{X(t)\}$ with large ω, which therefore oscillates rapidly, has a much magnified effect after differentiation, whereas a component with small ω, which therefore oscillates slowly, makes relatively little contribution to the differentiated process.

If the linear operation is a weighted averaging, we have that

$$Y(t) = \int_{-\infty}^{\infty} X(t-u)\,c(u)\,du.$$

Thus

$$\mathscr{L}\{e^{i\omega t}\} = e^{i\omega t} \int_{-\infty}^{\infty} e^{-i\omega u}\,c(u)\,du$$

and

$$l(\omega) = \int_{-\infty}^{\infty} e^{-i\omega u}\,c(u)\,du. \tag{51}$$

In particular for a simple averaging over $(0,k)$,

$$l(\omega) = \frac{1}{k}\int_{0}^{k} e^{-i\omega u}\,du = \frac{1-e^{-i\omega k}}{i\omega k}.$$

Thus we have that

$$|l(\omega)|^2 = \frac{2\{1-\cos(\omega k)\}}{\omega^2 k^2}.$$

That is, if

$$Y(t) = \frac{1}{k}\int_{t-k}^{t} X(u)\,du,$$

$$g_y(\omega) = \frac{2\{1-\cos(\omega k)\}}{\omega^2 k^2}\,g_x(\omega). \tag{52}$$

This should be compared with (39) for a simple moving average.

For processes in discrete time the definition of the transfer function is unchanged. In discrete time the most general stationary linear operation can be written as

$$Y_n = \mathscr{L}\{X_n\} = \sum_{u=-\infty}^{\infty} c_u X_{n-u}, \tag{53}$$

so that

$$\mathscr{L}\{e^{i\omega n}\} = e^{i\omega n}\sum_{u=-\infty}^{\infty} c_u e^{-i\omega u}.$$

Thus we have that if

$$l(\omega) = \sum_{u=-\infty}^{\infty} c_u e^{-i\omega u}, \qquad L(\omega) = |l(\omega)|^2, \tag{54}$$

the properties of $\{Y_n\}$ are given by

$$g_y(\omega) = L(\omega) g_x(\omega).$$

On using the relation (35) connecting the spectral function and the auto-covariance generating function, we have that

$$\Gamma_y(e^{i\omega}) = L(-\omega) \Gamma_x(e^{i\omega}). \tag{55}$$

If we write $\zeta = e^{i\omega}$, (55) becomes

$$\Gamma_y(\zeta) = L^{\dagger}(1/\zeta) \Gamma_x(\zeta), \tag{56}$$

where

$$L^{\dagger}(\zeta) = \left| \sum_{u=-\infty}^{\infty} c_u \zeta^{-u} \right|^2. \tag{57}$$

Equation (56) is an important general result; an alternative proof can be obtained by first finding the effect of the linear operation (53) on an individual autocovariance.

8.4. Derivation of the spectral representation

So far we have looked at the properties of processes having a spectral representation

$$X_n = \int_{-\pi}^{\pi} e^{i\omega n} dS(\omega) \quad \text{or} \quad X(t) = \int_{-\infty}^{\infty} e^{i\omega t} dS(\omega), \tag{58}$$

where $\{S(\omega)\}$ is a process with orthogonal increments. In Section 8.1 we saw that if a process is defined by an integral (58) for some process $\{S(\omega)\}$, the requirement that $\{S(\omega)\}$ has orthogonal increments is necessary for $\{X_n\}$ or $\{X(t)\}$ to be stationary. An important general issue, however, concerns whether all stationary processes have a spectral representation.

To sketch briefly the answer to this, we first consider how, given a process with a spectral representation, we can isolate the increments of the function $S(\omega)$. Then we apply the same operations to an arbitrary stationary process and verify that there result the components of a genuine spectral representation.

Suppose then that we are given the process $\{X(t)\}$,

$$X(t) = \int_{-\infty}^{\infty} e^{i\omega t} dS(\omega).$$

We require first to operate on $\{X(t)\}$ to produce, for given (ω_1, ω_2), the contribution arising from the interval $\omega_1 < \omega < \omega_2$, namely

$$\int_{\omega_1}^{\omega_2} e^{i\omega t}\, dS(\omega). \tag{59}$$

To produce this by the linear operation

$$\int_{-\infty}^{\infty} X(t-u)\, c_{\omega_1, \omega_2}(u)\, du, \tag{60}$$

we must have

$$\int_{-\infty}^{\infty} e^{-i\omega u} c_{\omega_1, \omega_2}(u)\, du = \begin{array}{ll} 1 & (\omega_1 < \omega < \omega_2), \\ 0 & \text{(otherwise).} \end{array} \tag{61}$$

Applying formally the Fourier inversion formula to (61), we have that

$$c_{\omega_1, \omega_2}(u) = \frac{1}{2\pi} \int_{\omega_1}^{\omega_2} e^{i\omega u}\, d\omega$$

$$= \frac{e^{i\omega_2 u} - e^{i\omega_1 u}}{2\pi i u}. \tag{62}$$

In fact it is easily shown directly that (62) satisfies (61) except at $\omega = \omega_1, \omega_2$, where the integral on the left-hand side is equal to $\frac{1}{2}$.

That is, the linear operation defined by (60) and (62) produces from the process $\{X(t)\}$ the process (59), provided that ω_1, ω_2 are points of continuity of the spectrum of $\{X(t)\}$. We call this linear operation a *band-pass filter*. The isolation of (59) is never possible exactly in practice, since (60) represents an averaging operation of infinite extent in both directions.

It follows on taking $t = 0$ in (59) that if ω_1 and ω_2 are continuity points of the spectrum

$$S(\omega_2) - S(\omega_1) = \int_{\omega_1}^{\omega_2} dS(\omega) = \int_{-\infty}^{\infty} X(-u)\, c_{\omega_1, \omega_2}(u)\, du. \tag{63}$$

In a rigorous treatment the right-hand side of (63) is considered as a limit in mean square as $u_0 \to \infty$ of the finite integral taken over the interval $(-u_0, u_0)$. Equation (63) determines $\{S(\omega)\}$ to within an additive constant, at the continuity points of the spectrum. At the discontinuity points we can define $S(\omega)$ by convention to be continuous on the right.

The argument so far shows how to isolate the process $\{S(\omega)\}$ given a process $\{X(t)\}$ with a spectral representation. We now reverse the argument. Given an arbitrary stationary process $\{X(t)\}$, with mean zero and autocovariance function $\gamma(h)$, define $\{S(\omega)\}$ by (63).

It is now fairly easy to show that $\{S(\omega)\}$ is a process with orthogonal increments and that for all t

$$E\left|X(t) - \int_{-\infty}^{\infty} e^{i\omega t} dS(\omega)\right|^2 = 0, \tag{64}$$

so that it does give a spectral resolution. To prove the first point, let (ω_1, ω_2) and (ω_3, ω_4) be non-overlapping intervals. Then

$$E\left\{\int_{\omega_1}^{\omega_2} dS(\omega) \int_{\omega_3}^{\omega_4} d\bar{S}(\omega)\right\} = \int_{-\infty}^{\infty}\int c_{\omega_1,\omega_2}(u)\, \bar{c}_{\omega_3,\omega_4}(v)\, E\{X(-u)\,\bar{X}(-v)\}\, du\, dv$$

$$= \int_{-\infty}^{\infty}\int c_{\omega_1,\omega_2}(u)\, \bar{c}_{\omega_3,\omega_4}(v)\, \gamma(v-u)\, du\, dv.$$

This is easily shown to be zero by taking u and $v-u$ as new variables and integrating with respect to u. The proof of (64) is similar.

Essentially the same proof applies to processes in discrete time. Yaglom (1962, p. 39) gives a detailed account of the proof.

8.5. Prediction and filtering theory

In Section 7.5 we described the problems of prediction, filtering and regulation and indicated how the prediction problem can be tackled by a time domain analysis. We now look at the corresponding frequency domain analysis, dealing mostly with processes in discrete time.

Consider then the stationary process $\{X_n\}$ of mean zero. Given values $\ldots, X_{n-2}, X_{n-1}$, we require to predict X_{n+k}. We shall, for simplicity, consider only the problem where the whole past of the process is given. Only processes with a continuous spectrum will be considered, since the prediction of any component of the process corresponding to a discrete spectrum offers little difficulty, in principle.

We first use the relation (37) connecting the spectrum and the generating function for the moving average representation of the process. In fact, $B(e^{i\omega})$ is a power series in $\zeta = e^{i\omega}$, with unit leading term; similarly $B(e^{-i\omega})$ is a series in inverse powers of ζ with unit leading term. Thus $\log g(\omega)$ has a Laurent expansion in ζ with constant term $\log\{\sigma_z^2/(2\pi)\}$. Thus, by the integral giving the constant term of a Laurent expansion,

$$\frac{\sigma_z^2}{2\pi} = \exp\left\{\frac{1}{2\pi}\int_{-\pi}^{\pi} \log g(\omega)\, d\omega\right\}; \tag{65}$$

it is assumed there that the annulus of convergence of the Laurent expansion includes the unit circle $|\zeta| = 1$, i.e. the ω axis from $-\pi$ to π.

In terms of the spectral density function $f(\omega) = g(\omega)/\sigma_x^2$, we have that

$$\frac{\sigma_z^2}{\sigma_x^2} = 2\pi \exp\left\{\frac{1}{2\pi} \int\limits_{-\pi}^{\pi} \log f(\omega)\, d\omega\right\}. \tag{66}$$

Now σ_x^2 is the variance of the process being predicted and σ_z^2 is the variance of the corresponding innovation process. Further, the discussion of Section 7.5 shows that σ_z^2 is the minimum mean square error attainable by a linear predictor for one-step ahead prediction. One consequence of (66) is that $\sigma_z^2 = 0$, i.e. the process can be predicted with zero error, if and only if

$$\int\limits_{-\pi}^{\pi} \log f(\omega)\, d\omega = -\infty. \tag{67}$$

The corresponding condition for processes in continuous time can be shown to be

$$\int\limits_{-\infty}^{\infty} \frac{\log f(\omega)}{1+\omega^2}\, d\omega = -\infty. \tag{68}$$

A process satisfying (68) is obtained by taking a Gaussian autocorrelation function,

$$\gamma(h) = e^{-\lambda h^2}, \tag{69}$$

for which, by (31),

$$f(\omega) = (\pi\lambda)^{\frac{1}{2}} \exp(-\tfrac{1}{4}\omega^2/\lambda), \tag{70}$$

and (68) holds. Autocorrelation functions closely approximated by (69) arise in the study of turbulence (Batchelor, 1953). One way of proving directly that this process can be predicted with zero error depends on the fact that (69) is differentiable any number of times at $h = 0$. This can be shown to imply the existence of all derivatives $d^m X(t)/dt^m$ of the process itself and that as m tends to infinity

$$E\left|X(t+k) - \sum_{r=0}^{m} \frac{k^r}{r!} \frac{d^r X(t)}{dt^r}\right|^2 \tag{71}$$

tends to zero. That is, a Taylor expansion of the process itself gives the required predictor.

We now consider briefly the form of the optimum predictor in discrete time. In the notation of (7.76), the predictor of X_{n+k} is

$$c_1^{(k)} X_{n-1} + c_2^{(k)} X_{n-2} + \dots$$

To minimize the expected mean squared modulus of error, we have, generalizing (7.73) to complex-valued processes,

$$\sum_{l=1}^{\infty} c_l^{(k)} \rho_x(m-l) = \rho_x(m+k) \quad (m = 1, 2, \ldots). \tag{72}$$

Assuming that there is a spectral density function, we introduce the representation (30) of the autocorrelation function. We have that

$$\int_{-\pi}^{\pi} e^{im\omega} f(\omega) \left\{ \sum_{l=1}^{\infty} c_l^{(k)} e^{-il\omega} - e^{ik\omega} \right\} d\omega = 0 \quad (m = 1, 2, \ldots). \tag{73}$$

Thus the choice of the coefficients $c_l^{(k)}$ requires our finding a function $C_k(e^{-i\omega})$,

$$C_k(\zeta) = \sum_{l=1}^{\infty} c_l^{(k)} \zeta^l,$$

of positive powers in ζ, such that the positive Fourier coefficients of

$$f(\omega)\{C_k(e^{-i\omega}) - e^{ik\omega}\}$$

are all zero. This can be shown to be equivalent to the approach of Section 7.5.

Since the details of this were explored in Section 7.5, we pass now to apply the same ideas to the filtering problem.

Suppose then that we observe the process $\{X_n\}$, where

$$X_n = W_n + Z_n. \tag{74}$$

Here the processes $\{W_n\}$ and $\{Z_n\}$, called the signal and the noise, are assumed to be mutually uncorrelated with zero means and with autocovariance functions $\gamma_w(h)$, $\gamma_z(h)$. It is easily seen that the autocovariance function of $\{X_n\}$ is

$$\gamma_x(h) = \gamma_w(h) + \gamma_z(h). \tag{75}$$

We use the previous notation for the spectral functions, assuming in all cases that the spectra are continuous.

We require to estimate the value W_n given observations on the process $\{X_n\}$. The simplest form of the problem, and the only one we consider here, arises when the whole doubly infinite sequence $\{X_n\}$ $(n = \ldots, -1, 0, 1, \ldots)$ is observed. Consider as an estimate of W_n

$$\sum_{r=-\infty}^{\infty} d_r X_{n-r}. \tag{76}$$

The squared modulus of error is

$$\left| W_n - \sum_{r=-\infty}^{\infty} d_r X_{n-r} \right|^2 = W_n \overline{W}_n - 2\mathscr{R}\{\sum \overline{d}_r \overline{X}_{n-r} W_n\} + \sum \overline{d}_r d_u \overline{X}_{n-r} X_{n-u}.$$

The expected value is

$$\sigma_w^2 - 2\mathscr{R}\{\sum \overline{d}_r \gamma_w(r)\} + \sum \overline{d}_r d_u \gamma_x(r-u). \tag{77}$$

Thus, on differentiating with respect to the $\overline{d}_r$, we have that the optimum filter is defined by

$$\gamma_w(r) = \sum d_u \gamma_x(r-u). \tag{78}$$

Further the corresponding expected squared modulus of error is

$$\sigma_w^2 - \sum \sum \overline{d}_r d_u \gamma_x(r-u), \tag{79}$$

where the d_r's are given by (78).

We now introduce into (78) the spectral form for the autocovariances, obtaining

$$\int_{-\pi}^{\pi} e^{i\omega r}\{g_w(\omega) - \sum d_u e^{-i\omega u} g_x(\omega)\} d\omega = 0. \tag{80}$$

Thus if

$$D(\zeta) = \sum_{u=-\infty}^{\infty} d_u \zeta^u,$$

equation (80) shows that all Fourier coefficients of

$$g_w(\omega) - D(e^{-i\omega}) g_x(\omega)$$

vanish and hence that the function itself vanishes. Note that *all* Fourier coefficients enter because of the doubly infinite limits for r in (76), i.e. because the filtering is to be based both on the past and on the future of the process $\{X_n\}$.

Thus the equations

$$D(e^{-i\omega}) = \frac{g_w(\omega)}{g_x(\omega)}, \tag{81}$$

$$d_r = \frac{1}{2\pi} \int_{-\pi}^{\pi} e^{i\omega r} \frac{g_w(\omega)}{g_x(\omega)} d\omega, \tag{82}$$

define the optimum filter. The quantities d_r can often most easily be found by writing $\zeta = e^{i\omega}$ and obtaining the coefficient of ζ^r in the Laurent expansion of (81).

Finally, the expected squared modulus error is by (79)

$$\sigma_w^2 - \sum \sum \bar{d}_r d_u \int_{-\pi}^{\pi} e^{i\omega(r-u)} g_x(\omega)\, d\omega$$

$$= \int_{-\pi}^{\pi} g_w(\omega)\, d\omega - \int_{-\pi}^{\pi} |D(e^{-i\omega})|^2 g_x(\omega)\, d\omega$$

$$= \int_{-\pi}^{\pi} \left\{ g_w(\omega) - \frac{g_w^2(\omega)}{g_x(\omega)} \right\} d\omega$$

$$= \int_{-\pi}^{\pi} \frac{g_w(\omega)\, g_z(\omega)}{g_x(\omega)}\, d\omega, \tag{83}$$

since $g_w(\omega) + g_z(\omega) = g_x(\omega)$.

Now the spectral functions are non-negative and hence (83) is zero if and only if there is no set of ω values (of positive measure) for which $g_w(\omega)$ and $g_z(\omega)$ are simultaneously non-zero, i.e. if the spectra of signal and noise are non-overlapping. When there is no overlap, the theoretical possibility of exact separation of signal and noise follows from the existence of band-pass filters.

A simple non-trivial example of a filtering problem occurs when the noise is an uncorrelated process

$$g_z(\omega) = \frac{\sigma_z^2}{2\pi},$$

and when the signal has a spectrum corresponding to a first-order auto-regressive process,

$$g_w(\omega) = \frac{\sigma_w^2(1-\lambda^2)}{2\pi(1-\lambda e^{i\omega})(1-\lambda e^{-i\omega})}.$$

Then

$$\frac{g_w(\omega)}{g_x(\omega)} = \frac{\sigma_w^2(1-\lambda^2)}{\sigma_w^2(1-\lambda^2)+\sigma_z^2(1-\lambda e^{i\omega})(1-\lambda e^{-i\omega})}.$$

Thus if $\zeta = e^{i\omega}$

$$D(1/\zeta) = \frac{\sigma_w^2(1-\lambda^2)}{\sigma_w^2(1-\lambda^2)+\sigma_z^2(1-\lambda\zeta)(1-\lambda/\zeta)}$$

$$= \frac{\sigma_w^2(1-\lambda^2)\zeta}{\sigma_w^2(1-\lambda^2)\zeta + \sigma_z^2(1-\lambda\zeta)(\zeta-\lambda)}. \tag{84}$$

The denominator of (84) has two real zeros ζ_0, $1/\zeta_0$, where $-1 < \zeta_0 < 1$

and in fact ζ_0 has the same sign as λ. Hence splitting (84) into partial fractions, we have that

$$D(1/\zeta) = A\left(\frac{\zeta_0}{\zeta - \zeta_0} + \frac{1}{1 - \zeta\zeta_0}\right),$$

$$= A\left\{\frac{\zeta_0}{\zeta}\left(1 - \frac{\zeta_0}{\zeta}\right)^{-1} + (1 - \zeta\zeta_0)^{-1}\right\},$$

where

$$A = \frac{\sigma_w^2(1 - \lambda^2)\,\zeta_0}{\lambda\sigma_z^2(1 - \zeta_0^2)}.$$

We can now expand in a Laurent series in $|\zeta_0| < |\zeta| < 1/|\zeta_0|$ obtaining

$$d_r = d_{-r} = A\zeta_0^r \quad (r = 0, 1, 2, \ldots). \tag{85}$$

Further by (83), the mean square error is

$$\frac{1}{2\pi}\int_{-\pi}^{\pi} \frac{\sigma_z^2\sigma_w^2(1 - \lambda^2)\,e^{i\omega}}{\sigma_w^2(1 - \lambda^2)\,e^{i\omega} + \sigma_z^2(1 - \lambda e^{i\omega})\,(e^{i\omega} - \lambda)}\,d\omega$$

$$= \frac{1}{2\pi i}\int \frac{\sigma_z^2\sigma_w^2(1 - \lambda^2)\,d\zeta}{-\lambda\sigma_z^2(\zeta - \zeta_0)\,(\zeta - 1/\zeta_0)},$$

where the last integral is taken round the unit circle $|\zeta| = 1$. Therefore the mean square error is $\sigma_w^2\zeta_0(1 - \lambda^2)/(1 - \zeta_0^2)$.

Thus the optimum filter attaches weights to the observations that decrease in geometric progression away from the point at which the signal is to be estimated. Note that we have considered a situation where both signal and noise have zero mean. If the signal had known mean μ_w, and the noise mean zero, we would write the filter (76)

$$\mu_w + \sum_{r = -\infty}^{\infty} d_r(X_{n-r} - \mu_w) \tag{86}$$

and, since, in ge. .a., $\sum d_r < 1$ this is not the same as applying (76) directly to the observations.

It is possible to develop much further the ideas outlined in this section. One general point is that the argument leading to (85) goes through in essentially the same way whenever the spectral functions $g_w(\omega)$ and $g_z(\omega)$ are rational functions of $e^{i\omega}$.

If the series $\{X_n\}$ is observed only say up to the time n at which W_n is to be estimated, the sum (76) is for r from 0 to ∞. Hence we have from (80) that the rth Fourier coefficient of

$$g_s(\omega) - D(e^{-i\omega})\,g_x(\omega) \tag{87}$$

vanishes for $r = 0, 1, \ldots$ We shall not consider here the determination of the appropriate $D(1/\zeta)$.

8.6. Multivariate processes

The methods and results of this and the previous chapter can be extended to multivariate stationary processes. All the essential ideas can be illustrated with bivariate processes in discrete time and so we suppose that for each integer n there is a random vector (X_n, Y_n) whose components are in general complex-valued. We assume that

$$E(X_n) = E(Y_n) = 0$$

and define not only the separate autocovariance functions

$$C(X_{n+h}, X_n) \quad \text{and} \quad C(Y_{n+h}, Y_n),$$

but also a cross-covariance function

$$C(X_{n+h}, Y_n) = E(X_{n+h} \overline{Y}_n). \tag{88}$$

If these three covariance functions are independent of n, the process is said to be stationary to the second order.

It is then convenient to use the notation

$$\gamma_{xx}(h) = C(X_{n+h}, X_n), \quad \gamma_{xy}(h) = C(X_{n+h}, Y_n), \quad \gamma_{yy}(h) = C(Y_{n+h}, Y_n). \tag{89}$$

Note that $\gamma_{xx}(h)$ is what we previously called $\gamma_x(h)$; the slightly more elaborate notation is useful for multivariate processes but is unnecessary otherwise. The covariance properties are thus specified by the matrix

$$\boldsymbol{\gamma}(h) = \begin{bmatrix} \gamma_{xx}(h) & \gamma_{xy}(h) \\ \gamma_{yx}(h) & \gamma_{yy}(h) \end{bmatrix}. \tag{90}$$

We have that $\gamma_{yx}(h) = E(Y_{n+h}\overline{X}_n) = E(Y_n \overline{X}_{n-h}) = \bar{\gamma}_{xy}(-h)$; in general $\gamma_{xy}(h)$ is not directly related to $\gamma_{xy}(-h)$.

It can be shown that the spectral representation of the process is

$$X_n = \int_{-\pi}^{\pi} e^{i\omega n}\, dS_x(\omega), \qquad Y_n = \int_{-\pi}^{\pi} e^{i\omega n}\, dS_y(\omega). \tag{91}$$

Here the processes $\{S_x(\omega)\}$, $\{S_y(\omega)\}$ have separately the properties discussed in Sections 8.1 and 8.2 for univariate processes; we write

$$V\left\{ \int_{-\pi}^{\omega_1} dS_x(\omega) \right\} = G_{xx}(\omega_1), \qquad V\left\{ \int_{-\pi}^{\omega_1} dS_y(\omega) \right\} = G_{yy}(\omega_1). \tag{92}$$

In addition, the bivariate process $\{S_x(\omega), S_y(\omega)\}$ is such that for all ω_1, ω_2, ω_3, ω_4 such that the intervals (ω_1, ω_2) and (ω_3, ω_4) do not overlap,

$$C\{S_x(\omega_2) - S_x(\omega_1), S_y(\omega_4) - S_y(\omega_3)\} = 0. \tag{93}$$

Further there exists a function $G_{xy}(\omega)$ such that for all $\omega_1 < \omega_2$

$$C\{S_x(\omega_2)-S_x(\omega_1),S_y(\omega_2)-S_y(\omega_1)\} = G_{xy}(\omega_2)-G_{xy}(\omega_1). \qquad (94)$$

Clearly $G_{yx}(\omega) = \bar{G}_{xy}(\omega)$, so that the properties of the spectral resolution are defined by the Hermitian matrix

$$\mathbf{G}(\omega) = \begin{bmatrix} G_{xx}(\omega) & G_{xy}(\omega) \\ G_{yx}(\omega) & G_{yy}(\omega) \end{bmatrix}. \qquad (95)$$

For processes with continuous spectra it is more convenient to work with the matrix of derivatives $\mathbf{g}(\omega) = \mathbf{G}'(\omega)$.

Now for univariate processes the function $G(\omega)$ corresponding to the matrix $\mathbf{G}(\omega)$ is non-decreasing. The analogous property here is that for all $\omega_1 < \omega_2$ the matrix $\mathbf{G}(\omega_2) - \mathbf{G}(\omega_1)$ is positive semi-definite. In particular, if $\mathbf{g}(\omega)$ exists, it follows on taking ω_2 near to ω_1 that the *coherency*, defined as

$$\frac{|g_{xy}(\omega)|^2}{g_{xx}(\omega)\,g_{yy}(\omega)}, \qquad (96)$$

lies in $[0,1]$.

The interpretation of (91)–(94) is that both X_n and Y_n can be regarded as the sum of harmonic oscillations with random coefficients. The coefficients in X_n and Y_n referring to different values of ω are orthogonal, but corresponding coefficients of the same value of ω may be correlated. In fact the coherency is a measure of the correlation at a particular value of ω.

Given the spectral representation (91), we have by direct calculation that

$$\gamma(h) = \int_{-\pi}^{\pi} e^{i\omega h}\,d\mathbf{G}(\omega). \qquad (97)$$

We can work equivalently with the correlation functions, for example with

$$\rho_{xy}(h) = \frac{\gamma_{xy}(h)}{\{V(X_n)\,V(Y_n)\}^{\frac{1}{2}}}$$

and with the corresponding generalization of the spectral distribution function

$$F_{xy}(\omega) = \frac{G_{xy}(\omega)}{\{V(X_n)\,V(Y_n)\}^{\frac{1}{2}}}.$$

They are related by

$$\rho(h) = \int_{-\pi}^{\pi} e^{i\omega h}\,d\mathbf{F}(\omega).$$

Sometimes we may wish to obtain the covariances given the spectral properties. This is usually most neatly done by generalizing (35). If we define a generating function by

$$\mathbf{\Gamma}(\zeta) = \sum_{h=-\infty}^{\infty} \gamma(h)\zeta^h,$$

we have, on inverting (97) in the case of continuous spectra, that

$$g(\omega) = \frac{1}{2\pi} \sum_{h=-\infty}^{\infty} e^{-i\omega h} \gamma(h).$$

Thus

$$g(\omega) = \frac{\mathbf{\Gamma}(e^{-i\omega})}{2\pi}. \tag{98}$$

Hence we compute $2\pi g(\omega)$ and then write $\zeta = e^{-i\omega}$ in the answer to obtain $\mathbf{\Gamma}(\zeta)$.

We now discuss two simple examples.

Example 8.1. *Input and output of a linear device.* Suppose that we have a linear electrical or mechanical device which converts an input process $\{X_n\}$ into an output process $\{Y_n\}$, where

$$Y_n = \sum_{u=-\infty}^{\infty} c_u X_{n-u}. \tag{99}$$

If $\{X_n\}$ is stationary to the second order and of zero mean, so is $\{Y_n\}$ provided that $\sum |c_u|^2 < \infty$.

In Section 8.3, equation (54), we obtained the spectral properties of $\{Y_n\}$ from those of $\{X_n\}$. The fundamental formal equation is in fact the discrete time version of (43), namely

$$dS_y(\omega) = l(\omega)\, dS_x(\omega), \tag{100}$$

where

$$l(\omega) = \sum_{u=-\infty}^{\infty} c_u e^{-i\omega u}.$$

It follows from (100) that since

$$dG_{yy}(\omega) = E\{dS_y(\omega)\, d\bar{S}_y(\omega)\},$$

we have, as before, that

$$g_{yy}(\omega) = |l(\omega)|^2 g_{xx}(\omega).$$

Further

$$dG_{xy}(\omega) = E\{dS_x(\omega)\, d\bar{S}_y(\omega)\},$$
$$= E\{\bar{l}(\omega)\, dS_x(\omega)\, d\bar{S}_x(\omega)\}.$$

Thus

$$g_{xy}(\omega) = \bar{l}(\omega)\, g_{xx}(\omega)$$

and

$$g(\omega) = g_{xx}(\omega) \begin{bmatrix} 1 & \bar{l}(\omega) \\ l(\omega) & |l(\omega)|^2 \end{bmatrix}. \tag{101}$$

The coherency is one for all ω, essentially because the output is a deterministic linear function of the input.

Suppose that we wished from observations on the input and output to infer the nature of the linear mechanism. If we are able to obtain only the separate spectral densities of input and output, only the function $|l(\omega)|$ can be found, i.e. the amplitude change in (100) can be found but not the phase change. If, however, the full matrix (101) is known, $l(\omega)$ can be found and the sequence $\{c_u\}$ is in principle determined. Consistency with (99) can be examined by computing the coherency.

These remarks are illustrated by considering the special case

$$Y_n = X_{n-k}. \tag{102}$$

Examination of the input and output separately would establish $|l(\omega)| = 1$, from which it follows only that (102) holds for some unknown value of k.

Example 8.2. *A simple econometric model.* In some fields, such as econometrics, multivariate stochastic difference equations are important. As an extremely simple example, suppose that X'_n and Y'_n are respectively the price and quantity of a certain crop in year n. Consider a model in which the amount sown in year n is influenced by the price obtained in year $n-1$, say

$$Y'_n = \alpha_1 + \beta_1 X'_{n-1} + U_n,$$

and in which the price in year n is influenced by the quantity available in that year, say

$$\dot{X}'_n = \alpha_2 - \beta_2 Y'_n + V_n.$$

Here the simplest assumption is that $\{U_n\}$ and $\{V_n\}$ are mutually uncorrelated sequences of uncorrelated random variables of zero mean and variances σ_u^2 and σ_v^2. Usually we shall be interested in the behaviour for $\beta_1, \beta_2 > 0$.

It can be shown that if $|\beta_1|, |\beta_2| < 1$ the process is stationary and if X_n and Y_n denote the deviations of X'_n and Y'_n from their expectations we have the simpler equations

$$Y_n = \beta_1 X_{n-1} + U_n, \qquad X_n = -\beta_2 Y_n + V_n. \tag{103}$$

There are several ways of analysing (103). For an analysis in the frequency domain, we introduce the spectral resolutions of all the processes and obtain, essentially as in (100), that the equations (103) become

$$dS_y(\omega) = \beta_1 e^{-i\omega} dS_x(\omega) + dS_u(\omega),$$

$$dS_x(\omega) = -\beta_2 dS_y(\omega) + dS_v(\omega).$$

Thus

$$dS_x(\omega) = \frac{-\beta_2\,dS_u(\omega)+dS_v(\omega)}{1+\beta_1\beta_2\,e^{-i\omega}},$$

$$dS_y(\omega) = \frac{dS_u(\omega)+\beta_1\,e^{-i\omega}\,dS_v(\omega)}{1+\beta_1\beta_2\,e^{-i\omega}}. \tag{104}$$

We can now obtain the spectral properties of $\{X_n, Y_n\}$, in fact for an arbitrary stationary process $\{U_n, V_n\}$. In our special case where the disturbance processes are random,

$$g_{uu}(\omega) = \frac{\sigma_u^2}{2\pi}, \qquad g_{uv}(\omega) = 0, \qquad g_{vv}(\omega) = \frac{\sigma_v^2}{2\pi}.$$

We obtain from (104), using results like $dG_{xy}(\omega) = E\{dS_x(\omega)\,d\bar{S}_y(\omega)\}$, that

$$2\pi|1+\beta_1\beta_2\,e^{-i\omega}|^2\,\mathbf{g}(\omega) = \begin{bmatrix} \beta_2^2\,\sigma_u^2+\sigma_v^2 & -\beta_2\,\sigma_u^2+\beta_1\,\sigma_v^2\,e^{i\omega} \\ -\beta_2\,\sigma_u^2+\beta_1\,\sigma_v^2\,e^{-i\omega} & \sigma_u^2+\beta_1^2\,\sigma_v^2 \end{bmatrix}. \tag{105}$$

It follows by comparison with (33) that the $\{X_n\}$ and $\{Y_n\}$ processes individually are first-order autoregressive processes, with autocorrelation functions $(-\beta_1\beta_2)^{|h|}$. The detailed interpretation of the off-diagonal term in (105) is left as an exercise. The term $e^{i\omega}$ indicates a connexion between Y_n and X_{n-1}, the constant term indicates a connexion between Y_n and X_n and the fact that the coherency is less than one indicates that Y_n is not a deterministic linear function of the sequence $\{X_n\}$. To obtain the cross-covariance function we can follow the procedure of (98).

Alternatively we can analyse (103) in the time domain by the methods of Example 7.5. Thus it would be possible to solve (103) obtaining X_n and Y_n as functions of the sequences $\{U_n\}$ and $\{V_n\}$; all properties would follow directly from this representation. Or again, we have directly from (103) that

$$X_n = -\beta_1\beta_2\,X_{n-1}+(V_n-\beta_2\,U_n),$$

$$Y_n = -\beta_1\beta_2\,Y_{n-1}+(U_n-\beta_1\,V_{n-1}); \tag{106}$$

further $V_n-\beta_2\,U_n$ is uncorrelated with X_{n-1} and $U_n-\beta_1\,V_{n-1}$ is uncorrelated with Y_{n-1}, so that (106) represents two first-order autoregressive schemes. To obtain the cross-covariance function, we multiply the two equations (103) by respectively X_{n-h} $(h>0)$ and Y_{n-h} $(h\geqslant 0)$ and take expectations. (The series are real-valued.) Thus

$$E(Y_n X_{n-h}) = \beta_1(-\beta_1\beta_2)^{h-1}\sigma_x^2 \quad (h>0)$$

$$E(X_n Y_{n-h}) = \beta_2(-\beta_1\beta_2)^{h-1}\sigma_y^2 \quad (h\geqslant 0).$$

Thus

$$\gamma_{xy}(h) = E(X_{n+h}\,Y_n) = \begin{array}{l} \beta_2(-\beta_1\beta_2)^{h-1}\sigma_y^2 \quad (h\geqslant 0), \\[2mm] \beta_1(-\beta_1\beta_2)^{-h-1}\sigma_x^2 \quad (h<0). \end{array} \tag{107}$$

In the symmetrical case $\beta_1 = \beta_2$, $\sigma_u^2 = \sigma_v^2$, $\sigma_x^2 = \sigma_y^2$, the cross-covariance has a maximum at $h = -1$, i.e. when X_{n-1} is correlated with Y_n. This might have been anticipated from the form of the original difference equations (103).

Bibliographic Notes

The main references are given in the text and in the Bibliographic Notes for Chapter 7. Two important early papers connected with the Wiener–Khintchine theorem are those of Wiener (1930) and Khintchine (1934).

Exercises

1. Show, by means of the following construction, that given a spectral distribution function $F(\omega)$ defined over $[-\pi, \pi)$, there exists a stationary process having that spectrum and yet being, in a non-linear sense, completely deterministic. Let U be uniformly distributed over $(-\pi, \pi)$ and let V be an independent random variable having distribution function $F(\omega)$. Let a process be defined by

$$X_n = e^{i(Vn+U)}.$$

Prove that this is a stationary process and show, by computing the autocorrelation function, that the spectral distribution function is $F(\omega)$. What calculations are necessary for perfect prediction of the process?

2. Obtain the autocorrelation functions of the continuous-time processes having as spectral density function

   (a) a normal density of zero mean;
   (b) the two-sided exponential distribution of zero mean;
   (c) an Edgeworth series of even terms.

3. Express the variance functions of Exercises 7.1 and 7.2 in terms of the spectral density function.

4. Prove that for a real-valued process in continuous time with a bounded spectral density, the variance function $v(l)$ of Exercise 7.2 is determined for large l by the spectral density at $\omega = 0$.

5. Prove directly that if $\gamma_y(h)$ and $\gamma_x(h)$ are the autocovariance functions of the process $\{Y_n\}$ and $\{X_n\}$ connected by (53), then

$$\gamma_y(h) = \sum c_u \bar{c}_v \gamma_x(h+v-u).$$

Hence prove (56).

6. Suppose that the process $\{Y_n\}$ is defined by

$$Y_n = \alpha Y_{n-1} + X_n,$$

where $\{X_n\}$ is a first-order autoregressive process of zero mean. Obtain the spectral density function of $\{Y_n\}$ and hence the autocorrelation function. Check the answer by obtaining the autocorrelation function directly by the methods of Chapter 7.

7. Obtain the functions $\gamma_{xy}(h)$ and $g_{xy}(\omega)$ for the system of Exercise 6.

8. Suppose that for a discrete state Markov process in continuous time, $G_i(z,t)$ is the probability generating function of the state occupied at t, given that state i is occupied at $t = 0$. Let there be an equilibrium distribution $\{p_i\}$, with mean μ and variance σ^2. Prove that the spectral density function of the process, considered as a stationary process in continuous time, is

$$\frac{1}{2\pi\sigma^2}\left[\frac{\partial}{\partial z}\{G^*(z,i\omega)+G^*(z,-i\omega)\}_{z=1}-\mu^2\right],$$

where $G(z,t) = \sum_i p_i G_i(z,t)$, with Laplace transform $G^*(z,s)$.

9. Use the result of Exercise 8 to obtain the spectral density function of the queueing process, Section 4.6(iii), in statistical equilibrium.

10. Assuming that an equilibrium distribution exists, express the spectral density function of the generalized birth–death process in terms of the Karlin–McGregor representation (4.100).

CHAPTER NINE

Point Processes

9.1. Introduction

In this chapter we consider processes whose realizations consist of a series of point events. We consider the events as occurring in continuous one-dimensional time. Roughly speaking, we consider a process as a point process when interest is concentrated on the individual occurrences of the events themselves, events being distinguished only by their positions in time. Thus, although the linear birth–death process is characterized by point events, namely the births and deaths, we would not normally think of it as a point process, since we are usually interested in the number of individuals alive at a particular time rather than directly in the instants at which births and deaths occur.

The Poisson process (Section 4.1) is in many ways the most important point process and most of the special processes considered in this chapter are fairly immediate generalizations of the Poisson process.

Some practical examples of point processes are the following: accidents occurring in time, stops of a machine taking running time of the machine as the time scale, mutations, fibre entanglements occurring along the length of a textile yarn, emissions from a radioactive source, etc.

Before dealing with particular processes, it is useful to introduce some general definitions. Consider an arbitrary process of point events. The time origin may, or may not, be a special point in the process. For example, an event may have occurred immediately before $t = 0$. More often, we deal with stationary processes in which the time origin is an arbitrary point. For simplicity, we shall suppose, unless explicitly stated otherwise, that there is zero probability that two or more events occur simultaneously. It is, nevertheless, easy to deal with problems in which the number of events, W_i, at the ith occurrence point is such that $\{W_i\}$ is a sequence of independent identically distributed random variables, independent of the positions of the occurrence points.

The following quantities connected with the process are of particular interest:

(a) the number, N_t, of events in the interval $(0,t]$ or, more generally, the number of events in $(t_1,t_2]$, equal to $N_{t_2} - N_{t_1}$;

(b) the expected number of events in $(0,t]$, denoted by $H(t) = E(N_t)$ or, more generally, the higher moments of N_t;

338

(c) the density of events, $h(t)$, defined by

$$h(t) = \lim_{\Delta t \to 0+} \frac{\text{prob}\{\text{event in } (t, t+\Delta t]\}}{\Delta t}; \qquad (1)$$

(d) the time, S_r, at which the rth event occurs;

(e) the backward recurrence-time, U_t, which is the length of time measured backwards from t to the last event at or before t (see Fig. 9.1);

(f) the forward recurrence-time, V_t, defined as the length of time from t to the next event at or after t.

There are broadly two ways of looking at point processes, in terms of the numbers of events occurring in fixed time intervals or in terms of intervals between events. Properties (a)–(c) are of the first type, properties (d)–(f) of the second type.

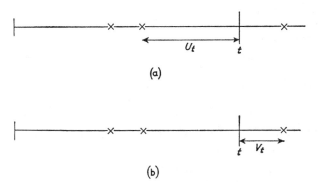

(a)

(b)

Fig. 9.1. Definition of recurrence-times U_t, V_t. (a) Backward recurrence-time, U_t; (b) forward recurrence-time, V_t.

For the Poisson process, properties (a)–(f) are given in Section 4.1. If the rate is ρ, we have in particular that the density of events is constant and equal to ρ. The p.d.f. of S_r is of the gamma type with index $r-1$. The forward recurrence-time has an exponential distribution with parameter ρ.

The formal definition of the stationarity of a point process is analogous to that given in Section 7.1. Let $A_1, A_2, \ldots$ be arbitrary sets on the real axis, and $T_h A_1, T_h A_2, \ldots$ be the sets obtained by translating through h. Let $N(A)$ be the number of events in A. The point process is stationary if the two sets of random variables

$$N(A_1), N(A_2), \ldots, N(A_k);$$

$$N(T_h A_1), N(T_h A_2), \ldots, N(T_h A_k)$$

have the same joint distribution for all initial sets $A_1, A_2, \ldots, A_k$ and all real h and all $k = 1, 2, \ldots$ An equivalent definition could be formulated in terms of intervals between successive events, although care is needed in this because the properties are in general quite different if the intervals are specified (a) starting from an arbitrary event and (b) starting from an arbitrary time point.

Of course the general definitions are not of much value in investigating particular processes, when we might, for instance, be concerned with $k = 2$ and with sets that are non-overlapping intervals. To examine the stationarity of particular processes, we normally look at the probability mechanism generating the process. If this is unaffected by time shifts, the point process will be stationary if the initial conditions are suitable. The Poisson process is stationary; numbers of events in non-overlapping intervals have independent Poisson distributions with means depending only on the lengths of the intervals.

The most important generalization of the Poisson process is the renewal process and we deal with this first in Section 9.2.

9.2. The renewal process

In the Poisson process, the intervals between successive events are independently exponentially distributed. An obvious and important generalization is to allow the intervals to be independently and identically distributed, say with p.d.f. $f(x)$. The resulting series of events is called a renewal process.

To allow for the possibility of different choices of time origin, it is convenient to introduce the following slightly more general process. Suppose that for $r = 1, 2, \ldots$ the rth event in the process occurs at time $X_1 + \ldots + X_r$, where $\{X_1, X_2, \ldots\}$ are independent positive random variables and $X_2, X_3, \ldots$ are identically distributed with p.d.f. $f(x)$, but X_1 has the possibly different p.d.f. $f_1(x)$. If

(a) $f_1(x) = f(x)$, so that all random variables are identically distributed, the process is an *ordinary renewal process*;
(b) $f_1(x)$ and $f(x)$ are not necessarily the same, the process is a *modified renewal process*;
(c) $f_1(x)$ has the special form,

$$f_1(x) = \frac{1 - F(x)}{\mu}, \qquad (2)$$

where $\mu = E(X_i)$ $(i = 2, \ldots)$ and $F(x)$ is the distribution function corresponding to $f(x)$, the process is an *equilibrium* (or *stationary*) *renewal process*.

Example 9.1. *Replacement of components.* As a concrete example, consider components subject to failure. Suppose that immediately on failure each component is replaced by a new one. Let the p.d.f. of component failure-time be $f(x)$. If we start with a new component at $t = 0$, all intervals X_i, including the first, have the distribution of failure-time and we have an ordinary renewal process. If, however, the component in use at $t = 0$ is not new, then X_1 is its remaining lifetime and will in general have a different distribution. We shall show later that if the time origin is at a fixed point a long time after the start of the process, X_1 has the p.d.f. (2) and an equilibrium renewal process is obtained.

There are many other examples of renewal processes. The following is one that has been extensively studied.

Example 9.2. Electronic counter. Consider point emissions, say, from a radioactive source, occurring in a Poisson process of rate ρ. Suppose that these are recorded in a counter which is subject to blockage according to

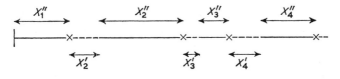

× Emissions recorded   – – – – Blocked time

Fig. 9.2. Electronic counter.

the following rules. The counter is initially free. After a time X_1'', exponentially distributed with parameter ρ, an emission occurs and is counted. The system is then blocked for a time X_2', during which no emissions can be counted. Then the counter becomes free and after a time X_2'', a further emission is recorded and so on. Figure 9.2 illustrates this.

Because emissions occur in a Poisson process, $\{X_1'', X_2'', \ldots\}$ are independently exponentially distributed with parameter ρ. Suppose we assume that the blocked times $\{X_2', X_3', \ldots\}$ are independently and identically distributed with p.d.f. $b(x)$, and that the sequences $\{X_i''\}$, $\{X_k'\}$ are independent.

The point events of interest are the emissions recorded. Under the above assumptions, these events form a modified renewal process in which

(a) $X_1 = X_1''$, the time to the first event, is exponentially distributed with parameter ρ;

(b) X_2, X_3, ..., the times between subsequent events, have as p.d.f. the convolution of $b(x)$ with the above exponential distribution.

Note also that we can produce an ordinary renewal process by changing the initial conditions, in fact by requiring that a blocked period starts at $t = 0$.

The main interest in counter problems is in relating the mean and variance of the number of emissions recorded to ρ and to the parameters of the distribution of blocked time.

The theory of the renewal process in continuous time is dealt with in the monograph by Cox (1962) and only a brief sketch of the main results is given here.

The time S_r up to the rth event is $X_1 + \ldots + X_r$ and in a renewal process is the sum of independent random variables. Familiar methods in the theory of probability are available for studying the exact or asymptotic distribution. In particular an application of the central limit theorem shows that S_r is for large r asymptotically normally distributed with mean $r\mu$ and variance $r\sigma^2$, where μ and σ^2 are the common mean and variance of X_2, X_3, ..., assumed here to be finite. For this asymptotic result the distribution of X_1 is irrelevant, provided that $V(X_1) < \infty$.

Consider now N_t, the number of events in $(0,t]$. Its distribution is related to that of S_r because

$$N_t < r \text{ if and only if } S_r > t. \tag{3}$$

Thus

$$\text{prob}(N_t < r) = \text{prob}(S_r > t) \tag{4}$$

$$= 1 - F_r(t),$$

say. Hence

$$\text{prob}(N_t = r) = F_r(t) - F_{r+1}(t), \tag{5}$$

with $F_0(t) = 1$. Hence the exact distribution of N_t can be obtained wherever the distribution functions of X_1, $X_1 + X_2$, ... can be calculated. An important special case is when the underlying distribution is of the gamma type. The distribution functions of the S_r are then of the gamma type also and (5) is very convenient for numerical work, especially for small r. For a Poisson process, the result (4) takes the special form given in (4.19).

We can use (4) in order to investigate the asymptotic distribution of N for large t. The argument depends on the asymptotic normality of S_r, for large r. In fact let $r_t = t/\mu + y_t \sigma \sqrt{(t/\mu^3)}$. Then

$$\text{prob}(N_t < r_t) = \text{prob}(S_{r_t} > t)$$

$$= \text{prob}\left\{\frac{S_{r_t} - r_t\mu}{\sigma\sqrt{r_t}} > -y_t\left(1 + \frac{y_t\sigma}{\sqrt{(t\mu)}}\right)^{-\frac{1}{2}}\right\}.$$

Fix $y_t = y$ and let $t \to \infty$. Then

$$\lim_{t \to \infty} \mathrm{prob}(N_t < r_t) = \lim_{t \to \infty} \mathrm{prob}\left\{ \frac{S_{r_t} - r_t \mu}{\sigma \sqrt{r_t}} > -y \right\}$$

$$= \Phi(y), \tag{6}$$

where $\Phi(y)$ is the standard normal probability integral. This proves that N_t is asymptotically normal with mean t/μ and variance $\sigma^2 t/\mu^3$. There is a minor difficulty in the above argument in that, since r_t must be an integer, we ought really to consider a limiting process in which $y_t = y + \eta_t$, where η_t is the smallest value such that r_t is an integer.

Since only the asymptotic distribution of S_r is required in the proof, the result holds for all three types of renewal process, ordinary, modified and equilibrium. In fact the method of proof of (6) requires only the asymptotic normality of S_r with a mean and variance proportional to r and can be adapted to much more general processes. If, however, we attempted to find further terms in an asymptotic expansion for the distribution, account would have to be taken of the possibly different distribution of X_1.

The simplest special case of the result is for the Poisson process of rate ρ, for which $\sigma = \mu = 1/\rho$. The limiting normal distribution has mean and variance equal to ρt and this is just the limiting normal form of the Poisson distribution, familiar from elementary probability theory.

It follows from the form of the asymptotic mean and variance that for the limiting distribution

$$\frac{\text{variance}}{\text{mean}} \sim \frac{\sigma^2 t}{\mu^3} \cdot \frac{\mu}{t} = \frac{\sigma^2}{\mu^2}. \tag{7}$$

This generalizes to an arbitrary renewal process the familiar property of the Poisson distribution that

$$\frac{\text{variance}}{\text{mean}} = 1. \tag{8}$$

Of course (7) is a limiting result only, whereas (8) is exact.

It is instructive to apply the general limiting result to Example 9.2, the electronic counter subject to blockage. The mean and variance of the interval between successive renewals are respectively $\mu_b + 1/\rho$ and $\sigma_b^2 + 1/\rho^2$, where μ_b, σ_b^2 refer to the distribution of blocked time and ρ is the rate parameter of the Poisson process of events. Hence N_t, the number of emissions recorded in time t, is asymptotically normally distributed with mean and variance respectively

$$\frac{\rho t}{1 + \mu_b \rho} \quad \text{and} \quad \frac{(1 + \sigma_b^2 \rho^2)\rho t}{(1 + \mu_b \rho)^3}.$$

The ratio of variance to mean is $(1 + \sigma_b^2 \rho^2)/(1 + \mu_b \rho)^2$, so that if σ_b/μ_b is

12

small, which would usually be the case, there is apparent underdispersion relative to the Poisson distribution.

Consider now the exact form of $H(t) = E(N_t)$; this is sometimes called the *renewal function*. We have from (5) that

$$H(t) = \sum_{r=0}^{\infty} r\{F_r(t) - F_{r+1}(t)\},$$

$$= \sum_{r=1}^{\infty} F_r(t), \qquad (9)$$

where $F_0(t) = 1$. Thus on taking Laplace transforms we have that

$$H^*(s) = \frac{1}{s} \sum_{r=1}^{\infty} f_r^*(s), \qquad (10)$$

where $f_r^*(s) = E\{e^{-s(X_1 + \ldots + X_r)}\}$. We now have to consider separately ordinary, modified and equilibrium renewal processes. The difference affects only the contribution of X_1 to $f_r^*(s)$.

First for ordinary renewal processes, $X_1, \ldots, X_r$ are identically distributed and $f_r^*(s) = \{f^*(s)\}^r$. Thus, on summing (10), we have for the ordinary renewal function $H_0(t)$,

$$H_0^*(s) = \frac{f^*(s)}{s\{1 - f^*(s)\}}. \qquad (11)$$

For the modified process for which X_1 has the p.d.f. $f_1(x)$, we have similarly $f_r^*(s) = f_1^*(s)\{f^*(s)\}^{r-1}$, so that the renewal function is

$$H_m^*(s) = \frac{f_1^*(s)}{s\{1 - f^*(s)\}}. \qquad (12)$$

For the equilibrium renewal process $f_1(x)$ is given by (2) and the corresponding Laplace transform is

$$\frac{1 - f^*(s)}{\mu s}. \qquad (13)$$

Hence for the equilibrium renewal function, $H_e(t)$, we have that

$$H_e^*(s) = \frac{1}{\mu s^2},$$

whence

$$H_e(t) = t/\mu. \qquad (14)$$

Thus for the equilibrium renewal process the expected number of events in $(t_1, t_2), H_e(t_2) - H_e(t_1)$, is simply $(t_2 - t_1)/\mu$.

We now study (11), for the ordinary renewal process, in more detail. If $f^*(s)$ is a rational function of s, so too is $H_0^*(s)$, and inversion in terms of combinations of exponential functions is in principle straightforward. The simplest non-trivial special case is

$$f(x) = \rho^2 x e^{-\rho x}, \qquad f^*(s) = \frac{\rho^2}{(\rho+s)^2}. \tag{15}$$

Then

$$H_0^*(s) = \frac{\rho^2}{s^2(2\rho+s)}$$

$$= \frac{\rho}{2s^2} - \frac{1}{4s} + \frac{1}{4(2\rho+s)}.$$

Therefore

$$H_0(t) = \tfrac{1}{2}\rho t - \tfrac{1}{4} + \tfrac{1}{4}e^{-2\rho t}. \tag{16}$$

We relate this to a general result below.

To examine the general limiting form of $H_0(t)$ as $t \to \infty$ requires an analysis of $H_0^*(s)$ as $s \to 0$. Now for small s we can write formally

$$f^*(s) = E(e^{-sX}) = 1 - s\mu + \tfrac{1}{2}s^2(\mu^2+\sigma^2) + \ldots;$$

and substitution into (11) gives the limiting result

$$H_0^*(s) = \frac{1}{s^2\mu} + \frac{\sigma^2-\mu^2}{2\mu^2 s} + o\left(\frac{1}{s}\right).$$

Formal inversion of this suggests that as $t \to \infty$

$$H_0(t) = \frac{t}{\mu} + \frac{\sigma^2-\mu^2}{2\mu^2} + o(1). \tag{17}$$

A rigorous proof of (17) is possible under very weak assumptions about the distribution $f(x)$, but requires difficult Tauberian arguments and will not be attempted here. If $f^*(s)$ is rational, a rigorous proof based on a partial fraction expansion of $H_0^*(s)$ is straightforward. For the special distribution (15), $\mu = 2/\rho$, $\sigma^2 = 2/\rho^2$, and so the first two terms of (16) agree with the first two terms of (17). For the distribution (15) the rate of approach to the asymptotic form is exponentially fast. This will be true in general when $f^*(s)$ is rational, the time constant in the exponential term being determined by the non-zero root of the equation $f^*(s) = 1$ with largest real part.

Similar results can be obtained for $V(N_t)$ and indeed for the higher moments. It can be shown that for an ordinary renewal process

$$V(N_t) = \frac{\sigma^2 t}{\mu^3} + \left(\frac{1}{12} + \frac{5\sigma^4}{4\mu^4} - \frac{2\mu_3}{3\mu^3}\right) + o(1), \tag{18}$$

whereas for an equilibrium renewal process

$$V(N_t) = \frac{\sigma^2 t}{\mu^3} + \left(\frac{1}{6} + \frac{\sigma^4}{2\mu^4} - \frac{\mu_3}{3\mu^3}\right) + o(1); \tag{19}$$

in these formulae $\mu_3 = E\{(X-\mu)^3\}$.

The leading term in the expansions for the higher cumulants of N_t is most easily obtained by applying Wald's identity (Section 2.4). Consider a random walk in discrete time in which X_i is the jump at the ith step. (This random walk has the very special property that the steps are all in the same direction.) The random walk is first across a barrier at t after $N_t + 1$ steps. Hence, ignoring the overshoot across the barrier, we have from Wald's identity that

$$E[e^{-st}\{f^*(s)\}^{-N_t-1}] = 1,$$

since $f^*(s)$ refers to the distribution of a single step. Therefore, if N_t is large, so that $N_t + 1$ can be replaced by N_t, we have that

$$\log E(e^{-N_t p}) = st, \tag{20}$$

where

$$p = \log f^*(s). \tag{21}$$

The left-hand side of (20), considered as a function of p, is the cumulant generating function of N_t and hence according to (20) the cumulant generating functions of N_t and of X_i are inverse functions. If we invert (21), we have that

$$s = -\frac{p}{\kappa_1} + \frac{\kappa_2 p^2}{\kappa_1^3 2!} - \left(-\frac{\kappa_3}{\kappa_1^4} + \frac{3\kappa_2^2}{\kappa_1^5}\right)\frac{p^3}{3!} + \left(\frac{\kappa_4}{\kappa_1^5} + \frac{15\kappa_2^3}{\kappa_1^7} - \frac{10\kappa_2\kappa_3}{\kappa_1^6}\right)\frac{p^4}{4!} - \dots,$$

where $\kappa_1, \kappa_2, \dots$ are the cumulants of X_i, from which it follows that the first four cumulants of N_t are asymptotically

$$\frac{t}{\kappa_1}, \qquad \frac{t\kappa_2}{\kappa_1^3}, \qquad t\left(\frac{3\kappa_2^2}{\kappa_1^5} - \frac{\kappa_3}{\kappa_1^4}\right) \quad \text{and} \quad t\left(\frac{\kappa_4}{\kappa_1^5} + \frac{15\kappa_2^3}{\kappa_1^7} - \frac{10\kappa_2\kappa_3}{\kappa_1^6}\right).$$

We turn now to the renewal density, $h(t)$, defined by (1). The probability that the rth renewal occurs in $(t, t + \Delta t)$ is $f_r(t)\,\Delta t + o(\Delta t)$. Since $h(t)\,\Delta t$ is as $\Delta t \to 0$, the probability of a renewal in $(t, t + \Delta t)$ irrespective of serial order, we have that

$$h(t) = \sum_{r=1}^{\infty} f_r(t) \tag{22}$$

$$= H'(t).$$

We have that for the ordinary renewal process

$$h_0^*(s) = \frac{f^*(s)}{1 - f^*(s)}. \tag{23}$$

It is easily shown, for example by differentiating $H_e(t) = t/\mu$, that for the equilibrium renewal process, $h_e(t) = 1/\mu$.

The limiting result for renewal densities, corresponding to (17) for the renewal function, is that

$$\lim_{t \to \infty} h(t) = \frac{1}{\mu}. \tag{24}$$

The behaviour of $h(t)$ for small t depends on the p.d.f. $f(x)$. If $f(x)$ is exponential, $f(x) = \rho e^{-\rho x}$, the renewal density $h(t)$ is constant and equal to ρ; this follows from (23) or from the defining property of the Poisson process. If the p.d.f. $f(x)$ is closely concentrated around its mean μ, the function $h(t)$ will have maxima near μ, 2μ, ..., the height of the maxima decreasing and the spread around the maxima increasing as t increases. For very small values of t, $h(t)$ is best calculated from a few terms of (22).

Finally note that (23) can be rewritten

$$h_0^*(s) = f^*(s) + h_0^*(s)f^*(s)$$

from which, on inversion, we have that

$$h_0(t) = f(t) + \int_0^t h_0(t-u)f(u)\,du. \tag{25}$$

This is often called the integral equation of renewal theory. It can be derived directly from probabilistic considerations. On the right-hand side, the first term refers to the possibility that the first renewal occurs near t, and the second term to the possibility that a renewal of some order occurs near $t-u$ and is followed by an interval of length u.

Consider now the backward and forward recurrence-times U_t and V_t. By convention we define $U_t = t$ if there have been no renewals in $(0,t)$, so that

$$\operatorname{prob}(U_t = t) = 1 - F_1(t).$$

Further if $q_t(x)$ $(0 < x < t)$ is the p.d.f. of U_t, we have

$$q_t(x) = h(t-x)\{1-F(x)\}. \tag{26}$$

For (26), multiplied by Δx, is, as $\Delta x \to 0$, the limiting probability that there is a renewal in $(t-x, t-x+\Delta x)$ followed by an interval long enough to ensure that the next event is after t. Of particular interest is the distribution of U_t for large t. By (24) and (26), the limiting p.d.f. of U_t is

$$\frac{1-F(x)}{\mu}. \tag{27}$$

Similarly the p.d.f. of V_t, denoted by $r_t(x)$, is given in general by

$$r_t(x) = f_1(t+x) + \int_0^t h(t-y)f(y+x)\,dy. \tag{28}$$

As $t \to \infty$, this tends to

$$\frac{1}{\mu}\int_0^\infty f(y+x)\,dy = \frac{1-F(x)}{\mu}. \tag{29}$$

The identity of (27) and (29) is plausible because the process is reversible in time when viewed from a time point sufficiently far from the special conditions introduced at $t = 0$. The p.d.f. (27) is a decreasing function of x and has Laplace transform (13), from which the mean and variance are found to be

$$\frac{1}{2}\left(\mu + \frac{\sigma^2}{\mu}\right) \quad \text{and} \quad \frac{\mu_3}{3\mu} + \frac{\sigma^2}{2}\left(1 - \frac{\sigma^2}{2\mu^2}\right) + \frac{\mu^2}{12}. \tag{30}$$

The result (29) motivates our definition of an equilibrium renewal process as one for which the p.d.f. of the first interval, X_1, is (27). We can consider first an ordinary renewal process and then transfer the time origin to a point a long time after the start of the process. The interval from the new origin will then have the p.d.f. (27) and an equilibrium renewal process results. It can be shown from (28) that in the equilibrium process V_t has the p.d.f. (27) for all t.

The following is an example of the direct application of the recurrence-time distribution (27).

Example 9.3. *Survey of component failure properties.* Consider a system with a large number of independent components all with the same distribution of failure-time and suppose that components are replaced immediately on failure, so that each component position generates a renewal process. Imagine that in order to investigate the distribution of component failure-time, observations are made, beginning a long time after the start of the process, of either

(a) the age of the components currently in use; or
(b) the time from the start of observation to failure of the components currently in use.

The frequency distribution of these observations, Y, will be (27). It may be required to estimate from the Y's the underlying distribution of failure-time, $f(x)$. There are several ways of doing this. First the height of the p.d.f. of Y at the origin gives an estimate of $1/\mu$ and then (30) can be used to estimate the first r moments of $f(x)$ from the first $r - 1$ moments of Y. Alternatively $f(x)$ can be estimated by numerical differentiation of a smoothed frequency curve of Y. These methods do not require an assumption about the functional form of $f(x)$.

An alternative procedure is to assume a suitable functional form either for $\{1 - F(x)\}/\mu$ or for $f(x)$ and to estimate the parameters by maximum likelihood.

Example 9.4. *Road traffic.* We can illustrate the results and techniques of renewal theory by a simple problem concerning road traffic; see Haight (1963) for much fuller discussion of the applications of stochastic processes to road-traffic problems, and Weiss and Maradudin (1962) for this particular problem. Consider a single lane of cars going along a major road and consider the sequence of point events defined by the times at which the cars pass a fixed reference point. A first approximate model is to treat the sequence as a Poisson process; this is a good approximation for light traffic. A second approximation is to treat it as a renewal process, with p.d.f. of intervals $f(x)$.

Suppose now that at the reference point there is a minor road and that at $t = 0$ a car arrives there and wishes to merge with the stream of traffic on the major road. (Alternatively consider a pedestrian wishing to cross the single stream of traffic.) We assume that there is a gap acceptance function $\alpha(x)$, such that independently for each gap between successive cars

$$\alpha(x) = \text{prob\{minor road vehicles merges} | \text{gap to next major road car is } x\}.$$

Now if the arrival of the minor road car is independent of the major road traffic, the major road traffic has to be treated as an equilibrium renewal process, the p.d.f. of the first interval being (29). Hence the probability that the first gap is acceptable, i.e. that the minor road vehicle has zero waiting time is

$$\int_0^\infty \left\{ \frac{1 - F(x)}{\mu} \right\} \alpha(x)\, dx.$$

To find for $t > 0$ the p.d.f. $w(t)$ of the delay to the minor road vehicle, note that

$$w(t) = k(t) \int_0^\infty f(x)\, \alpha(x)\, dx,$$

where $k(t)\, \Delta t$ is, as $\Delta t \to 0$, the conditional probability that a car on the major road passes the reference point in $(t, t + \Delta t)$, given that no merging action has yet been taken. That is, $k(t)$ is a type of conditional renewal density for the major road traffic. We can build up an integral equation for $k(t)$ by the same type of probabilistic argument used to develop the integral equation of renewal theory, (25). In fact, either the vehicle arriving at t is the first, or one arrived previously at say $t - u$ and the gap u was rejected. These two possibilities lead to the equation

$$k(t) = \frac{\{1 - F(t)\}\{1 - \alpha(t)\}}{\mu} + \int_0^t k(t - u) f(u) \{1 - \alpha(u)\}\, du.$$

If we write $\beta(t) = \{1 - F(t)\}\{1 - \alpha(t)\}/\mu$ and $\gamma(t) = f(t)\{1 - \alpha(t)\}$, the integral equation

$$k(t) = \beta(t) + \int_0^t k(t - u)\, \gamma(u)\, du$$

can be solved formally by Laplace transforms. In particular the moments of the distribution of delay can be found. Explicit inversion is possible when the major road traffic is a Poisson process and the gap acceptance function a unit step function.

It is possible to make a much more thorough study of the renewal process including, for example, investigation of the joint distribution of such random variables as U_t and N_t. Instead we shall in the remaining sections examine briefly more general forms of point process.

9.3. Renewal processes with more than one type of interval

One natural generalization of the renewal process is to have intervals of different types characterized by different probability distributions, the sequence of intervals of different types being possibly deterministic or possibly itself being specified stochastically.

Example 9.5. *Alternating renewal process*. Suppose that we have two independent sequences of random variables $\{X_1', X_2', \ldots\}$ and $\{X_1'', X_2'', \ldots\}$. Let each sequence consist of mutually independent and identically distributed random variables, the two p.d.f.'s being $f_1(x)$ and $f_2(x)$. Suppose also that we take random variables alternately from the two series. Figure 9.3 illustrates the resulting process. We start with the type 1 interval X_1', ending with a type 1 event. This is followed by the type 2

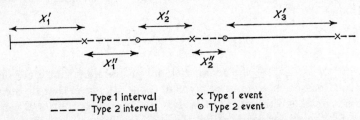

Fig. 9.3. Alternating renewal process.

interval X_1'', ending with a type 2 event after time $X_1' + X_1''$, and so on. The $(2r-1)$th event is a type 1 event and occurs after time

$$X_1' + \ldots + X_r' + X_1'' + \ldots + X_{r-1}''$$

and the $2r$th event is a type 2 event and occurs after time

$$X_1' + \ldots + X_r' + X_1'' + \ldots + X_r''.$$

This system is called an *alternating renewal process*. If the two p.d.f.'s are exponential, say with parameters ρ_1 and ρ_2, the process is called an *alternating Poisson process*.

As one application, consider a machine subject to stoppages and call the time necessary to start a stopped machine a repair time. There is thus an alternating sequence of running times and repair times. If it can be assumed that these are two sequences of independent random variables, each sequence having its own p.d.f., we have an alternating renewal process. We might often assume stoppages to occur in a Poisson process, i.e. take $f_1(x)$ to be an exponential p.d.f. Note that if we are concerned solely to study the pattern of incidence of stops, we consider the stops as points occurring in the running time of the machine.

Another application occurs in the comparison of the mean failure-times μ_1, μ_2 of alternative types of component under conditions of routine use, without keeping detailed records of failure-times. Goodman (1953b) suggested that this can be done, provided that a reasonable number of components are in use simultaneously, by replacing each component on failure by a component of the opposite type. It is clear that after some time the proportion of type 1 components in use will estimate $\mu_1/(\mu_1 + \mu_2)$. The full discussion of this process requires, however, the analysis of an alternating renewal process.

Example 9.6. *Simple semi-Markov process.* In Example 9.4 the two types of interval occur in strict alternation. Another possibility is that the sequence of interval types is stochastic and the simplest such situation is when a two-state Markov chain, say with transition matrix

$$
\begin{bmatrix} \alpha_{11} & \alpha_{12} \\ \alpha_{21} & \alpha_{22} \end{bmatrix} = \begin{bmatrix} \alpha_1 & 1-\alpha_1 \\ 1-\alpha_2 & \alpha_2 \end{bmatrix}, \tag{31}
$$

describes the occurrence of types. That is, following an interval of type 1 there is a probability $\alpha_{11} = \alpha_1$ of another interval of type 1. Similarly $\alpha_{22} = \alpha_2$ is the probability that a type 2 interval is followed by another type 2 interval. If $\alpha_1 = \alpha_2 = 0$ we have the alternating renewal process. If $\alpha_1 + \alpha_2 = 1$, the sequence of interval types is completely random. Then the sequence of events forms an ordinary renewal process with p.d.f. $\alpha_1 f_1(x) + \alpha_2 f_2(x)$ and each successive event has independently probabilities α_1 and α_2 of being of type 1 and type 2. Note that the matrix (α_{ij}) is exactly the transition probability matrix (p_{ij}) of Chapter 3.

In Example 9.4, we have already mentioned a renewal process as an approximate model for a sequence of cars passing along a one-lane road. The present type of process provides a more flexible, but still quite empirical model. We identify type 1 intervals with the times between cars travelling 'independently' and in particular we might take $f_1(x)$ to be exponential with a large mean. We identify type 2 intervals with the times between cars travelling in a bunch. Thus $f_2(x)$ has small mean and might be expected to have considerably less dispersion than an exponential distribution, although for simplicity we might sometimes take it to be exponential too. The parameter α_2 determines the geometric distribution of the number of cars per bunch, whereas α_1 and α_2 together determine the proportion of cars travelling singly. We suggest that the reader constructs a realization of such a process with $f_1(x)$, $f_2(x)$ exponential with means 5 and 1 and with $\alpha_1 = \frac{1}{5}$, $\alpha_2 = \frac{1}{2}$.

The practical difficulty of the model is that even with $f_1(x)$ and $f_2(x)$ exponential, four mathematically independent parameters are involved and quite extensive data are necessary for estimating the parameters and examining the adequacy of the model. Note that in this application the two types of event would not usually be directly distinguished in empirical data; we would simply record a single series of events. Of course, a full stochastic specification of road traffic would be complicated, requiring a statement of the position coordinates of all cars as functions of time.

A direct generalization of Example 9.6 is to have k p.d.f.'s $f_1(x)$, ..., $f_k(x)$ associated with k states of a Markov chain. In fact, however, it is possible to deal with a more general process still, in which the time spent in a transition from state i to state j of the chain has p.d.f. $f_{ij}(x)$, depending possibly on both initial and final state.

More explicitly, consider a process constructed as follows. First take a realization of the k-state Markov chain with transition matrix $\mathbf{A} = (\alpha_{ij})$. Now construct a process in continuous time by making the time spent in a transition from i to j have p.d.f. $f_{ij}(x)$, all such times being mutually independent. At the end of the interval we imagine a point event of type j. Sometimes we call this an interval of type j.

We call such a process a *semi-Markov process*, or a *Markov renewal process*. These processes are of general interest in that they join together the theory of two different types of process, the renewal process and the Markov chain. In the brief treatment to be given here we follow the same methods as used for renewal processes. It would, however, be possible to base the treatment much more heavily on the theory of Markov chains, regarding the lengths of the intervals as random functions defined on a Markov chain; see Section 3.12.

For simplicity we shall write out formulae at length only for processes with $k = 2$. When expressed in matrix form, the results do, however, extend immediately to general k. We say that an event is of type j if it comes at the end of an interval with p.d.f. $f_{ij}(x)$, for some i.

Suppose first that an event of type i has just occurred at $t = 0$. Let $h_{ij}(t)$ be the renewal density for events of type j. That is

$$h_{ij}(t) = \lim_{\Delta t \to 0} \frac{\text{prob\{event of type } j \text{ in } (t, t + \Delta t)|\text{event of type } i \text{ at } 0\}}{\Delta t}. \quad (32)$$

The conditional expected number of type j events in $(0, t)$ is then

$$\int_0^t h_{ij}(u)\, du. \tag{33}$$

To determine the renewal densities (32), we have four integral equations of which a typical one is

$$h_{11}(t) = \alpha_{11} f_{11}(t) + \alpha_{11} \int_0^t h_{11}(t-u) f_{11}(u)\, du$$

$$+ \alpha_{21} \int_0^t h_{12}(t-u) f_{21}(u)\, du. \tag{34}$$

To prove (34), note that in order to have a type 1 event in $(t, t+\varDelta t)$, we must have one of the following three things happen:

(a) a transition to state 1, occurring with probability α_{11}, associated with an interval of length nearly t;

(b) a sequence of events leading to a type 1 event near $t-u$, followed by a transition to state 1, associated with an interval of length nearly u;

(c) a sequence of events leading to a type 2 event near $t-u$, followed by a transition to state 1, associated with an interval of length nearly u.

These three possibilities account for the three terms of (34).

If we apply a Laplace transformation to (34), we have that

$$h_{11}^*(s)\, \{1 - \alpha_{11} f_{11}^*(s)\} - h_{12}^*(s)\, \alpha_{21} f_{21}^*(s) = \alpha_{11} f_{11}^*(s), \tag{35}$$

the corresponding equation based on $h_{12}(t)$ being

$$- h_{11}^*(s)\, \alpha_{12} f_{12}^*(s) + h_{12}^*(s)\, \{1 - \alpha_{22} f_{22}^*(s)\} = \alpha_{12} f_{12}^*(s). \tag{36}$$

There are two more equations starting from state 2, and the four equations can be put in matrix form as

$$\mathbf{h}^*(s)\, \{\mathbf{I} - \mathbf{g}^*(s)\} = \mathbf{g}^*(s), \tag{37}$$

where

$$\mathbf{h}^*(s) = \begin{bmatrix} h_{11}^*(s) & h_{12}^*(s) \\ h_{21}^*(s) & h_{22}^*(s) \end{bmatrix}, \qquad \mathbf{g}^*(s) = \begin{bmatrix} \alpha_{11} f_{11}^*(s) & \alpha_{12} f_{12}^*(s) \\ \alpha_{21} f_{21}^*(s) & \alpha_{22} f_{22}^*(s) \end{bmatrix}. \tag{38}$$

Equation (37) is a matrix generalization of (23), the corresponding result for a renewal process. The equation holds for any number of states in the Markov chain. In the special case when $f_{ij}(x) = f_j(x)$, i.e. when the distribution of interval length depends only on the final state,

$$\mathbf{g}^*(s) = \begin{bmatrix} \alpha_{11} & \alpha_{12} \\ \alpha_{21} & \alpha_{22} \end{bmatrix} \begin{bmatrix} f_1^*(s) & 0 \\ 0 & f_2^*(s) \end{bmatrix}. \tag{39}$$

In some contexts we may require the analogue of a modified renewal process. Suppose that with probability ϖ_i the first event occurs after a time distributed with p.d.f. $g_i^{(1)}(x)$ and is of type i. Now the time to the first event contributes directly in (35)–(37) only to the term on the right-hand side. Hence the equation analogous to (37) for the row vector of renewal densities is

$$\mathbf{h}_m^*(s) \{\mathbf{I} - \mathbf{g}^*(s)\} = \mathbf{g}^{*(1)}(s), \tag{40}$$

where

$$\mathbf{g}^{(1)}(t) = (\varpi_1 g_1^{(1)}(t), \varpi_2 g_2^{(1)}(t)). \tag{41}$$

We say that the system is in state j at time t if the next event after t is of type j. Given that an event of type i occurs at $t = 0$, let $p_{ij}(t)$ be the probability that the system is in state j at time t. Further let $\mathscr{F}_{ij}(x)$ denote the survivor function corresponding to $f_{ij}(x)$,

$$\mathscr{F}_{ij}(x) = \int_x^\infty f_{ij}(u)\, du.$$

Then we can set up integral equations of which the following is a typical one:

$$p_{11}(t) = \alpha_{11} \mathscr{F}_{11}(t) + \alpha_{11} \int_0^t h_{11}(t-u) \mathscr{F}_{11}(u)\, du$$

$$+ \alpha_{21} \int_0^t h_{12}(t-u) \mathscr{F}_{21}(u)\, du. \tag{42}$$

If we apply a Laplace transformation and write the resulting equations in matrix form, we have that

$$\mathbf{p}^*(s) = \mathbf{G}^*(s) + \mathbf{h}^*(s)\, \mathbf{G}^*(s), \tag{43}$$

where

$$\mathbf{G}^*(s) = \begin{bmatrix} \alpha_{11} \mathscr{F}_{11}^*(s) & \alpha_{12} \mathscr{F}_{12}^*(s) \\ \alpha_{21} \mathscr{F}_{21}^*(s) & \alpha_{22} \mathscr{F}_{22}^*(s) \end{bmatrix}. \tag{44}$$

Since $s\mathscr{F}_{ij}^*(s) = 1 - f_{ij}^*(s)$, it follows that

$$s\mathbf{G}^*(s) = \mathbf{A} - \mathbf{g}^*(s), \tag{45}$$

where $\mathbf{A} = (\alpha_{ij})$. Combining (37), (43) and (45), we have that

$$s\mathbf{p}^*(s) = (\mathbf{I} - \mathbf{g}^*)^{-1}(\mathbf{A} - \mathbf{g}^*). \tag{46}$$

Reasonably simple results can be extracted from these formulae when the process is an alternating renewal process, or when the p.d.f.'s $f_{ij}(x)$ have simple rational Laplace transforms. We deal here only with the simplest special case of the alternating process. Equation (39) applies with $\alpha_{11} = \alpha_{22} = 0$. Thus, from (37)

$$
\mathbf{h}^*(s) = \begin{bmatrix} 0 & f_2^*(s) \\ f_1^*(s) & 0 \end{bmatrix} \begin{bmatrix} 1 & -f_2^*(s) \\ -f_1^*(s) & 1 \end{bmatrix}^{-1}
$$

$$
= \frac{1}{\{1 - f_1^*(s) f_2^*(s)\}} \begin{bmatrix} f_1^*(s) f_2^*(s) & f_2^*(s) \\ f_1^*(s) & f_1^*(s) f_2^*(s) \end{bmatrix}. \tag{47}
$$

Of course this formula can be obtained directly from the results for the renewal density in ordinary and modified renewal processes. For example, the diagonal terms in (47) refer to an ordinary renewal process in which the distribution is the convolution of $f_1(x)$ and $f_2(x)$.

Similarly from (46), we have for the probability distribution of the state occupied at t

$$
s\mathbf{p}^*(s) = \frac{1}{\{1 - f_1^*(s) f_2^*(s)\}} \begin{bmatrix} f_2^*(s)\{1 - f_1^*(s)\} & 1 - f_2^*(s) \\ 1 - f_1^*(s) & f_1^*(s)\{1 - f_2^*(s)\} \end{bmatrix}. \tag{48}
$$

In the special case of the alternating Poisson process, $f_i^*(s) = \rho_i/(\rho_i + s)$, and

$$
\mathbf{p}^*(s) = \begin{bmatrix} \dfrac{\rho_2}{s(s+\rho_1+\rho_2)} & \dfrac{s+\rho_1}{s(s+\rho_1+\rho_2)} \\ \dfrac{s+\rho_2}{s(s+\rho_1+\rho_2)} & \dfrac{\rho_1}{s(s+\rho_1+\rho_2)} \end{bmatrix}. \tag{49}
$$

Thus

$$
\mathbf{p}(t) = \begin{bmatrix} \dfrac{\rho_2}{\rho_1+\rho_2}\{1 - e^{-(\rho_1+\rho_2)t}\} & \dfrac{\rho_1}{\rho_1+\rho_2} + \dfrac{\rho_2}{\rho_1+\rho_2}e^{-(\rho_1+\rho_2)t} \\ \dfrac{\rho_2}{\rho_1+\rho_2} + \dfrac{\rho_1}{\rho_1+\rho_2}e^{-(\rho_1+\rho_2)t} & \dfrac{\rho_1}{\rho_1+\rho_2}\{1 - e^{-(\rho_1+\rho_2)t}\} \end{bmatrix}. \tag{50}
$$

For example, the leading element is the probability of being in state 1 at time t given that a type 1 event occurred at $t = 0$, which in this particular process means that state 2 is occupied immediately after $t = 0$. Thus the term in question is zero at $t = 0$ and increases exponentially to its limit $\rho_2/(\rho_1+\rho_2)$.

In a general alternating renewal process, in which μ_i is the mean of the p.d.f. $f_i(x)$, it can be shown by examining the form of (48) as $s \to 0$, that the limiting probabilities are $\mu_1/(\mu_1+\mu_2)$, $\mu_2/(\mu_1+\mu_2)$. More generally still, in a semi-Markov process associated with an ergodic Markov chain, the limiting probabilities can be evaluated by finding from (46)

$$
\lim_{s \to 0} \{s\mathbf{p}^*(s)\}.
$$

The p.d.f.'s $f_{ij}(x)$ enter only through their means μ_{ij}. More directly, if (π_1, π_2) is the equilibrium distribution of the discrete time Markov chain, a proportion $\pi_1 \alpha_{11}$ of a very large number of transitions are, with probability one, from state 1 to state 1. The duration in continuous time of such transitions is on the average μ_{11}. Hence of a very long time period a proportion

$$\frac{\pi_1 \alpha_{11} \mu_{11}}{\pi_1 \alpha_{11} \mu_{11} + \pi_1 \alpha_{12} \mu_{12} + \pi_2 \alpha_{21} \mu_{21} + \pi_2 \alpha_{22} \mu_{22}}$$

is, with probability one, spent in passing from state 1 to state 1. Thus the limiting probability of state 1, as evaluated from (46) must be

$$\frac{\pi_1 \alpha_{11} \mu_{11} + \pi_2 \alpha_{21} \mu_{21}}{\pi_1 \alpha_{11} \mu_{11} + \pi_1 \alpha_{12} \mu_{12} + \pi_2 \alpha_{21} \mu_{21} + \pi_2 \alpha_{22} \mu_{22}}.$$

9.4. Stationary point processes

We next deal with a few general properties of stationary point processes.

First consider the distribution of the forward recurrence-time V_t, i.e. the time measured from t forward to the next event after t. Now in a stationary process the distribution of V_t cannot involve t. Hence the distribution of V_t for any fixed t is the same as that for the recurrence-time V measured forward from a time point chosen at random over a long time interval. Let $f(x)$ be the p.d.f. of the interval between successive events. Then if a time point is selected at random, the probability that it falls between two events separated by an interval of length between x and $x + \Delta x$ is

$$xf(x)\, \Delta x / \mu \tag{51}$$

where μ is the mean interval. The justification of (51) is that the chance that the sampling point falls in such an interval is proportional to the total length of such intervals, which is proportional to $xf(x)$; the denominator μ is a normalizing constant.

Now given that the sampling point falls in an interval of length x, it is equally likely to fall anywhere in the interval and hence the conditional p.d.f. of V is rectangular over $(0, x)$. Hence the unconditional p.d.f. of V is $q(v)$, where

$$q(v) = \int_v^\infty \frac{1}{x} \cdot \frac{xf(x)}{\mu}\, dx = \frac{1 - F(v)}{\mu}, \tag{52}$$

the integration being over (v, ∞), since only intervals longer than v can produce forward recurrence-times equal to v. The argument applies equally to the backward recurrence-time. Now (52) is exactly the result (27) and (29) obtained for renewal processes by a quite different argument. Thus the discussion in Section 9.2 of the connexion between the recurrence-time distribution and the distribution of intervals between successive events applies perfectly generally to stationary point processes. We shall use this result in particular in Section 9.5 in the discussion of the superposition of processes.

We now turn to the specification of the properties of stationary point processes by methods analogous to those developed for real and complex valued stationary processes in Chapters 7 and 8. We can describe a stationary point process by the sequence of intervals $(X_1, X_2, ...)$ between successive events. Alternatively, we may divide the time axis into a large number of narrow intervals of width Δt and count the number of events in each interval. The two specifications are equivalent, but lead to different second-order functions. We shall use suffices x and n to distinguish functions referring respectively to the intervals and to the numbers.

If there is a non-zero probability of coincident events, we can apply directly the general formulae that follow. Alternatively, it may be preferable to deal first with the process governing 'occurrence points' and to examine separately the distribution of the number of events per occurrence point.

First we consider the sequence of intervals $\{X_1, X_2, ...\}$. In order that the sequence should be stationary, we suppose that an event occurs at $t = 0$. Let $\mu_x = E(X_i)$, $\sigma_x^2 = V(X_i)$ be the mean and variance of intervals. We can consider this sequence as a stationary real-valued process in discrete time, the time parameter being the serial number of the event. Therefore we define the autocovariance function

$$\gamma_x(h) = C(X_{i+h}, X_i) \quad (h = 0, \pm 1, ...),\tag{53}$$

and the corresponding spectral density function

$$\phi_x(\omega) = \frac{1}{2\pi} \sum_{h=-\infty}^{\infty} e^{-ih\omega} \gamma_x(h),\tag{54}$$

with

$$\gamma_x(h) = \int_{-\pi}^{\pi} e^{ih\omega} \phi_x(\omega)\, d\omega.\tag{55}$$

Since $\gamma_x(h)$ is an even real function of h, so too is $\phi_x(\omega)$.

Further we can introduce a variance function $\psi_x(h)$ specifying the variance of the time interval from an arbitrary event to the hth successive event. We have that

$$\psi_x(h) = V(X_1 + ... + X_h) = h\gamma_x(0) + 2\sum_{j=1}^{h-1}\sum_{i=1}^{j} \gamma_x(i).\tag{56}$$

Sometimes it is useful to express the variance as a ratio to its value for the Poisson process by means of the index of dispersion

$$\tau_x(h) = \frac{\psi_x(h)}{h\mu_x^2}. \tag{57}$$

The functions $\gamma_x(h)$, $\phi_x(\omega)$, $\psi_x(h)$ are mutually equivalent. For a renewal process, $\gamma_x(h) = 0$ ($h \neq 0$) and $\tau_x(h)$ is constant.

Consider now the formal process $\{\Delta N(t)\}$, where $\Delta N(t)$ is the number of events in $(t, t + \Delta t)$. Let

$$E\{\Delta N(t)\} = \mu_n \Delta t, \qquad V\{\Delta N(t)\} = \sigma_n^2 \Delta t, \tag{58}$$

neglecting terms that are $o(\Delta t)$. In the important special case when multiple occurrences are excluded, $\Delta N(t)$ takes values 0 and 1 and

$$E\{\Delta N(t)\} = V\{\Delta N(t)\} = \mu_n \Delta t. \tag{59}$$

The number of events in a long time interval t_0 is asymptotically $t_0 \mu_n$, whereas the length of time up to the occurrence of the n_0th event is, for large n_0, asymptotically $n_0 \mu_x$. Hence

$$\mu_n \mu_x = 1. \tag{60}$$

To examine the covariance properties, we define for $v > 0$ a function $\gamma_n(v)$ by

$$C\{\Delta N(t+v), \Delta N(t)\} = \gamma_n(v)(\Delta t)^2. \tag{61}$$

We call $\gamma_n(v)$ the covariance density. To obtain an alternative form for $\gamma_n(v)$ we consider first a process in which multiple occurrences arise with probability zero. Then

$$\begin{aligned} E\{\Delta N(t+v)\,\Delta N(t)\} &= \text{prob}\{\Delta N(t) = \Delta N(t+v) = 1\} \\ &= \mu_n \Delta t \,\text{prob}\{\Delta N(t+v) = 1 | \Delta N(t) = 1\} \\ &= \mu_n h_n(v)(\Delta t)^2, \end{aligned} \tag{62}$$

say, where

$$h_n(v) = \lim_{\Delta v \to 0} \frac{\text{prob}\{\text{event in } (t+v, t+v+\Delta v)|\text{event at } t\}}{\Delta v}$$

is a function generalizing the renewal density, (22). Converting (62) into a covariance, we have that for $v > 0$

$$\gamma_n(v) = \mu_n\{h_n(v) - \mu_n\}. \tag{63}$$

If multiple occurrences do arise it is convenient to define $h_n(v)$ so that (63) still holds. For this, let

$$h_n(v; i) = \lim_{\Delta v \to 0} \frac{E\{\text{no. events in } (t+v, t+v+\Delta v)|i \text{ events at } t\}}{\Delta v}, \tag{64}$$

and
$$h_n(v) = \sum_{i=1}^{\infty} i\varpi_i h_n(v;i)/\mu_n, \tag{65}$$

where $\varpi_i \Delta t$ is the probability that there are i events in $(t, t+\Delta t)$ and $\mu_n = \sum i\varpi_i$.

Furthermore it is useful to extend the definition of $\gamma_n(v)$ so that for $v = 0$, (58) holds. We therefore add to $\gamma_n(v)$, as previously defined, $\sigma_n^2 \delta(v)$, where $\delta(v)$ denotes a Dirac delta function. That is

$$\gamma_n(v) = \mu_n\{h_n(v) - \mu_n\} + \sigma_n^2 \delta(v). \tag{66}$$

We can define a formal spectral density function, $\phi_n(\omega)$, by taking Fourier transforms of (66); thus

$$\phi_n(\omega) = \frac{\mu_n}{2\pi} \int_{-\infty}^{\infty} \{h_n(v) - \mu_n\} e^{-i\omega v} dv + \frac{\sigma_n^2}{2\pi}. \tag{67}$$

If, as in studying the renewal process, it is convenient to find the Laplace transform $h_n^*(s)$, we can use (67) in the form

$$\phi_n(\omega) = \frac{\mu_n}{2\pi} \{h_n^*(i\omega) + h_n^*(-i\omega)\} + \frac{\sigma_n^2}{2\pi}. \tag{68}$$

Thus a Poisson process has a 'white' spectrum, $\phi_n(\omega) = \mu_n/(2\pi)$. A renewal process with $f(x)$ as the p.d.f. for the intervals between successive events has $h_n^*(s) = f^*(s)/\{1 - f^*(s)\}$ and hence, again assuming there are no multiple occurrences,

$$\phi_n(\omega) = \frac{\mu_n}{2\pi} \left\{ \frac{f^*(i\omega)}{1 - f^*(i\omega)} + \frac{f^*(-i\omega)}{1 - f^*(-i\omega)} + 1 \right\}. \tag{69}$$

It can be shown from (69) that, if $f(x)$ is a gamma distribution, $\phi_n(\omega)$ increases or decreases as $\omega \to \infty$ to its limiting value $\mu_n/(2\pi)$, depending on whether the coefficient of variation is less than or greater than one.

We can further define a variance function

$$\psi_n(v) = V\left\{ \int_0^v dN(t) \right\}$$

$$= \sigma_n^2 v + 2 \int_{0+}^v dx \int_{0+}^x dy\, \gamma_n(y) \tag{70}$$

$$= \sigma_n^2 v + 4 \int_0^\infty \left\{ \phi_n(\omega) - \frac{\sigma_n^2}{2\pi} \right\} \left\{ \frac{1 - \cos(v\omega)}{\omega^2} \right\} d\omega. \tag{71}$$

The corresponding index of dispersion is

$$\tau_n(v) = \frac{\psi_n(v)}{v\mu_n}. \tag{72}$$

тис general discussion of the relations between the autocovariance, the spectral density and the variance function is parallel to that for ordinary real or complex-valued processes.

Before continuing the general development it is worth while calculating for a special process the various functions that have been introduced so far.

Example 9.7. *Two-state semi-Markov process*. We consider in more detail the process introduced in Example 9.6 in which there are two types of interval, having p.d.f.'s $f_1(x), f_2(x)$ and the transitions between them are determined by the Markov chain with transition matrix (31)

$$\begin{bmatrix} \alpha_1 & 1-\alpha_1 \\ 1-\alpha_2 & \alpha_2 \end{bmatrix},$$

where we suppose $0 < \alpha_1, \alpha_2 < 1$. The equilibrium distribution in the chain is

$$\pi_1 = \frac{1-\alpha_2}{2-\alpha_1-\alpha_2}, \qquad \pi_2 = \frac{1-\alpha_1}{2-\alpha_1-\alpha_2}. \tag{73}$$

Note that this is the equilibrium distribution in the discrete 'time' of the imbedded Markov chain and is not the same as the probability distribution of the state of the process at time t.

First consider $\gamma_x(h)$, the autocovariance function when we work with intervals between successive events. Let μ_1, μ_2 denote the means, and σ_1^2, σ_2^2 the variances of the p.d.f.'s $f_1(x)$, $f_2(x)$. If the variances are zero, the intervals X_i take one of the values μ_1 and μ_2 and we are observing a simple Markov process, with autocovariance function of the form $\sigma_x^2 \lambda^h$; see equation (7.9). When the variances are non-zero, we are observing a Markov process with superimposed error, having therefore the form $k\sigma_x^2 \lambda^h$ ($k < 1$) for $h = 1, 2, \ldots$; see equation (7.30).

More explicitly, consider two intervals X_1, X_{h+1}. Let (i,j) denote that X_1 is sampled from $f_i(x)$ and X_{h+1} from $f_j(x)$. Then for $h = 1, 2, \ldots$

$$E\{X_1 X_{h+1} | (i,j)\} = \mu_i \mu_j. \tag{74}$$

Now it follows from the elementary theory of the two-state Markov chain that the four possible forms (i,j) occur with the following probabilities:

$$\begin{aligned} (1,1) &: \pi_1(\pi_1 + \pi_2 \eta^h), & (1,2) &: \pi_1(\pi_2 - \pi_2 \eta^h), \\ (2,1) &: \pi_2(\pi_1 - \pi_1 \eta^h), & (2,2) &: \pi_2(\pi_2 + \pi_1 \eta^h), \end{aligned} \tag{75}$$

where $\eta = \alpha_1 + \alpha_2 - 1$. Thus $E(X_1 X_{h+1})$, and hence $C(X_1, X_{h+1})$, are obtained by averaging (74) with respect to (75), thus giving

$$\gamma_x(h) = (\mu_1 - \mu_2)^2 \pi_1 \pi_2 \eta^h \quad (h = 1, 2, \ldots). \tag{76}$$

Also the variance is

$$\gamma_x(0) = \pi_1 \sigma_1^2 + \pi_2 \sigma_2^2 + \pi_1 \pi_2 (\mu_1 - \mu_2)^2. \tag{77}$$

Note that no assumptions are involved about the form of the distributions $f_1(x)$, $f_2(x)$. The spectral density and variance function can be obtained by transformation of (76) and (77).

If now we consider the complementary analysis in terms of numbers of events, it is easiest to examine the spectral density function. We need to calculate the function $h_n(v)$. This is given in terms of the functions $h_{ij}(v)$ of (32) by

$$h_n(v) = \pi_1\{h_{11}(v) + h_{12}(v)\} + \pi_2\{h_{21}(v) + h_{22}(v)\} = (\pi_1 \pi_2)\mathbf{h}(v)\binom{1}{1}.$$

After some calculation from (37) we have that

$$h_n^*(s) = \frac{\pi_1 f_1^*(s) + \pi_2 f_2^*(s) + (1 - \alpha_1 - \alpha_2) f_1^*(s) f_2^*(s)}{1 - \alpha_1 f_1^*(s) - \alpha_2 f_2^*(s) - (1 - \alpha_1 - \alpha_2) f_1^*(s) f_2^*(s)}. \tag{78}$$

This reduces to the corresponding formula (23) for a renewal process when either (a) $f_1^*(s)$ and $f_2^*(s)$ are identical, or (b) $\alpha_1 + \alpha_2 = 1$, so that the Markov chain is completely random.

We now obtain the spectrum of the process by substitution in (68). If $f_i(x) = \rho_i e^{-\rho_i x}$, we have that

$$h_n^*(s) = \frac{\rho_1 \rho_2 (2 - \alpha_1 - \alpha_2) + s(\pi_1 \rho_1 + \pi_2 \rho_2)}{s(s + \rho_1 + \rho_2 - \alpha_1 \rho_1 - \alpha_2 \rho_2)},$$

so that

$$h_n(t) = \frac{\rho_1 \rho_2 (2 - \alpha_1 - \alpha_2)}{\rho_1 + \rho_2 - \alpha_1 \rho_1 - \alpha_2 \rho_2}$$

$$+ \frac{\pi_1 \rho_1 + \pi_2 \rho_2 - \rho_1 \rho_2 (2 - \alpha_1 - \alpha_2)}{(\rho_1 + \rho_2 - \alpha_1 \rho_1 - \alpha_2 \rho_2)} \exp\{-(\rho_1 + \rho_2 - \alpha_1 \rho_1 - \alpha_2 \rho_2)t\}. \tag{79}$$

We now develop briefly some further general theory of the functions defined above. First suppose that the time S_r from an origin up to the rth subsequent event is for large r asymptotically normally distributed with mean $r\mu_x$ and with a variance proportional to r, so that, in particular, the index of dispersion $\tau_x(r)$ tends to a limit as $r \to \infty$. We know that this is true for renewal processes with $\sigma^2 < \infty$, but it is clearly true much more generally since S_r is the sum of a large number of random variables. We can now repeat the argument used in (6) for renewal processes to show that the number of events in a long time interval v is asymptotically normally distributed with mean $v/\mu_x = v\mu_n$ and with variance related to that of S_r. It follows that the index of dispersion $\tau_n(v)$ is asymptotically equal to $\tau_x(r)$; that is

$$\lim_{r \to \infty} \tau_x(r) = \lim_{v \to \infty} \tau_n(v). \tag{80}$$

Next we consider briefly the general relations that connect the covariance functions $\gamma_n(v)$, $\gamma_x(h)$ with respectively the distributional properties of intervals and numbers of events. We deal for simplicity with processes without multiple occurrences.

Suppose then that an event occurs at the origin and that X_1, X_2, ... are the subsequent intervals between successive events. Let $p_x(r;y)$ be the p.d.f. of $Y = X_1 + ... + X_r$. Then

$$h_n(v) = \sum_{r=1}^{\infty} p_x(r;v),$$

since both sides when multiplied by Δv give the probability of an event in $(v, v + \Delta v)$. Thus from (63) we have that

$$\gamma_n(v) = \mu_n \left\{ \sum_{r=1}^{\infty} p_x(r;v) - \mu_n \right\}. \tag{81}$$

The result obtained here is essentially a generalization to an arbitrary process of the result (22) for renewal processes.

The complementary relation expressing $\gamma_x(h)$ in terms of the distributional properties of numbers will be given without proof. Let $p_n(t;s)$ be the probability that there are s events in the interval $(0,t)$, where the origin is an arbitrary point, not in general a point at which an event occurs. Then (McFadden, 1962)

$$\gamma_x(h) = \mu_x \left\{ \int_0^{\infty} p_n(t;h)\, dt - \mu_x \right\}. \tag{82}$$

An important general point about (81) and (82) is that the functions of the second degree are expressed in terms of distributional properties of all orders.

9.5. Operations on point processes

We now consider two types of operation on point processes, the superposition of processes and random translation.

(i) SUPERPOSITION OF PROCESSES

First suppose that we have p independent point processes and that a new process is formed by superimposing the p separate processes. Figure 9.4 shows an example with $p = 3$.

For example, consider an industrial process in one stage of which p similar machines operate independently and in parallel. Then the sequence of instants at which items are produced will be a pooled output of the type under consideration. We may need to know the properties of the pooled sequence in order to plan the next stage of production.

Another application concerns the sequence of pulses arriving at a central nerve cell from a number of independent sources.

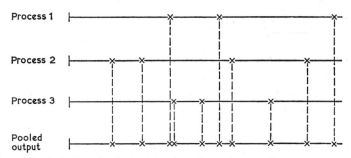

Fig. 9.4. Process formed by superimposing $p = 3$ processes.

Let $N_{t(p)}$ be the number of events in the interval $(0, t)$ in the pooled process and $N_t^{(1)}$, ..., $N_t^{(p)}$ the corresponding numbers in the component processes, these being independent random variables. Since

$$N_{t(p)} = N_t^{(1)} + \ldots + N_t^{(p)},$$

the probability generating function and other properties of $N_{t(p)}$ are easily obtained. Further, the functions $\gamma_n(h)$, $\phi_n(\omega)$, $\psi_n(h)$ for the combined process are the sums of the corresponding functions for the component processes.

We may sometimes want the distribution of the interval between successive events in the pooled process. We shall suppose that the component processes have identical stationary structure and that multiple occurrences have zero probability. Denote by $f(x)$ the p.d.f. of intervals between successive events in a single process. Then, by (52), the p.d.f. of forward recurrence time for a single process is $\mathscr{F}(x)/\mu$, where $\mathscr{F}(x) = 1 - F(x)$ is the complement of the distribution function and μ is the mean corresponding to $f(x)$. But the forward recurrence time, V_p, in the combined process is the smallest of the p independent component recurrence times. Thus the distribution function of V_p is

$$1 - \left\{ \int_x^\infty \frac{\mathscr{F}(u)}{\mu} du \right\}^p$$

and hence the p.d.f. of V_p is

$$\frac{p\mathscr{F}(x)}{\mu} \left\{ \int_x^\infty \frac{\mathscr{F}(u)}{u} du \right\}^{p-1} \tag{83}$$

Now if $f_p(x)$ is the p.d.f. of intervals in the combined process, the p.d.f. of V_p is, on again applying (52),

$$\frac{\int_x^\infty f_p(u)\, du}{\mu/p}. \tag{84}$$

On equating (83) and (84) and differentiating, we have that

$$f_p(x) = -\frac{d}{dx}\left[\mathscr{F}(x)\left\{ \int_x^\infty \frac{\mathscr{F}(u)}{\mu}\, du \right\}^{p-1} \right]. \tag{85}$$

For example, if $f(x) = \rho e^{-\rho x}$, (85) gives $f_p(x) = p\rho e^{-p\rho x}$. This is obvious if the original processes are Poisson processes, but the result is more general in that it refers also to component processes in which intervals have correlated exponential distributions. As another example, suppose that in the component process $f(x)$ is a rectangular distribution over $(0, 2\mu)$. Then

$$f_p(x) = \frac{2p-1}{2\mu}\left(1 - \frac{x}{2\mu}\right)^{2p-2} \qquad (0 < x < 2\mu).$$

It follows easily that as $p \to \infty$ the p.d.f. of the normalized random variable pX/μ tends rapidly to e^{-x}.

In many ways the most important properties of the combined process refer to the 'local' behaviour for large p, i.e. to behaviour over time periods short compared with the time between events in the component processes. The general limiting result is that the local properties are those of a Poisson process, provided that multiple occurrences have zero probability. One important consequence is an explanation of the occurrence in practice of processes closely approximating to Poisson processes. For example, the sequence of calls received at a large telephone exchange is the superposition of the individual sequences arising from single subscribers. Hence the overall sequence can be expected to be locally a Poisson process more or less independently of the form of the individual sequences.

We shall illustrate the limiting result by examining two special properties of the combined process. First we examine the limiting distribution of the intervals, $X_{(p)}$, between successive events in the combined process. We first normalize $X_{(p)}$ by considering the random variable $X_{(p)}/(\mu/p)$. On integrating (85), we have that the survivor function of this random variable is

$$\mathscr{F}\left(\frac{x\mu}{p}\right)\left\{ 1 - \int_0^{x\mu/p} \frac{\mathscr{F}(u)}{\mu}\, du \right\}^{p-1}.$$

For fixed x, μ, the limit of this as $p \to \infty$ is e^{-x}, since $\mathscr{F}(u)$ is continuous as $u \to 0+$ and $\mathscr{F}(0) = 1$; this last follows because we have assumed that multiple occurrences have zero probability.

Secondly we consider the distribution of the number of events in a time period $t = v\mu/p$ as $p \to \infty$ with v and μ fixed. For this, we require the distribution of the number of events in an arbitrary short interval of a single component process. By (52), the probability that such an interval contains no events is

$$\int\limits_t^\infty \frac{\mathscr{F}(x)}{\mu} \, dx = 1 - \int\limits_0^t \frac{\mathscr{F}(x)}{\mu} \, dx$$

$$= 1 - \frac{t}{\mu} + o(t),$$

as $t \to 0$.

Further, under some weak restrictions on the type of process the probability of more than one event will be $o(t)$. Then the probability generating function of the number of events in a single process is

$$1 - \frac{t}{\mu} + \frac{tz}{\mu} + o(t)$$

and hence that for the p processes is as $p \to \infty$

$$\left\{ 1 - \frac{t}{\mu} + \frac{tz}{\mu} + o(t) \right\}^p \sim \exp(zv - v),$$

the generating function of the Poisson distribution of mean v. A thorough discussion of the asymptotic properties of superimposed processes is given by Khintchine (1960).

(*ii*) RANDOM DISPLACEMENT OF EVENTS

We now consider briefly a second type of operation on a stationary point process. Consider again a process without multiple occurrences and suppose that each point is displaced by a random amount, the displacements of different points being independent identically distributed random variables. Let $h_n^{(1)}(v)$ and $h_n^{(2)}(v)$ be the renewal densities of the process before and after this random displacement of points.

Consider a particular event O_2 in the final process. Let O_1 be its original position. In order that there should in the final process be an event at distance $(v, v + \varDelta v)$ from O_2, it is necessary that in the original process, there should have been an event at distance $(v + u, v + u + \delta u + \varDelta v)$ from O_1, where the relative displacement is $(u, u + \delta u)$, for some u. (In this argument, we suppose that $\delta u \ll \varDelta v$.) Thus

$$h_n^{(2)}(v) \, \varDelta v = \int\limits_{-\infty}^{\infty} h_n^{(1)}(v + u) \, g(u) \, du \varDelta v,$$

where $g(u)$ is the p.d.f. of the difference of two independent displacements. We can write this in the slightly more convenient form

$$h_n^{(2)}(v) - \mu_n = \int\limits_{-\infty}^{\infty} \{h_n^{(1)}(v+u) - \mu_n\} g(u)\, du. \tag{86}$$

This is most readily expressed in terms of the spectral functions $\phi_n^{(1)}(\omega)$, $\phi_n^{(2)}(\omega)$ before and after displacement. In equation (67), we can take $\sigma_n^2 = \mu_n$ since we are dealing with processes without multiple occurrences; further μ_n is clearly unaltered by random displacement. Therefore

$$\phi_n^{(2)}(\omega) - \frac{\mu_n}{2\pi} = \left\{\phi_n^{(1)}(\omega) - \frac{\mu_n}{2\pi}\right\} g^\dagger(\omega), \tag{87}$$

where $g^\dagger(\omega)$ is the Fourier transform of $g(u)$ and is thus equal to $E(e^{iZ\omega}) E(e^{-iZ\omega})$, where Z is a typical displacement.

Thus $g^\dagger(\omega)$ plays the role of the function $L(\omega)$ of Section 8.3 applying to the non-Poisson part of the spectrum. Note that if the original process is a Poisson process, so too is the final process for all $g(x)$. A displacement distribution having a characteristic function vanishing identically in an interval can be used to transform very special types of non-Poisson process into processes having a 'white' spectrum.

As a simple example, suppose that the original process is a renewal process with the p.d.f. of intervals being xe^{-x}. Then $f^*(s) = 1/(1+s)^2$, so that by (69), with $\mu_n = \frac{1}{2}$,

$$\phi_n^{(1)}(\omega) = \frac{1}{2\pi} \left\{ -\frac{1}{(4+\omega^2)} + \frac{1}{2} \right\}.$$

If, to take an artificial special case, the displacements have an exponential p.d.f. $\rho e^{-\rho x}$, we have that $g^\dagger(\omega) = \rho^2/(\rho^2 + \omega^2)$, and hence by (87)

$$\phi_n^{(2)}(\omega) - \frac{1}{4\pi} = -\frac{\rho^2}{2\pi(4+\omega^2)(\rho^2+\omega^2)}. \tag{88}$$

It follows from (71) that the negative sign on the right-hand side of (88) implies that the variance of the number of events in any interval is always smaller than that for a Poisson distribution of the same mean. Further, it is easily shown that in this case the random displacement has made the process more nearly a Poisson process.

9.6. Real-valued processes associated with a point process

Example 9.8. *Idealized model of a textile yarn.* As a very idealized model of a textile yarn, consider a one-dimensional assembly of parallel fibres, each fibre being represented by an interval. The number of fibres, $Y(t)$, passing through t, is called the thickness at t. In practice the fibres are not straight and parallel; also an allowance for the variability of fibre cross-section has to be included. However, the present model is useful as a first approximation.

To describe the process, we need first to give the point process specifying the positions of the fibre left ends; the choice of left rather than right is made here for mathematical convenience only. Then we must give the distribution of fibre length and specify any stochastic dependency between the fluctuations of length and those of numbers of fibre left ends. The simplest case, which we shall consider first below, is when all fibres have the same length l. The next case is when fibre length is a random variable having the same distribution independently for all fibre left ends.

If $\Delta N(x)$ denotes the number of fibre left ends between x and $x + \Delta x$, and if all fibres are of length l,

$$Y(t) = \int_{t-l}^{t} dN(x). \tag{89}$$

Quite generally, if $\Delta N(x, l)$ is the number of left ends between x and $x + \Delta x$ associated with fibres of length between l and $l + \Delta l$, then

$$Y(t) = \int_{l=0}^{\infty} \int_{x=t-l}^{t} dN(x, l). \tag{90}$$

Example 9.9. *The shot effect.* Suppose that electrons arrive at the electrode of a valve and that each electron generates a signal (voltage or current), such that at time u after the arrival at the electrode the signal is $g(u)$. We assume that $g(u)$ tends to zero sufficiently rapidly as $u \to \infty$ for all integrals to be convergent. If the signals from different electrons are additive, the total signal $Y(t)$ at time t is

$$Y(t) = \int_{-\infty}^{t} g(t-x) dN(x), \tag{91}$$

where, as in the previous example, $\Delta N(x)$ is the number of point events in $(x, x + \Delta x)$. The random fluctuations of the process $\{Y(t)\}$ are called the shot effect.

An alternative notation for (91) is

$$Y(t) = \sum_{i=-\infty}^{\infty} g(t-T_i), \tag{92}$$

where $\{T_i\}$ are the times at which electrons arrive and $g(u)$ is defined to be zero for negative arguments.

If $g(u) = 1$ $(0 \leqslant u \leqslant l)$, and is zero otherwise, (91) reduces to (89), and we have a process equivalent to that of the previous Example.

More generally the function $g(u)$ may itself be randomly distributed. Denote the possible functions by $g(u; l)$, where l is index specifying the function. Thus if there were two possible functions l could be restricted to the values $(0,1)$. There is no real loss of generality, however, in supposing that l takes positive real values and denoting by $\Delta N(x,l)$ the number of events in $(x, x + \Delta x)$ being associated with values in $(l, l + \Delta l)$. Then

$$Y(t) = \int_{l=0}^{\infty} \int_{x=-\infty}^{\infty} g(t-x;l)\,dN(x,l). \tag{93}$$

Equation (90) of the previous example can be recovered by taking $g(u;l)$ to be a rectangular pulse of unit height and length l.

Example 9.10. *Immigration–death process.* There is an interesting connexion between the process considered in the two previous examples and that of Example 4.5. In that example, we considered individuals entering a system in a Poisson process of rate α; the times spent in the system by different individuals are independently exponentially distributed with parameter μ. The state of the system at time t is defined by the number of individuals present at that instant.

This is completely equivalent to the special case of (90) in which the fibre left ends occur in a Poisson process and in which the lengths of fibres are independently exponentially distributed. Thus equation (90), and even more so (93), can be regarded as defining immigration–death processes of a much more general type than was considered in Example 4.5.

We suppose first that the function $g(u)$ is fixed, so that $Y(t)$ is defined by (91), or in the special case where $g(u)$ is a rectangular pulse, by (89). Let the underlying point process be stationary and let its first- and second-moment properties be specified as in (59) and (66) by

$$E\{\Delta N(x)\} = \mu_n \Delta x, \qquad C\{\Delta N(x), \Delta N(x+v)\} = \gamma_n(v)\,(\Delta x)^2,$$

where $\gamma_n(v) = \gamma_n^0(v) + \sigma_n^2 \delta(v)$, say. Then directly from (91) we have that

$$E\{Y(t)\} = \int_{-\infty}^{t} g(t-x)\,\mu_n\,dx = \mu_n \int_{0}^{\infty} g(u)\,du, \tag{94}$$

and that

$$V\{Y(t)\} = \int_{-\infty}^{t} \int_{-\infty}^{\infty} g(t-x)\,g(t-y)\,\{\gamma_n^0(x-y) + \sigma_n^2 \delta(x-y)\}\,dx\,dy$$

$$= \sigma_n^2 \int_{0}^{\infty} \{g(u)\}^2\,du + \int_{0}^{\infty} \int_{0}^{\infty} g(u)\,g(w)\,\gamma_n^0(u-w)\,du\,dw. \tag{95}$$

An exactly similar argument gives the autocovariance function of $\{Y(t)\}$ as

$$C\{Y(t+v),\, Y(t)\} = \sigma_n^2 \int_0^\infty g(u)\,g(u+v)\,du$$

$$+ \int_0^\infty \int_0^\infty g(u+v)\,g(w)\,\gamma_n^0(u-w)\,du\,dw. \qquad (96)$$

Thus the second-degree properties of the process $\{Y(t)\}$ are in principle readily calculated. Equations (94)–(96) simplify appreciably when $g(u) = 1$ $(0 \leqslant u \leqslant l)$ and is zero otherwise, i.e. for the process (89).

If the underlying point process is a Poisson process,

$$C\{Y(t+v),\, Y(t)\} = \mu_n \int_0^\infty g(u)\,g(u+v)\,du. \qquad (97)$$

For the process (89)

$$C\{Y(t+v),\, Y(t)\} = \mu_n \max\{0, l-v\} \quad (v > 0). \qquad (98)$$

Thus the autocorrelation function is triangular, as is otherwise obvious since the process is a simple moving average of a purely random process.

When the point process is a Poisson process, the whole theory of (91) is much simplified. In fact, on considering a discrete approximation to the integral (91), we have that $Y(t)$ is the weighted sum of independent and identically distributed random variables. Since the rth cumulant of $\Delta N(x)$ is $\mu_n\,\Delta x$, we have from the additive properties of cumulants that

$$\kappa_r\{Y(t)\} = \mu_n \int_0^\infty \{g(u)\}^r\,du, \qquad (99)$$

with a similar result for joint cumulants. In particular, for the process (89), the integral equals l for all r, and all cumulants are $\mu_n l$. Hence $Y(t)$ has a Poisson distribution of mean $\mu_n l$, as can be seen also by direct arguments. Equation (99) for $r = 2$ is called Campbell's theorem.

We turn now to the situation where we allow a randomly distributed function $g(u; l)$, so that the process is specified by the process $\{\Delta N(x, l)\}$, giving the number of events in $(x, x+\Delta x)$ associated with values in $(l, l+\Delta l)$. If $q(l)$ is the p.d.f. of the random variable L, we have that

$$E\{\Delta N(x, l)\} = \mu_n q(l)\,\Delta x \Delta l. \qquad (100)$$

Now there would be no difficulty in principle in introducing a function of v, l, m to specify $C\{\Delta N(x, l), \Delta N(x+v, m)\}$. Here, however, we shall restrict attention to point processes without multiple occurrences and for which the values of L for different events are independent. Then

$$C\{\Delta N(x, l), \Delta N(x+v, m)\} = \{\gamma_n^0(v)\, q(l)\, q(m)$$

$$+ \mu_n q(l)\, \delta(v)\, \delta(l-m)\}\, (\Delta x)^2\, \Delta l \Delta m. \quad (101)$$

If now we take the process $Y(t)$ as defined in the general form, (93)

$$Y(t) = \int_{l=0}^{\infty} \int_{x=-\infty}^{t} g(t-x; l)\, dN(x, l),$$

we have from (100) that

$$E\{Y(t)\} = \mu_n \int_{l=0}^{\infty} \int_{x=-\infty}^{t} g(t-x; l)\, q(l)\, dx\, dl$$

$$= \mu_n \int_{0}^{\infty} E_l\{g(u; l)\}\, du, \quad (102)$$

where E_l denotes expectation over the p.d.f. $q(l)$. Further, by a similar calculation, we have that

$$C\{Y(t), Y(t+v)\} = \mu_n \int_{0}^{\infty} E_l\{g(u; l)\, g(u+v; l)\}\, du$$

$$+ \int_{0}^{\infty} \int_{0}^{\infty} \gamma_n^0(u-w)\, E_l\{g(u; l)\}\, E_l\{g(w+v; l)\}\, du\, dw.$$

$$(103)$$

If the point process is a Poisson process, the second term on the right-hand side is zero.

For Example 9.8, we identify $g(u; l)$ with a rectangular pulse of length l and unit height. If $\mathcal{Q}(l)$ is the survivor function corresponding to L, then in general for $v > 0$,

$$E_l\{g(u; l)\} = \mathcal{Q}(u), \qquad E_l\{g(u; l)\, g(u+v; l)\} = \mathcal{Q}(u+v). \quad (104)$$

Thus, for $v > 0$,

$$E\{Y(t)\} = \mu_n \int_{0}^{\infty} \mathcal{Q}(u)\, du = \mu_n E_l(L), \quad (105)$$

$$C\{Y(t+v), Y(t)\} = \mu_n \int_{v}^{\infty} \mathcal{Q}(u)\, du + \int_{0}^{\infty} \int_{0}^{\infty} \gamma_n^0(u-w)\, \mathcal{Q}(u)\, \mathcal{Q}(w+v)\, du\, dw.$$

$$(106)$$

The most interesting and important special case is when the underlying point process is a Poisson process. It then follows from (106) that for $v \geqslant 0$

$$C\{Y(t+v), Y(t)\} = \mu_n \int_v^\infty \mathscr{Q}(u)\, du \qquad (107)$$

and in particular that

$$V\{Y(t)\} = \mu_n E_l(L). \qquad (108)$$

Thus the autocorrelation function is

$$\int_v^\infty \mathscr{Q}(u)\, du / E_l(L). \qquad (109)$$

Therefore in Example 9.8 it is possible to deduce the distribution of fibre length by differentiating the autocovariance function of $\{Y(t)\}$; note that the derivative of (109) at $v = 0$ can be used to determine $E_l(L)$. This result has a number of other applications, for example to the estimation of the distribution of the speeds of moving points crossing an area, from counts of the numbers of individuals in the area at a series of time points.

In particular, results for the immigration–death process of Example 4.5 are obtained by taking $\mathscr{Q}(u) = e^{-\rho u}$. The autocorrelation function of $\{Y(t)\}$ is then $e^{-\rho v}$; this result can be deduced also by the methods of Chapter 4.

Now by (105) and (108) we have that when the underlying point process is a Poisson process

$$E\{Y(t)\} = V\{Y(t)\}.$$

This suggests that the marginal distribution of $Y(t)$ may be a Poisson distribution of mean (105).

One way of proving this that leads also to further results is to suppose first that the distribution of L is discrete with probabilities $q_1, q_2, \ldots$ at values $l_1, l_2, \ldots$ Then by the results of Section 4.1 the Poisson process of rate μ_n combined with a random choice of L for each point is equivalent to a set of independent Poisson processes with parameters $\mu_n q_1, \mu_n q_2, \ldots$ and associated with fixed lengths $l_1, l_2, \ldots$ Denote by $Y_i(t)$ the contribution to $Y(t)$ arising from the pulses of fixed length l_i. Then

$$Y(t) = \sum Y_i(t) \qquad (110)$$

and the processes referring to different lengths $l_i, l_j, \ldots$ are mutually independent. Now $Y_i(t)$ is the number of points in the Poisson process of rate $\mu_n q_i$ occurring in the interval $(t - l_i, t)$. Thus $Y_i(t)$ has a Poisson distribution of mean $\mu_n q_i l_i$ and therefore, by (110), $Y(t)$ has a Poisson distribution of mean $\mu_n \sum q_i l_i = \mu_n E_l(L)$. Since any distribution of L can be approximated by a discrete distribution the general result is proved. It is left as an exercise to prove (107) by the same type of argument.

Although the process $\{Y(t)\}$ is not a Markov process, we can by elementary methods obtain results about returns to the state zero. First, in a very long time, there is an alternation of periods in state 0 and 'busy' periods spent entirely away from state zero. Now the periods in state zero are exponentially distributed with mean $1/\mu_n$ and further the long-run proportion of time in state 0 is the zero term of the Poisson distribution and is $\exp\{-\mu_n E_l(L)\}$. Since in a long period, the number of 'busy' periods and the number of periods in state 0 are equal, it follows that the expected length of a 'busy' period is

$$\frac{\exp\{\mu_n E_l(L)\} - 1}{\mu_n}. \tag{111}$$

This is the expected recurrence-time of state 0, i.e. the expected time from first leaving state 0 to first returning to it.

To extend this result, suppose that state 0 is occupied at time zero, and consider the probability $p_{00}(t)$ that state 0 is occupied at time t. Now in terms of the discrete distribution of L, $Y(t) = 0$ if and only if $Y_i(t) = 0$ $(i = 1, 2, \ldots)$. Thus we consider first the simple problem: given that $Y_i(0) = 0$, i.e. that there were no events in the relevant Poisson process in $(-l_i, 0)$, to find $\mathrm{prob}\{Y_i(t) = 0\}$. We require the conditional probability that there are no events in $(t - l_i, t)$; this is

$$\exp(-\mu_n q_i l_i) \quad (l_i < t),$$
$$\exp(-\mu_n q_i t) \quad (l_i \geqslant t).$$

Thus if we combine the probabilities referring to different lengths, we have that if $Q(l)$ is the distribution function of length

$$p_{00}(t) = \exp\left(-\mu_n \sum_{l_i < t} q_i l_i - \mu_n t \sum_{l_i \geqslant t} q_i\right)$$

$$= \exp\left\{-\mu_n \int_0^t l\, dQ(l) - \mu_n t \int_t^\infty dQ(l)\right\}$$

$$= \exp\left\{-\mu_n \int_0^t \mathcal{Q}(v)\, dv\right\}. \tag{112}$$

Now let $r_{00}(t)\, \Delta t + o(\Delta t)$ be the probability, given that state 0 is occupied at zero time, that the state 0 is re-entered in $(t, t + \Delta t)$. Then, we have that

$$p_{00}(t + \Delta t) = p_{00}(t)(1 - \mu_n \Delta t) + r_{00}(t)\, \Delta t + o(\Delta t), \tag{113}$$

so that

$$r_{00}(t) = p_{00}'(t) + \mu_n p_{00}(t)$$

$$= \mu_n Q(t) \exp\left\{-\mu_n \int_0^t \mathcal{Q}(v)\, dv\right\}. \tag{114}$$

The justification of (113) is that, although the processes $\{Y(t)\}$ is not a Markov process, the transition probabilities out of state 0 are independent of the previous history of the process.

Finally let $f_{00}(t)\,\Delta t + o(\Delta t)$ be the probability, given that state 0 is occupied at zero time, that the *first* re-entry into state 0 occurs in $(t, t + \Delta t)$. Now the intervals between successive re-entries into state 0 are easily seen to be independent and identically distributed random variables and so these re-entries constitute a renewal process. We can apply the integral equation of renewal theory, (25), to give

$$r_{00}(t) = f_{00}(t) + \int_0^t f_{00}(t-u)\, r_{00}(u)\, du.$$

In terms of Laplace transforms

$$f_{00}^*(s) = \frac{r_{00}^*(s)}{1 + r_{00}^*(s)}. \tag{115}$$

Thus given the form of the distribution of L, the distribution of first recurrence-time can be studied.

The argument is parallel to that used in discrete time in Chapter 3. However in the present analysis the special properties of the zero state are vital. The method just given would not yield $r_{ij}(t), f_{ij}(t)$, referring to passages from a general state i to a general state j. Another way of looking at the analysis given here is that we are studying the imbedded process of entries into state 0, this imbedded process having the form of a renewal process. Particular properties of non-Markov processes can quite often be obtained by this technique.

Bibliographic Notes

The very extensive literature on the general limiting theorems of renewal theory is best approached through the paper of Smith (1958). A much more applied account of renewal processes is given by Cox (1962). The key papers on semi-Markov processes are by Smith (1955), Lévy (1956) and Pyke (1961a, b). Properties of general stationary point processes are discussed by Khintchine (1960), Ryll-Nardzewski (1961), Bartlett (1955, Chapter 3), Lewis (1961), McFadden (1962), McFadden and Weissblum (1963) and Cox and Lewis (1966). References on the superposition of processes are Khintchine (1960) and Cox and Smith (1953, 1954). Topics related to the last section are discussed by Doob (1953, Chapter 9), Bartlett (1955, Chapter 5), Middleton (1960, Chapter 4) and Riordan (1962, Chapter 3).

Exercises

1. Obtain an alternative derivation of equation (9) for the function $H(t)$ for a renewal process by introducing random variables

$$A_r(t) = \begin{array}{ll} 1 & (r\text{th event at or before } t), \\ 0 & (r\text{th event after } t), \end{array}$$

noting that

$$N_t = \sum_{r=1}^{\infty} A_r(t).$$

Does this derivation apply to arbitrary point processes ?

2. Derive the integral equation of renewal theory of Section 9.2 directly by probabilistic arguments. Extend it to modified renewal processes. Suggest how it could be used to approximate to the renewal density for small t.

3. In a renewal process, let the p.d.f. of intervals be the displaced exponential function, $\rho e^{-\rho(x-\gamma)}$ $(x > \gamma \geqslant 0)$. Prove from (4) that if N_t is the number of events in $(0, t)$

$$\text{prob}(N_t < r) = \sum_{k=0}^{r-1} e^{-\rho(t-r\gamma)} \frac{\{\rho(t-r\gamma)\}^k}{k!}.$$

4. Prove from (5) that $G_0(z, t)$, the probability generating function of N_t for an ordinary renewal process, satisfies

$$G_0^*(z, s) = \frac{1 - f^*(s)}{s\{1 - zf^*(s)\}},$$

whereas for the corresponding equilibrium renewal process, with mean interval μ,

$$G_e^*(z, s) = \frac{1}{s} + \frac{z-1}{\mu s} G_0^*(z, s).$$

Hence show that if $N_t^{(o)}$, $N_t^{(e)}$ refer respectively to the ordinary and equilibrium renewal processes

$$\text{prob}\{N_t^{(e)} = 0\} = 1 - \frac{1}{\mu} \int_0^t \text{prob}\{N_u^{(o)} = 0\}\,du,$$

$$\text{prob}\{N_t^{(e)} = r\} = \frac{1}{\mu} \int_0^t [\text{prob}\{N_u^{(o)} = r-1\} - \text{prob}\{N_u^{(o)} = r\}]\,du$$

$$(r = 1, 2, \ldots).$$

Give a direct probabilistic proof of these last results.

5. Suppose that independently of an ordinary renewal process with p.d.f. of intervals $f(x)$, 'catastrophes' occur in a Poisson process of rate s. Prove that the probability is $f^*(s)$ that an event occurs in the renewal process before there is a catastrophe and that $h_0^*(s)$ is the expected number of events in the renewal process before the first catastrophe. Hence obtain a direct probabilistic proof of (23), in the form

$$h_0^*(s) = f^*(s) + h_0^*(s) f^*(s).$$

6. Find the renewal density and the renewal function for an ordinary renewal process for which the p.d.f. of intervals is $\varpi_1 \rho_1 e^{-\rho_1 x} + \varpi_2 \rho_2 e^{-\rho_2 x}$, with $\varpi_1 + \varpi_2 = 1$.

7. Examine the form of (46) as $s \to 0$ for a 2×2 system in which $\mathbf{A}$ has no zero elements and hence prove the limiting results stated at the end of Section 9.3.

8. Prove that for an arbitrary stationary point process, the origin being taken where one or more events occur, the expected number of events in $(0, t)$ is

$$\int\limits_0^t h_n(v) \, dv,$$

where $h_n(v)$ is defined by (65).

9. Obtain the functions $h_n(v; i)$ and $h_n(v)$ of (64) and (65) for a process in which occurrence points occur in a Poisson process and in which the numbers of events at occurrence points $\{1, 2, \ldots\}$ form a stationary process.

10. Obtain and plot the general form of the functions $h_n(v)$, $\gamma_n(v)$, $\phi_n(\omega)$ for the renewal processes with p.d.f.'s (a) xe^{-x}, (b) $\frac{1}{2}e^{-x} + \frac{1}{6}e^{-\frac{1}{3}x}$.

11. Give examples of stationary point processes for which (a) $\tau_n(v) \to 0$ as $v \to \infty$, (b) $\tau_n(v)$ increases monotonically to a finite limit as $v \to \infty$, (c) $\tau_n(v)$ decreases monotonically to a non-zero limit as $v \to \infty$.

12. For an ordinary renewal process, show that if $\mathscr{F}(x) = 1 - F(x)$, then

$$h_0^{(1)}(t) = f(t) + \frac{F^2(t)}{\displaystyle\int\limits_0^t \mathscr{F}(x) \, dx}$$

is such that

(a) $h_0^{(1)}(t) = h_0(t)$ for the Poisson process;
(b) $h_0^{(1)}(t)$ and $h_0(t)$ have the same limits as $t \to \infty$;
(c) $h_0^{(1)}(t)$ and $h_0(t)$ have the same value and first two derivatives at $t = 0$.

(Bartholomew, 1963)

13. Let $\{X_1, X_2, \ldots\}$ be a sequence of independent and identically distributed random variables with p.d.f. $f(x)$, and let R be an integer-valued random variable with $\text{prob}(R = r) = p_r$, the X_i's and R being independent. A finite renewal process is one in which events occur at $X_1, X_1 + X_2, \ldots, X_1 + \ldots + X_R$. Calculate

   (a) the mean, variance and Laplace transform of the distribution of the time of the last event;
   (b) the expected number of events in $(0, t)$.

14. Suppose events occur in a renewal process $X_1, X_2, \ldots$, where the intervals X_i have p.d.f. $f(x)$. Let $Y_1, Y_2, \ldots$ be the intervals of a Poisson process of unit rate. The number of events N_n of the renewal process occurring in the interval Y_n of the Poisson process may be termed the Poisson count of the renewal process. Let

$$a_k = \int\limits_0^\infty \frac{e^{-t} t^k}{k!} f(t) \, dt.$$

Show that the Poisson count of the renewal process is given by $N_n = E_n V_n$, where

   (a) $\{E_n\}$ and $\{V_n\}$ are independent;
   (b) E_n takes the value 1 at the occurrences of a modified renewal process in discrete time $(n = 1, 2, \ldots)$ whose first interval has the distribution a_{k-1} $(k = 1, 2, \ldots)$ and subsequent intervals have the distribution $a_k(1 - a_0)$ $(k = 1, 2, \ldots)$ and E_n is zero otherwise;
   (c) the V_n are independent and identically distributed with

$$\text{prob}(V_n = r) = (1 - a_0) a_0^{r-1} \quad (r = 1, 2, \ldots).$$

(Kingman, 1963)

Table of Exponentially Distributed Random Quantities

This appendix gives a short table of exponentially distributed random numbers representing realizations of independent random variables having the probability density function e^{-x} ($x > 0$). The table is suitable for use only in short trial simulations. For extensive simulations by hand calculation, the tables of Clark and Holz (1960) can be used. With an electronic computer, the standard procedure is to generate pseudo-random quantities using a suitable recurrence equation.

0·22	2·53	1·16	0·55	0·14	0·19	0·92	0·94
2·46	0·20	1·19	0·19	0·12	0·25	0·91	1·40
0·56	2·31	0·98	0·51	1·49	2·40	5·60	1·01
1·01	0·36	3·11	0·01	0·72	0·57	1·82	1·53
0·73	1·17	2·65	3·80	0·07	0·89	2·06	0·12
0·01	1·52	5·41	0·61	1·02	3·25	0·07	0·39
0·25	5·78	0·17	1·34	0·69	0·04	2·40	1·05
0·58	0·17	1·98	0·77	0·25	2·15	1·48	0·36
0·09	1·71	0·52	0·29	1·15	0·25	0·64	0·22
1·31	0·16	1·61	0·17	0·51	1·80	1·97	2·40
0·31	0·07	0·26	0·42	2·21	0·01	0·85	1·01
0·89	0·08	0·08	0·27	0·58	0·33	1·76	0·47
0·05	1·87	0·03	0·77	0·07	2·64	0·12	1·37
1·68	1·83	0·54	0·74	0·31	0·15	0·04	0·46
0·38	2·20	1·99	0·51	1·99	0·66	0·08	0·29
4·51	3·01	1·65	0·90	0·64	0·58	1·53	0·63
0·68	4·05	2·20	0·65	1·98	0·38	2·47	0·96
0·76	0·26	0·81	0·92	0·05	0·04	1·81	0·98
0·70	0·45	0·09	3·75	2·11	0·32	0·25	1·46
2·24	0·64	1·05	2·40	1·10	2·52	1·05	1·26
1·90	0·02	0·26	0·29	6·30	6·00	0·10	0·79
0·13	0·36	0·13	0·40	0·32	0·75	2·72	1·03
1·43	1·86	0·11	2·00	4·03	0·05	0·02	0·77
0·86	0·73	1·49	1·54	0·31	0·29	0·08	0·14
0·55	3·29	2·32	0·46	0·20	0·29	1·35	2·55

APPENDIX 2

Bibliography

We give below first a selected list of general books on stochastic processes, then a list of books on special applications, and finally a list of books and papers referred to in the present book. Current work in the subject is best traced through *Mathematical Reviews* and *Statistical Theory and Method Abstracts*.

(i) General books

BAILEY, N. T. J. (1964). *The elements of stochastic processes with applications to the natural sciences.* New York: Wiley.

BARTLETT, M. S. (1955). *An introduction to stochastic processes.* Cambridge University Press.

BENDAT, J. S. (1958). *Principles and application of random noise theory.* New York: Wiley.

BHARUCHA-REID, A. T. (1960). *Elements of the theory of Markov processes and their applications.* New York: McGraw-Hill.

BLANC-LAPIERRE, A. and FORTET, R. (1953). *Théorie des fonctions aléatoires.* Paris: Masson.

CHUNG, K. L. (1960). *Markov chains with stationary transition probabilities.* Berlin: Springer.

DOOB, J. L. (1953). *Stochastic processes.* New York: Wiley.

DYNKIN, E. B. (1960). *Theory of Markov processes.* London: Pergamon. (English translation.)

FELLER, W. (1957). *Introduction to probability theory and its applications,* 2nd ed. New York: Wiley.

HARRIS, T. E. (1963). *The theory of branching processes.* Berlin: Springer.

KEMENY, J. G. and SNELL, J. L. (1960). *Finite Markov chains.* New York: Van Nostrand.

KEMPERMAN, J. H. B. (1961). *The passage problem for a stationary Markov chain.* Chicago University Press.

LANING, J. H. and BATTIN, R. H. (1956). *Random processes in automatic control.* New York: McGraw-Hill.

LÉVY, P. (1948). *Processus stochastiques et mouvement Brownien.* Paris: Gauthier-Villars.

LOÈVE, M. M. (1963). *Probability theory,* 3rd ed. New York: Van Nostrand.

PARZEN, E. (1962). *Stochastic processes.* San Francisco: Holden-Day.

ROSENBLATT, M. (1962). *Random processes.* New York: Oxford University Press.

SPITZER, F. (1964). *Principles of random walk.* New York: Van Nostrand.

TAKÁCS, L. (1960). *Stochastic processes; problems and solutions.* London: Methuen.

WAX, N. (editor) (1954). *Selected papers on noise and stochastic processes.* New York: Dover.

WOLD, H. (1954). *A study in the analysis of stationary time series,* 2nd ed. Stockholm: Almqvist and Wiksell.

YAGLOM, A. M. (1962). *An introduction to the theory of stationary random functions.* Englewood Cliffs, N.J.: Prentice-Hall. (English translation.)

378

(ii) Books on special applications

ARROW, K. J., KARLIN, S. and SCARF, H. (editors) (1958). *Studies in the mathematical theory of inventory and production.* Stanford University Press.

ARROW, K. J., KARLIN, S. and SCARF, H. (editors) (1962). *Studies in applied probability and management science.* Stanford University Press.

BAILEY, N. T. J. (1957). *Mathematical theory of epidemics.* London: Griffin.

BARTLETT, M. S. (1960). *Stochastic population models.* London: Methuen.

BENES, V. E. (1963). *General stochastic processes in the theory of queues.* Cambridge, Mass.: Addison Wesley.

BLACKMAN, R. B. and TUKEY, J. W. (1959). *The measurement of power spectra.* New York: Dover.

BOGDANOFF, J. L. and KOZIN, F. (editors) (1963). *Proceedings of first symposium on engineering applications of random function theory and probability.* New York: Wiley.

BUSH, R. R. and MOSTELLER, F. (1955). *Stochastic models for learning.* New York: Wiley.

COLEMAN, J. S. (1964). *Introduction to mathematical sociology.* New York: Free Press of Glencoe.

COX, D. R. (1962). *Renewal theory.* London: Methuen.

COX, D. R. and SMITH, W. L. (1961). *Queues.* London: Methuen.

HAIGHT, F. A. (1963). *Mathematical theories of traffic flow.* New York: Academic Press.

KHINTCHINE, A. J. (1960). *Mathematical methods in the theory of queueing.* London: Griffin. (English translation.)

MIDDLETON, D. (1960). *Introduction to statistical communication theory.* New York: McGraw-Hill.

MORAN, P. A. P. (1959). *The theory of storage.* London: Methuen.

MORSE, P. M. (1958). *Queues, inventories and maintenance.* New York: Wiley.

RIORDAN, J. (1962). *Stochastic service systems.* New York: Wiley.

SAATY, T. L. (1961). *Elements of queueing theory, with applications.* New York: McGraw-Hill.

SOLODOVNIKOV, V. V. (1960). *Introduction to the statistical dynamics of automatic control systems.* New York: Dover. (English translation.)

SYSKI, R. (1960). *Congestion theory in telephone systems.* Edinburgh: Oliver and Boyd.

TAKÁCS, L. (1962). *Introduction to the theory of queues.* New York: Oxford University Press.

WHITTLE, P. (1963). *Prediction and regulation.* London: English Universities Press.

WIENER, N. (1949). *Extrapolation, interpolation and smoothing of stationary time series.* New York: Wiley.

(iii) References

BAILEY, N. T. J. (1954). 'A continuous time treatment of a simple queue using generating functions', *J. R. statist. Soc.*, B, **16**, 288–91.

BAILEY, N. T. J. (1957). *Mathematical theory of epidemics.* London: Griffin.

BARNARD, G. A. (1959). 'Control charts and stochastic processes', *J. R. statist. Soc.*, B, **21**, 239–71.

BARTHOLOMEW, D. J. (1963). 'An approximate solution of the integral equation of renewal theory', *J. R. statist. Soc.*, B, **25**, 432–41.

BARTLETT, M. S. (1953). 'Stochastic processes or the statistics of change', *Appl. Statistics*, **2**, 44–64.

BARTLETT, M. S. (1955). *An introduction to stochastic processes*. Cambridge University Press.

BATCHELOR, G. K. (1953). *Theory of homogeneous turbulence*. Cambridge University Press.

BATHER, J. A. (1963). 'Control charts and the minimization of costs', *J. R. statist. Soc.*, B, **25**, 49–80.

BELLMAN, R. (1960). *Introduction to matrix analysis*. New York: McGraw-Hill.

BENDAT, J. S. (1958). *Principles and application of random noise theory*. New York: Wiley.

BHARUCHA-REID, A. T. (1960). *Elements of the theory of Markov processes and their applications*. New York: McGraw-Hill.

BLUMEN, I., KOGAN, M. and MCCARTHY, P. J. (1955). *The industrial mobility of labor as a probability process*. Vol. VI of Cornell studies of industrial and labor relations. The New York State School of Industrial and Labor Relations, Cornell University, Ithaca, New York.

BOX, G. E. P. and JENKINS, G. M. (1962). 'Some statistical aspects of adaptive optimization and control', *J. R. statist. Soc.*, B, **24**, 297–343.

BOX, G. E. P. and JENKINS, G. M. (1963). 'Further contributions to adaptive quality control: simultaneous estimation of dynamics: non-zero costs', *Bull. int. statist. Inst.*, **40**, 943–74.

BROADBENT, S. R. and KENDALL, D. G. (1953). 'The random walk of *trichostrongylus retortaeformis*', *Biometrics*, **9**, 460–66.

CHUNG, K. L. (1960). *Markov chains with stationary transition probabilities*. Berlin: Springer.

CHUNG, K. L. and FUCHS, W. H. J. (1951). 'On the distribution of values of sums of random variables', *Mem. Amer. math. Soc.*, No. 6.

CLARK, C. E. and HOLZ, B. W. (1960). *Exponentially distributed random numbers*. Baltimore: Johns Hopkins Press.

COX, D. R. (1955a). 'The analysis of non-Markovian stochastic processes by the inclusion of supplementary variables', *Proc. Camb. phil. Soc.*, **51**, 433–41.

COX, D. R. (1955b). 'A use of complex probabilities in the theory of stochastic processes', *Proc. Camb. phil. Soc.*, **51**, 313–19.

COX, D. R. (1962). *Renewal theory*. London: Methuen.

COX, D. R. and LEWIS, P. A. W. (1966). *Statistical analysis of series of events*. London: Methuen.

COX, D. R. and SMITH, W. L. (1953). 'The superposition of several strictly periodic sequences of events', *Biometrika*, **40**, 1–11.

COX, D. R. and SMITH, W. L. (1954). 'On the superposition of renewal processes', *Biometrika*, **41**, 91–9.

CRAMÉR, H. (1946). *Mathematical methods of statistics*. Princeton University Press.

CRAMÉR, H. (1954). 'On some questions connected with mathematical risk', *Univ. California Publ. Statist.*, **2**, 99–123.

CRAMÉR, H. (1961). 'On some classes of non-stationary stochastic processes', *Proc. 4th Berkeley Symposium*, **2**, 57–78.

CRAMÉR, H. (1964). 'Model building with the aid of stochastic processes', *Technometrics*, **6**, 133–60.

DANIELS, H. E. (1960). 'Approximate solutions of Green's type for univariate stochastic processes', *J. R. statist. Soc.*, B, **22**, 376–401.

DARLING, D. A. and SIEGERT, A. J. F. (1953). 'The first passage problem for a continuous Markov process', *Ann. math. Statist.*, **24**, 624–39.

DEBREU, G. and HERSTEIN, I. N. (1953). 'Nonnegative square matrices', *Econometrica*, **21**, 597–607.

DERMAN, C. (1954). 'A solution to a set of fundamental equations in Markov chains', *Proc. Amer. math. Soc.*, **5**, 332–4.

DOOB, J. L. (1942). 'The Brownian movement and stochastic equations', *Ann. Math.*, **43**, 351–69. (Also in Wax (1954).)

DOOB, J. L. (1953). *Stochastic processes.* New York: Wiley.

EINSTEIN, A. (1956). *Investigations on the theory of Brownian movement.* New York: Dover.

ERDÖS, P., FELLER, W. and POLLARD, H. (1949). 'A property of power series with positive coefficients', *Bull. Amer. math. Soc.*, **55**, 201–4.

FELLER, W. (1951). 'Diffusion processes in genetics', *Proc. 2nd Berkeley Symposium*, 227–46.

FELLER, W. (1952). 'The parabolic differential equations and the associated semi-groups of transformations', *Ann. Math.*, **55**, 468–519.

FELLER, W. (1957). *An introduction to probability theory and its applications*, 2nd ed. New York: Wiley.

FOSTER, F. G. (1953). 'On the stochastic matrices associated with certain queueing processes', *Ann. math. Statist.*, **24**, 355–60.

FRÉCHET, M. (1938). *Recherches theoriques moderns sur le calcul des probabilités*, Vol. II. Paris: Hermann.

GABRIEL, K. R. (1959). 'The distribution of the number of successes in a sequence of dependent trials', *Biometrika*, **46**, 454–60.

GABRIEL, K. R. and NEUMANN, J. (1957). 'On a distribution of weather cycles by length', *Quart. J. R. met. Soc.*, **83**, 375–80.

GABRIEL, K. R. and NEUMANN, J. (1962). 'A Markov chain model for daily rainfall occurrence at Tel Aviv', *Quart. J. R. met. Soc.*, **88**, 90–5.

GANI, J. (1957). 'Problems in the probability theory of storage systems', *J. R. statist. Soc.*, B, **19**, 181–206.

GANTMACHER, F. R. (1959). *Applications of the theory of matrices.* New York: Interscience. (English translation.)

GAVER, D. P. (1963). 'Time to failure and availability of paralleled systems with repair', *I.E.E.E. Trans. on Reliability*, R12, 30–8.

GNEDENKO, B. V. (1962). *The theory of probability.* New York: Chelsea. (English translation.)

GNEDENKO, B. V. and KOLMOGOROV, A. N. (1954). *Limit distributions for sums of independent random variables.* Cambridge, Mass.: Addison-Wesley. (English translation.)

GOODMAN, L. A. (1953a). 'Population growth of the sexes', *Biometrics*, **9**, 212–25.

GOODMAN, L. A. (1953b). 'Methods of measuring useful life of equipment under operational conditions', *J. Amer. statist. Ass.*, **48**, 503–30.

GOODMAN, L. A. (1961). 'Statistical methods for the mover-stayer model', *J. Amer. statist. Ass.*, **56**, 841–68.

HAIGHT, F. A. (1963). *Mathematical theories of traffic flow.* New York: Academic Press.

HAJNAL, J. (1956). 'The ergodic properties of non-homogeneous finite Markov chains', *Proc. Camb. phil. Soc.*, **52**, 67–77.

HAJNAL, J. (1958). 'Weak ergodicity in non-homogeneous Markov chains', *Proc. Camb. phil. Soc.*, **54**, 233–46.

HAMMERSLEY, J. M. and HANDSCOMB, D. C. (1964). *Monte Carlo methods.* London: Methuen.

HARRIS, T. E. (1963). *The theory of branching processes.* Berlin: Springer.

ITÔ, K. (1951). 'On stochastic differential equations', *Mem. Amer. math. Soc.*, No. 4.

JENSEN, A. (1948). 'An elucidation of Erlang's statistical works through the theory of stochastic processes', *The life and works of A. K. Erlang*, 23–100. Copenhagen Telephone Company.

JENSEN, A. (1954). *A distribution model applicable to economics*. Copenhagen: Munksgaard.

JOSHI, D. D. (1954). 'Les processus stochastiques en démographie', *Publ. Inst. statist. Univ. Paris*, 3, 153–77.

KAC, M. (1947). 'Random walk and the theory of Brownian motion', *Amer. math. Monthly*, 54, 369–91. (Also in Wax (1954).)

KARLIN, S. and MCGREGOR, J. (1957a). 'The differential equations of birth and death processes and the Stieltjes moment problem', *Trans. Amer. math. Soc.*, 85, 489–546.

KARLIN, S. and MCGREGOR, J. (1957b). 'The classification of birth and death processes', *Trans. Amer. math. Soc.*, 86, 366–400.

KARLIN, S. and MCGREGOR, J. (1958a). 'Linear growth, birth and death processes', *J. Math. Mech.*, 7, 643–62.

KARLIN, S. and MCGREGOR, J. (1958b). 'Many server queueing processes with Poisson input and exponential service times', *Pacific J. Math.*, 8, 87–118.

KEILSON, J. (1963). 'The first passage time density for homogeneous skip-free walks on the continuum', *Ann. math. Statist.*, 34, 1003–11.

KEILSON, J. (1965). *Green's function methods in probability theory*. London: Griffin.

KEILSON, J. and KOOHARIAN, A. (1960). 'On time dependent queueing processes', *Ann. math. Statist.*, 31, 104–12.

KEILSON, J. and WISHART, D. M. G. (1964). 'A central limit theorem for processes defined on a finite Markov chain', *Proc. Camb. phil. Soc.*, 60, 547–67.

KEMENY, J. G. and SNELL, J. L. (1960). *Finite Markov chains*. New York: Van Nostrand.

KEMPERMAN, J. H. B. (1961). *The passage problem for a stationary Markov chain*. University of Chicago Press.

KENDALL, D. G. (1949). 'Stochastic processes and population growth', *J. R. statist. Soc.* B, 11, 230–64.

KENDALL, D. G. (1951). 'Some problems in the theory of queues', *J. R. statist. Soc.*, B, 13, 151–85.

KENDALL, D. G. (1953). 'Stochastic processes occurring in the theory of queues and their analysis by the method of the imbedded Markov chain', *Ann. math. Statist.*, 24, 338–54.

KENDALL, D. G. (1959). 'Unitary dilations of one-parameter semigroups of Markov transition operators and the corresponding integral representations for Markov processes with a countable infinity of states', *Proc. Lond. math. Soc.*, 9, 417–31.

KENDALL, D. G. (1960). 'Geometric ergodicity and the theory of queues', in K. J. Arrow, S. Karlin, P. Suppes (editors), *Mathematical methods in the social sciences*. Stanford University Press.

KENDALL, D. G. (1961a). 'The distribution of energy perturbations for Halley's and some other comets', *Proc. 4th Berkeley Symposium*, 3, 87–98.

KENDALL, D. G. (1961b). 'Some problems in the theory of comets, I', *Proc. 4th Berkeley Symposium*, 3, 99–120.

KENDALL, D. G. (1961c). 'Some problems in the theory of comets, II', *Proc. 4th Berkeley Symposium*, 3, 121–47.

KHINTCHINE, A. J. (1932). 'Mathematisches über die Erwartung vor einem öffertlichen Schalter', *Mat. Sbornik*, 39, 73–84.

KHINTCHINE, A. J. (1934). 'Korrelationstheorie der stationären stochastischen Prozesse', *Math. Ann.*, 109, 604–15.

KHINTCHINE, A. J. (1960). *Mathematical methods in the theory of queueing.* London: Griffin. (English translation.)

KINGMAN, J. F. C. (1963). 'Poisson counts for random sequences of events', *Ann. math. Statist.*, **34**, 1217–32.

KOLMOGOROV, A. N. (1936). 'Anfangsgründe der Theorie der Markoffschen ketten mit unendlich vielen möglichen Zuständen', *Mat. Sbornik* (N.S.), **1**, 607–10.

LANING, J. H. and BATTIN, R. H. (1956). *Random processes in automatic control.* New York: McGraw-Hill.

LÉVY, P. (1948). *Processus stochastiques et mouvement Brownien.* Paris: Gauthier-Villars.

LÉVY, P. (1956). 'Processus semi-markoviens', *Proc. Int. Congr. Math. (Amsterdam)*, **3**, 416–26.

LEWIS, T. (1961). 'The intervals between regular events displaced in time by independent random deviations of large dispersion', *J. R. statist. Soc.*, B, **23**, 476–83.

LINDLEY, D. V. (1952). 'The theory of queues with a single server', *Proc. Camb. phil. Soc.*, **48**, 277–89.

LINDLEY, D. V. (1959). Discussion following C. B. Winsten's paper: 'Geometric distributions in the theory of queues', *J. R. statist. Soc.*, B, **21**, 22–3.

LLOYD, E. H. (1963). 'Reservoirs with serially correlated inflows', *Technometrics*, **5**, 85–93.

LOTKA, A. J. (1931). 'The extinction of families', *J. Wash. Acad. Sci.*, **21**, 377–80 and 453–9.

MCFADDEN, J. A. (1962). 'On the lengths of intervals in a stationary point process', *J. R. statist. Soc.*, B, **24**, 364–82.

MCFADDEN, J. A. and WEISSBLUM, W. (1963). 'Higher-order properties of a stationary point process', *J. R. statist. Soc.*, B, **25**, 413–31.

MCMILLAN, B. and RIORDAN, J. (1957). 'A moving single server problem', *Ann. math. Statist.*, **28**, 471–8.

MEYER, H. A. (editor) (1956). *Symposium on Monte Carlo methods.* New York: Wiley.

MIDDLETON, D. (1960). *Introduction to statistical communication theory.* New York: McGraw-Hill.

MILLER, H. D. (1961a). 'A generalization of Wald's identity with applications to random walks', *Ann. math. Statist.*, **32**, 549–60.

MILLER, H. D. (1961b). 'A convexity property in the theory of random variables defined on a finite Markov chain', *Ann. math. Statist.*, **32**, 1260–70.

MILLER, H. D. (1962a). 'A matrix factorization problem in the theory of random variables defined on a finite Markov chain', *Proc. Camb. phil. Soc.*, **58**, 268–85.

MILLER, H. D. (1962b). 'Absorption probabilities for sums of random variables defined on a finite Markov chain', *Proc. Camb. phil. Soc.*, **58**, 286–98.

MORAN, P. A. P. (1956). 'A probability theory of a dam with a continuous release', *Quart. J. Math.*, **7**, 130–7.

MORAN, P. A. P. (1959). *The theory of storage.* London: Methuen.

MORSE, P. M. (1958). *Queues inventories and maintenance.* New York: Wiley.

NEYMAN, J. (1960). 'Indeterminism in science and new demands on statisticians', *J. Amer. statist. Ass.*, **55**, 625–39.

NOBLE, B. (1958). *The Wiener-Hopf technique.* London: Pergamon.

PARZEN, E. (1960). *Modern probability theory and its applications.* New York: Wiley.

PARZEN, E. (1961). 'An approach to time series analysis', *Ann. math. Statist.*, **32**, 951–89.

384 · *The Theory of Stochastic Processes*

PARZEN, E. (1962). *Stochastic processes.* San Francisco: Holden-Day.
PIAGGIO, H. T. H. (1942). *Differential equations.* London: Bell.
PITT, H. R. (1963). *Integration, measure and probability.* Edinburgh: Oliver and Boyd.
PRABHU, N. U. (1958). 'Some exact results for the finite dam', *Ann. math. Statist.,* **29,** 1234–43.
PRABHU, N. U. (1964). 'Time dependent results in storage theory', *J. appl. Prob.,* **1,** 1–46.
PYKE, R. (1961a). 'Markov renewal processes: definitions and preliminary properties', *Ann. math. Statist.,* **32,** 1231–42.
PYKE, R. (1961b). 'Markov renewal processes with finitely many states', *Ann. math. Statist.,* **32,** 1243–59.
REUTER, G. E. H. (1957). 'Denumerable Markov processes and the associated contraction semigroups on *l*', *Acta Math.,* **97,** 1–46.
REUTER, G. E. H. and LEDERMANN, W. (1953). 'On the differential equations for the transition probabilities of Markov processes with enumerably many states', *Proc. Camb. phil. Soc.,* **49,** 247–62.
RICE, S. O. (1944). 'Mathematical analysis of random noise, I', *Bell syst. tech. J.,* **23,** 282–332. (Also in Wax (1954).)
RICE, S. O. (1945). 'Mathematical analysis of random noise, II', *Bell syst. tech. J.,* **24,** 46–156. (Also in Wax (1954).)
RIORDAN, J. (1962). *Stochastic service systems.* New York: Wiley.
ROBERTS, S. W. (1959). 'Control chart tests based on geometric moving averages', *Technometrics,* **1,** 239–50.
ROSENBLATT, M. (1962). *Random processes.* New York: Oxford University Press.
RYLL-NARDZEWSKI, C. (1961). 'Remarks on processes of calls', *Proc. 4th Berkeley Symposium,* **2,** 455–66.
SAATY, T. L. (1961a). 'Some stochastic processes with absorbing barriers', *J. R. statist. Soc.,* B, **23,** 319–34.
SAATY, T. L. (1961b). *Elements of queueing theory with applications.* New York: McGraw-Hill.
SMITH, W. L. (1953). 'On the distribution of queueing times', *Proc. Camb. phil. Soc.,* **49,** 449–61.
SMITH, W. L. (1955). 'Regenerative stochastic processes', *Proc. Roy. Soc.,* A, **232,** 6–31.
SMITH, W. L. (1958). 'Renewal theory and its ramifications', *J. R. statist. Soc.,* B, **20,** 243–302.
SOLODOVNIKOV, V. V. (1960). *Introduction to the statistical dynamics of automatic control systems.* New York: Dover. (English translation.)
SOMMERFELD, A. (1949). *Partial differential equations in physics.* New York: Academic Press.
SPITZER, F. (1956). 'A combinatorial lemma and its applications to probability theory', *Trans. Amer. math. Soc.,* **82,** 323–39.
SPITZER, F. (1957). 'The Wiener-Hopf equation whose kernel is a probability density', *Duke math. J.,* **24,** 327–43.
SPITZER, F. (1964). *Principles of random walk.* New York: Van Nostrand.
TAKÁCS, L. (1955). 'Investigation of waiting time problems by reduction to Markov processes', *Acta Math. Acad. Sci. Hungar.,* **6,** 101–29.
TAKÁCS, L. (1960). *Stochastic processes, problems and solutions.* London: Methuen.
TITCHMARSH, E. C. (1939). *Theory of functions,* 2nd ed. Oxford University Press.
TITCHMARSH, E. C. (1948). *Introduction to the theory of Fourier integrals.* 2nd ed. Oxford University Press.
TOCHER, K. D. (1963). *The art of simulation.* London: English Universities Press.

UHLENBECK, G. E. and ORNSTEIN, L. S. (1930). 'On the theory of Brownian motion', *Phys. Rev.*, **36**, 823–41. (Also in Wax (1954).)

VERE-JONES, D. (1962). 'Geometric ergodicity in denumerable Markov chains', *Quart. J. Math.*, **13**, 7–28.

WALD, A. (1947). *Sequential analysis*. New York: Wiley.

WATTERSON, G. A. (1961). 'Markov chains with absorbing states: a genetic example', *Ann. math. Statist.*, **32**, 716–29.

WAX, N. (editor) (1954). *Selected papers on noise and stochastic processes*. New York: Dover.

WEESAKUL, B. (1961). 'The random walk between a reflecting and an absorbing barrier', *Ann. math. Statist.*, **32**, 765–9.

WEISS, G. H. and MARADUDIN, A. A. (1962). 'Some problems in traffic delay', *Operations Research*, **10**, 74–104.

WENDEL, J. G. (1958). 'Spitzers' formula: a short proof', *Proc. Amer. math. Soc.*, **9**, 905–8.

WHITTAKER, E. T. and WATSON, G. N. (1952). *A course of modern analysis*, 5th ed. Cambridge University Press.

WHITTLE, P. (1957). 'On the use of the normal approximation in the treatment of stochastic processes', *J. R. statist. Soc.*, B, **19**, 268–81.

WHITTLE, P. (1963). *Prediction and regulation*. London: English Universities Press.

WIENER, N. (1923). 'Differential space', *J. Math. and Phys.*, **2**, 131–74.

WIENER, N. (1930). 'Generalized harmonic analysis', *Acta Math.*, **55**, 117–258.

WIENER, N. (1949). *Extrapolation, interpolation and smoothing of stationary time series*. New York: Wiley.

WIENER, N. and MASANI, P. (1957). 'The prediction theory of multivariate stochastic processes, I', *Acta Math.*, **98**, 111–50.

WIENER, N. and MASANI, P. (1958). 'The prediction theory of multivariate stochastic processes, II', *Acta Math.*, **99**, 93–137.

WOLD, H. (1954). *A study in the analysis of stationary time series*, 2nd ed. Stockholm: Almqvist and Wiksell.

YAGLOM, A. M. (1962). *An introduction to the theory of stationary random functions*. Englewood Cliffs, N.J.: Prentice Hall. (English translation.)

YULE, G. U. (1927). 'On a method of investigating periodicities in disturbed series with special reference to Wolfer's sunspot numbers', *Phil. Trans. Roy. Soc.*, A, **226**, 267–98.

Author Index

Subject Index

389